Instructor's Manual

J. Richard Christman
Professor Emeritus
U. S. Coast Guard Academy

with the assistance of

Stanley A. Williams
Iowa State University

Walter Eppenstein
Rensselaer Polytechnic Institute

FUNDAMENTALS OF PHYSICS

Seventh Edition

David Halliday
University of Pittsburgh

Robert Resnick
Rensselaer Polytechnic Institute

Jearl Walker
Cleveland State University

WILEY

John Wiley & Sons, Inc.

To order books or for customer service call 1-800-CALL-WILEY (225-5945).

ISBN 0-471-47058-9

Printed in the United States of America

10 9 8 7 6 5 4 3 2 1

Printed and bound by Courier Kendallville, Inc.

PREFACE

This manual contains material designed to be useful in the design of an introductory physics course based on the text *FUNDAMENTALS OF PHYSICS*, seventh edition, by David Halliday, Robert Resnick, and Jearl Walker. It may be used with either the extended or regular versions of the text. Section One includes material to help instructors choose topics and design courses. Section Two contains a discussion of sources for ancillary material that might be helpful in designing a course or obtaining lab and demonstration apparatus and audio/visual material. Section Three contains lecture notes outlining the important topics of each chapter, suggested demonstration and laboratory experiments, computer software, video cassettes, and DVDs.

Sections Four, Five, and Six contain answers to checkpoints, end-of-chapter questions, and end-of-chapter problems. To help ease the transition from the sixth to the seventh edition of the text, Section Seven of the manual cross references end-of-chapter problems between the two editions. Because some instructors avoid assigning problems that are discussed in *A Student's Companion*, in the *Student Solution Manual* or on the Wiley website, while others desire to include a few of these in many assignments, Section Eight of the manual contains a list of these problems.

The principal author is grateful to Stanley Williams, who co-authored the first edition of the instructor manual for *Fundamentals of Physics*. Much of his material has been retained in this manual. He is also grateful to Walter Eppenstein, who helped with suggestions for demonstration and laboratory experiments. Jearl Walker helped significantly by supplying answers to checkpoint questions, end-of-chapter questions, and end-of-chapter problems.

The author is indebted to the Project Editor Geraldine Osnato, who managed many aspects of this project. Special thanks go to Sharon Prendergast, the Production Editor. Karen Christman carefully read earlier editions of the manuscript and made many useful suggestions. Her fine work is gratefully noted. The unfailing support of Mary Ellen Christman is joyfully acknowledged.

J. Richard Christman
Professor Emeritus
U.S. Coast Guard Academy
New London, Connecticut 06320

TABLE OF CONTENTS

SECTION ONE
ABOUT THE TEXT

Fundamentals of Physics, seventh edition, follows the sequence of topics found in most introductory courses. In fact, earlier editions of this text were instrumental in establishing that sequence. It is, however, extremely flexible in regard to both the range of topics and the depth of coverage. As a result, it can be used for a two, three, or four term course along traditional lines. It can also be used with many of the innovative courses that are presently being designed and taught. In many instances sections that discuss fundamental principles and give applications are followed by other sections that go deeper into the physics. Some instructors prefer to cover fewer topics than others but treat the topics they do cover in great depth. Others prefer to cover more topics with less depth. Courses of both types can easily be accommodated by selecting appropriate sections of the text.

By carefully choosing sections of the text to be included, your course might be a two-term, in-depth study of the fundamentals of classical mechanics and electromagnetism. With the addition of another term you might include more applications and the thermodynamics and optics chapters. In a three-term course, you might also forgo thermodynamics and optics but include Chapter 37 (Relativity) and some of the quantum mechanics chapters added in the extended version.

When designing the course, some care must be taken in the selection of topics because many discussions in later chapters presume coverage of prior material. Here are some comments you might find useful in designing your course. Also refer to the *Lecture Notes* section of this manual.

Mechanics. The central concepts of classical mechanics are covered in Chapters 1 through 11. Some minor changes that are possible, chiefly in the nature of postponements, are mentioned in the Lecture Notes. For example, the scalar product can be postponed until the discussion of work in Chapter 7 and the vector product can be postponed until the discussion of torque in Chapter 11.

Coverage of Chapter 5 can be shortened to two lectures or elongated to over four, depending on the time spent on applications. Sections 9–8, 9–9, 9–10, and 9–11, on collisions, can be covered as part of laboratory exercises. Other sections in the first twelve chapters that can be used to adjust the length of the course are 2–10, 3–7, 4–8, 4–9, 6–4, 7–8, 9–10, 9–11, 9–12, 11–5, and 11–12. Section 10–7, which deals with the calculation of the rotational inertias of extended bodies, can be covered in detail or can be shortened by simply stating results once the definition as a sum over particles has been discussed. The parallel axis theorem is needed to solve some end-of-chapter problems in this chapter and in Chapter 16 and it should be covered if those problems are assigned.

The order of the chapters should be retained. For example, difficulties arise if you precede dynamics with statics as is sometimes done in other texts. To do so, you would need to discuss torque, introduced in Chapter 10, and explain its relation to angular acceleration. This involves considerable effort and is of questionable value.

Chapters 12 through 18 apply the fundamental principles of the first 11 chapters to special systems and, in many cases, lay the groundwork for what is to come. Many courses omit one or more of Chapters 12 (Equilibrium and Elasticity), 13 (Gravitation), 14 (Fluids), and 17 (Waves — II). There is some peril in these omissions, however. Chapter 13, for example, is pedagogically important. The central idea of the chapter is a force law and the discussions of many of its ramifications show by example how physics works. Since the chapter brings together many previously discussed ideas it can be used as a review. In addition, Newton's law of gravity is used later to introduce Coulomb's law and the proof that the electrostatic force is conservative relies on the analogy. The basis of Gauss' law is laid in Chapter 13 and inclusion of this chapter makes teaching of the law easier.

The idea of a velocity field is first discussed in Chapter 14 and is used to introduce electric flux in Chapter 23 (Gauss' Law). The concepts of pressure and density are explained in Chapter 14 and are used again in the thermodynamics chapters. If Chapter 14 is omitted, you should be prepared to make up for the loss of material by presenting definitions and discussions of velocity fields, pressure, and density when they are first used in your course.

Chapter 12 (Equilibrium and Elasticity) can be safely omitted. If it is, a brief description of the equilibrium conditions might be included in the discussion of Chapter 10 or 11. The few problems in later chapters that depend on material in this chapter can be passed over. If Chapter 12 is included, be sure you have already covered torque and have explained its relation to angular acceleration.

Chapters 15 (Oscillations) and 16 (Waves — I) are important parts of an introductory course and should be covered except when time constraints are severe. Chapter 15 is required for Chapter 16 and both are required for Chapter 17 (Waves — II). Chapter 15 is also required for Chapter 31 (Electromagnetic Oscillations and Alternating Current) and parts of Chapter 16 are required for Chapters 33 (Electromagnetic Waves), 35 (Interference), 36 (Diffraction), 38 (Photons and Matter Waves), and 39 (More About Matter Waves). Chapters 15 and 16 may be covered in the mechanics part of the course or may be delayed until electromagnetic waves are covered.

Sections of Chapters 12 through 17 that can be used to adjust the length of the course are 12–6, 12–7, 13–7, 13–8, 13–9, 14–5, 15–6, 15–7, 15–8, 15–9, 16–8, 17–7, 17–9, and 17–10.

Thermodynamics. Chapters 18 through 20 cover the ideas of thermodynamics. Most two-term courses and some three-term courses omit these chapters entirely. If they are covered, they can be placed as a unit almost anywhere after the mechanics chapters. The idea of temperature is used in Chapter 26 (Current and Resistance) and in some of the modern physics chapters, as well as in the other thermodynamics chapters. If Chapter 18 is not covered prior to Chapter 26, you should plan to discuss the idea of temperature in connection with that chapter or else omit the section that deals with the temperature dependence of the resistivity. Sections of these chapters that can be used to adjust the length of the course are 18–6, 18–12, 19–6, 19–10, 20–5, 20–6, 20–7, and 20–8.

Electromagnetism. The fundamentals of electricity and magnetism are covered in Chapters 21 through 33. Chapter 33 (Electromagnetic Waves) may be considered a capstone to the electromagnetism chapters or as an introduction to the optics chapters. Sections that might be omitted to adjust the length of the course are 21–5, 24–8, 25–6, 25–7, 25–8, 26–6, 26–8, 26–9, 27–8, 27–9, 28–7, 30–9, 30–12, 31–11, 32–6, 32–7, 32–8, 32–9, 32–10, 32–11, and 33–7. Sections 33-8, 33–9, and 33–10 can be om8itted if the optics chapters are not covered. Otherwise, they must be included.

Sections 25–6, 25–7, and 25–8, on dielectrics, should be included in an in-depth course but may be omitted in other courses to make room for other topics. Similarly, coverage of Chapters 27 (Circuits) and 31 (Electromagnetic Oscillations and Alternating Currents) may be adjusted considerably, depending on the extent to which the course emphasizes practical applications. They may also be covered as laboratory exercises. Section 26–6 is required if Chapter 41 is covered although the material can be shorted and presented in conjunction with Chapter 41 rather than at an earlier time.

Section 32–2 contains a discussion of Gauss' law for magnetism, one of Maxwell's equations, and should be included in every course, as should sections 32–3, 32–4, and 32–5, on the displacement current, the Ampere-Maxwell law, and the complete set of Maxwell's equations. The last portion of the chapter deals with magnetic properties of materials and some of ramifications of those properties. It nicely complements the previous sections on dielectrics. These parts of the chapter might be omitted or passed over swiftly to gain time for other sections. On the other hand, they should be included if you intend to emphasize properties of materials.

Optics. Chapters 34 through 36 are the optics chapters. You might wish to precede them with Chapter 33 (Electromagnetic Waves) or you might wish to replace Chapter 33 with a short qualitative discussion. You can be somewhat selective in your coverage of Chapter 34 (Images). It can be covered as lightly or as deeply as desired. Much of the material in this chapter can be covered as laboratory exercises.

Chapters 35 (Interference) and 36 (Diffraction) are important in their own right and are quite useful for the discussion of photons and matter waves in Chapter 38. Chapter 36 cannot be included without Chapter 35 but coverage of both chapters can be reduced somewhat to make room for other topics. The fundamentals of interference and diffraction are contained in Sections 35–1 through 35–6 and 36–1 through 36–5. Other sections of these chapters can be included or excluded, as desired.

Modern Physics. Chapter 37 (Relativity) may be used as a capstone to the mechanics section of the course, as a capstone to the entire course, or as an introduction to the modern physics included in the extended version of the text. Some results of relativity theory are needed for the chapters that follow. If you do not wish to cover Chapter 37 in detail you can describe these results as they are needed. However, it is probably more satisfying to present a more complete and logically connected description of relativity theory. If you plan to cover some of the other modern physics chapters you should consider including Chapter 37.

The fundamentals of the quantum theory are presented in Chapters 38 (Photons and Matter Waves) and 39 (More About Matter Waves). This material should be treated as a unit and must follow in the order written. If you include these chapters, be sure earlier parts of the course include discussions of uniform circular motion, angular momentum, Coulomb's law, electrostatic potential energy, electromagnetic waves, and diffraction. $E = mc^2$ and $E^2 = (pc)^2 + (mc^2)^2$, from relativity theory, are used in discussions of the Compton effect.

The introductory modern physics chapters are followed by application chapters: Chapters 40 (All About Atoms), 41 (Conduction of Electricity in Solids), 42 (Nuclear Physics), 43 (Energy from the Nucleus), and 44 (Quarks, Leptons, and the Big Bang). You may choose to end the course with Chapter 39 or you may choose to include one or more of the application chapters.

The ideas of temperature and the Kelvin scale are used in several places in the modern physics chapters: Sections 40–12 (How a Laser Works), 41–5 (Metals), 41–6 (Semiconductors), 43–6 (Thermonuclear Fusion: The Basic Process), and 44–12 (The Microwave Background Radiation). With a little supplementary material, these sections can be covered even if Chapter 18 is not.

Chapter 43 (Energy from the Nucleus) requires Chapter 42 (Nuclear Physics) for background material, but Chapter 42 need not be followed by Chapter 43. $E = mc^2$ and $E^2 = (pc)^2 + (mc^2)^2$ from relativity theory are also used. The discussion of thermonuclear fusion uses some of the ideas of kinetic theory, chiefly the distribution of molecular speeds. Either Chapter 19 (particularly Section 19–7) should be covered first or you should be prepared to supply a little supplementary material here.

Chapter 44 includes an introduction to high energy particle physics and tells how the ideas of physics are applied to cosmology. Both these topics fascinate many students. In addition, the chapter provides a nice overview of physics.

Some knowledge of the Pauli exclusion principle (from Chapter 40) and spin angular momentum (from Chapters 32 and 40) is required. Knowledge of the strong nuclear force (discussed in Chapters 42 and 43) is also required. In addition, beta decay (discussed in Chapter 42) is used several times as an illustrative example. Nevertheless, the chapter can be made to stand alone with the addition of only a small amount of supplementary material.

SUGGESTED COURSES

A bare bones two-semester course (about 90 meetings) can be constructed around Chapters 1 through 11, 15, 16, and 21 through 33, with the omission of Chapter 31 Sections 32–7 through 32–

11. The course can be adjusted to the proper length by the inclusion or omission of supplementary material and optional topics. If four to eight additional meetings are available each term, Chapter 13 or 14 (or perhaps both) can be inserted after Chapter 11 and one or more of the optics chapters can be inserted after Chapter 33. As an alternative, you might consider including sections on dielectrics, magnetic properties, semiconductors, and superconductors to emphasize properties of materials.

A three-term course (about 135 meetings) can be constructed by adding the thermodynamics chapters (18 through 20) and some or all of the modern physics chapters (37 through 44) to those mentioned above. If the needs of the class dictate a section on alternating current, some modern physics material can be replaced by Chapter 31.

ESTIMATES OF TIME

The following chart gives estimates of the time required to cover all of each chapter, in units of 50 minute periods. The second and fifth columns of the chart contain estimates of the number of lecture periods needed and includes the time needed to perform demonstrations and discuss the main points of the chapter. The third and sixth columns contain estimates of the number of recitation periods required and includes the time needed to go over problem solutions, answers to end-of-chapter questions, and points raised by students. If your course is organized differently, you may wish to add the two numbers to obtain the total estimated time for each chapter.

Use the chart as a rough guide when planning the syllabus for a semester, quarter, or year course. If you omit parts of chapters, reduce the estimated time accordingly.

Text Chapter	Number of Lectures	Number of Recitations	Text Chapter	Number of Lectures	Number of Recitations
1	0.3	0.2	23	1.8	1.8
2	2.0	2.0	24	1.8	1.8
3	1.0	1.0	25	1.5	2.0
4	2.0	2.5	26	1.0	1.0
5	2.0	2.0	27	2.0	2.3
6	2.0	2.0	28	2.0	1.8
7	1.8	1.5	29	2.0	1.2
8	2.0	2.0	30	2.5	2.5
9	2.5	2.0	31	1.5	1.8
10	2.0	1.5	32	2.5	2.5
11	2.0	2.0	33	2.9	2.7
12	1.0	2.0	34	2.5	2.5
13	2.3	2.0	35	2.0	2.0
14	2.0	2.0	36	2.0	2.0
15	2.5	1.8	37	2.5	2.0
16	2.5	2.0	38	2.0	2.0
17	2.5	2.0	39	2.0	2.0
18	2.5	2.5	40	2.2	2.0
19	2.0	1.4	41	2.0	2.0
20	1.5	1.6	42	1.8	2.0
21	1.0	1.0	43	2.0	2.0
22	1.6	1.3	44	2.0	2.0

SECTION TWO
SUGGESTIONS FOR THE COURSE

General Two excellent books that deal with teaching the introductory calculus-based course are

Teaching Introductory Physics; Arnold B. Arons; John Wiley (1997); also available from the American Association of Physics Teachers (AAPT, One Physics Ellipse, College Park, MD 20740-3845; www.aapt.org).

Teaching Introductory Physics (A Sourcebook); Clifford E. Swartz and Thomas Miner; Springer-Verlag (1998); also available from the AAPT (see above for address).

Both of these provide well thought-out explanations of some of the concepts that perplex students and give help with teaching those concepts. They are also excellent sources of demonstration and laboratory experiments that illuminate the important ideas of the introductory physics course.

Over the past fifteen years or so the field of physics education research has grown tremendously. Many research projects focus on the troubles students have in learning physics and analyze proposed remedies. Lillian McDermot and Edward Redish have compiled an extensive resource letter that lists books and journal articles in the field. It appeared in the September 1999 issue of the American Journal of Physics and is highly recommended as a source of material for improvement of the course.

Also see *On Teaching Physics*; edited by Melba Phillips. An older but still valuable collection of *American Journal of Physics* articles dealing with physics education.

Class Participation Each chapter contains several semi-quantitative questions, called checkpoints. Encourage students to use them to check their understanding of the concepts and relationships discussed in the chapter. Go over some or all of them in recitation classes or lectures. Answers to the checkpoint questions are given in Section Four of this manual.

If funds are available, consider setting up an interactive class room or lecture hall in which students can be polled remotely. The checkpoints and end-of-chapter questions are excellent for this purpose. You might look into *Classroom Performance System* from Texas Instruments (www.einstruction.com).

Assessment. Several books deal with grading practices and the use of grading in effective teaching. See for example

Effective Grading: A Tool for Learning and Assessment; Barbara E. Walvoord and Virginia Johnson; Jossy-Bass, A Wiley Company; 250 pages. Available through the AAPT (see above for address).

Classroom Assessment Techniques: A Handbook for College Teachers; Thomas A. Angelo and K. Patricia Cross; Jossey-Bass, a Wiley Company; 427 pages. Available through the AAPT (see above for address).

Many schools now use computer submission and grading for homework, quizzes, and exams. *WebAssign* (Box 8202, NCSU, Raleigh NC 27695; www.webassign.net/info) and *mapleT.A.* (Waterloo Maple, 615 Kumpf Drive, Waterloo, Ontario, Canada N2V 1K8; www.maplesoft.com) are two such software products. Both allow you to generate assignments and exams containing your own problems.

Video. All of the video cassette and DVD items listed in the SUGGESTIONS sections of the Lecture Notes are short, well done, and highly pertinent to the chapter. It is not possible to review all available material and there are undoubtedly many other fine video cassettes and disks that are not listed. Video might be incorporated into the lectures, shown during laboratory periods, or set up in a special room for more informal viewing.

An excellent set of DVDs, *The Mechanical Universe*, can be obtained from The Annenberg CPB Collection (PO Box 2345, South Burlington, VT 05407–2345; www.learner.org) and from the AAPT (see above for address). The set consists of 52 half-hour segments dealing with nearly all the important concepts of introductory physics. Historical information and animated graphics are used to present the concepts in an imaginative and engaging fashion. Some physics departments run appropriate segments throughout the course in special viewing rooms. Accompanying textbooks, teacher manuals, and study guides are also available.

Many time-tested film loops originally from Project Physics have been transferred to DVD and are available under the title *Physics Single-Concept Film Collection* from Ztek Co. (PO Box 11768, Lexington, KY 40577–1768, www.ztek.com); and from the AAPT (see above for address). The films cover a host of topics in mechanics, thermodynamics, electricity and magnetism, optics, and modern physics. Other short films that have been transferred to video are the AAPT Collections 1 and 2 and the Miller Collection.

The following is available from Films for the Humanities and Sciences (PO Box 2053, Princeton, NJ 08543–2053; www.films.com): *Physics: Introductory Concepts*; VHS and DVD. A 29-part series of experiments covering a wide range of topics and using slow motion and high-speed photography to capture details. Also of interest from the same company are *The Physics of Sports* and *The Physics of Amusement Park Rides*.

Physics Demonstrations in Mechanics (two parts), *Physics Demonstrations in Heat* (three parts), *Physics Demonstrations in Sound and Waves* (three parts), *Physics Demonstrations in Light* (two parts), and *Physics Demonstrations in Electricity and Magnetism* (three parts) are available in VHS and DVD formats from Physics Curriculum & Instruction (22585 Woodhill Drive, Lakeville, MN 55044; www.physicscurriculum.com). Each is a collection of 3 to 4 minute demonstrations that can be incorporated into lecture demonstrations.

Computer Software. Computers have made significant contributions to the teaching of physics. They are widely used in lectures to provide animated illustrations, with parameters under the control of the user; they also provide tutorials and drills that students can work through on their own. Specialized programs are listed in appropriate SUGGESTION sections of the Lecture Notes. In addition, several available software packages cover large portions of an introductory course. Some of them are:

Core Concepts in Physics; Macintosh, Windows; Thomson Brooks/Cole (10 Davis Drive, Belmont, CA 94002; www.brookscole.com). A great many animations and live videos, laboratory demonstrations, and graphics. Most are interactive. Many step-by-step solutions are given to example problems.

Interactive Physics; MSC Working knowledge; available from Physics Curriculum & Instruction (see above for address); Windows, Macintosh. Animations and graphs for a wide variety of mechanical phenomena. The user can set up "experiments" with massive objects, strings, springs, dampers, and constant forces. Parameters can easily be changed. Reviewed in The Physics Teacher, September 1991.

Interactive Physics Player Workbook; tutorial oriented work book and CD-ROM; Macintosh, Windows; Cindy Schwartz and John Ertel; Prentice-Hall (240 Frisch Ct., Paramus, NJ 07652-5240; www.phptr.com). A large number of animations and simulations. Self-check quizzes are associated with the simulations.

Physics 4.2 CD; MCH Multimedia, Inc; available from the AAPT (see above for address); Windows and Macintosh. A collection of interactive demonstrations covering topics in introductory mechanics, with some quantum mechanics.

Exploration of Physics; Physics Curriculum & Instruction (see above for address); Windows, Macintosh; two-volume set. A comprehensive collection of highly interactive simulations. Useful for demonstrations and for student activities.

Physics of Sports; Physics Curriculum & Instruction (see above for address); Windows, Macintosh. Simulations of activities from basketball, baseball, gymnastics, diving, biking, skiing, car racing, weight lifting, high jumping, and hammer throwing, with graphical analysis. Can be used for demonstrations and for student activities. User supplies parameters.

Amusement park Physics; Physics Curriculum & Instruction (see above for address); Windows, Macintosh. Digitized video clips of amusement park rides, suitable for any analysis tool that can be used with AVI Video for Windows files.

Physlets. Physlets are small Java applets designed by Wolfgang Christian and others at Davidson College. They can be incorporated into interactive demonstrations, interactive problems for homework and exams, or student activities. The book *Physlet Physics: Interactive Illustrations, Explorations and Problems for Introductory Physics* by Wolfgang Christian and Mario Belloni and published by Prentice-Hall shows you how to use them. No programming experience is necessary. Physlets can be downloaded from the website webphysics.davidson.edu/Applets/Applets.html.

You might consider setting aside a room or portion of a lab, equip it with several computers, and make tutorial, drill, and simulation programs available to students. If you have sufficient hardware (and software), you might base some assignments on computer materials.

Computers and top-of-the line graphing calculators might also be used by students to perform calculations. Properly selected computer projects can add greatly to the students' understanding of physics. Projects involving the investigation of some physical system of interest might be assigned to individuals or might be carried out by a laboratory class. The PTRA workshop manual *Role of Graphing Calculators in Teaching Physics* by Cheri Bibo Lehman, Linda J. Armstrong, and John E. Gastineau is available from the AAPT (see above for address). A large number of suitable problems and projects can also be found in the book *Introduction to Computational Physics* by Marvin L. De Jong (Addison-Wesley, 1991).

Commercial spreadsheet programs can facilitate problem solving. PSI-Plot (Windows; Poly Software International, P.O. Box 60, Pearl River, NY 10965, www.polysoftware.com) and $f(g)$ Scholar (Macintosh, Windows; Future Graph, Inc., Suite 200, 538 Street Road75 James Way, Southampton, PA 18966, www.graduating engineer.com) are high-end spreadsheet programs that incorporate many science and engineering problem-solving and graphing capabilities. Commercial problem-solving programs such as *MathCAD* (Windows; MathSoft, Inc., 101 Main Street, Cambridge, MA 02142-1521; www.mathsoft.com), *DERIVE* (Windows; Texas Instruments; www.education.ti.com), *MAPLE* (Macintosh, Windows; Waterloo Maple, 615 Kumpf Drive, Waterloo, Ontario, Canada N2V 1K8; www.maplesoft.com), and *Mathematica* (Macintosh, Windows; Wolfram Research, Inc., 100 Trade Center Drive, Champaign, IL 61820–7237; www.wolfram.com) can easily be used by students to solve problems and graph results. All these programs allow students to set up a problem generically, then view solutions for various values of input parameters. For example, the range or maximum height of a projectile can be found as a function of initial speed or firing angle, even if air resistance is taken into account.

A number of computer programs allow you to view digitized video on a computer monitor and mark the position of an object in each frame. The coordinates of the object can be listed and plotted. They can then be used to find the velocity and acceleration of the object, either within the program itself or by exporting the data to a spreadsheet. Three of these are: *Videopoint* (Windows, CD-ROM; Pasco Scientific, 10101 Foothills Blvd., Roseville, CA 95747–7100; www.pasco.com), *VideoGraph* (Macintosh; Physics Academic Software, Centennial Campus, 940 Main Campus Drive,

Raleigh, NC 27606–5212; www.aip.org/pas), and *World-in-Motion* (Windows; Physics Curriculum and Instruction; see above for address). All of these come with an assortment of video clips. Home-made videos can also be used. The capabilities of the programs are different. Check carefully before purchasing.

Alberi's Window (304 Pleasant Street, Watertown, MA 02472; www. albertiswindow.com) makes *Motion Visualizer 2D* and *Motion Visualizer 3D*, which analyze input from video cameras to produce computer graphics displaying trajectories, velocity graphs, and acceleration graphs. Two objects can be followed, making the system amenable to collision studies. The systems are also available from Pasco Scientific (see above for address).

Demonstrations. Notes for most of the chapters are developed around demonstration experiments. Generally speaking, these use relatively inexpensive, readily available equipment, yet clearly demonstrate the main ideas of the chapter. The choice of demonstrations, however, is highly personal and you may wish to substitute others for those suggested here or you may wish to present the same ideas using chalkboard diagrams. Several excellent books give many other examples of demonstration experiments. The following are available from the AAPT (see above for address):

A Demonstration Handbook for Physics, G.D. Freier and F.J. Anderson, 320 pages (1981). Contains over 800 demonstrations, including many that use everyday materials and that can be constructed with minimal expense. Line drawings are used to illustrate the demonstrations.

String and Sticky Tape Experiments, Ronald Edge, 448 pages (1987). Contains a large number of illuminating experiments that can be constructed from inexpensive, readily available materials.

Apparatus for Teaching Physics, edited by Karl C. Mamola. A collection of articles from The Physics Teacher that describe laboratory and demonstration apparatus.

How Things Work, H. Richard Crane, 114 pages, 1992. A collection of 20 articles from The Physics Teacher.

Turning the World Inside Out and 174 Other Simple Physics Demonstrations, Robert Ehrlich, 216 pages. A collection of demonstration experiments using common, inexpensive materials.

Apparatus for Teaching Physics, edited by Karl C. Mamola, 247 pages. A collection of articles from *The Physics Teacher* dealing with laboratory and demonstration apparatus.

Interactive Physics Demonstrations, edited by Joe Pizza. Describes 46 interactive demonstrations suitable for hallway exhibits. From *The Physics Teacher*.

The following is currently out of print but is available in many college libraries and physics departments:

Physics Demonstration Experiments, H.F. Meiners, ed. An excellent source of ideas, information, and construction details on a large number of experiments, with over 2000 line drawings and photographs. It also contains some excellent articles on the philosophical aspects of lecture demonstrations, the use of shadow projectors, TV, films, overhead projectors, and stroboscopes.

Appropriate demonstrations described in Freier and Anderson are listed in the SUGGESTIONS sections of the notes. This book does not give any construction details, but more information about most demonstrations can be obtained from the book edited by Meiners.

The *Physics InfoMall* CD-ROM (The Learning Team, 84 Business Park Drive, Suite 307, Armonk, NY 10504; www.phys.ksu.edu/perg/infomall), a searchable database of over 1000 demonstrations, is another excellent source. There are both Windows and Macintosh versions. The CD also contains articles and abstracts, problems with solutions, whole reference books, and a physics calendar.

Monographs The following books, all available from the AAPT (see above for address), are also sources of ideas for demonstrations and examples:

Physics of Sports; edited by C. Frohlich. Contains reprints and a resource letter.

Amusement Park Physics; edited by Carole Escobar. In workbook form. The activities described are perhaps more appropriate for a high school class but some can be used in college level lectures as examples.

Potpourri of Physics Teaching Ideas; edited by Donna Berry; reprints of articles on apparatus from The Physics Teacher.

The Role of Toys in Teaching Physics; Jodi and Roy McCullough; AAPT (see above for address); 292 pages. A PTRA workshop manual.

Flying Circus of Physics; Jearl Walker; John Wiley and Sons. A collection of problems and questions about every day phenomena.

A computer can also be used for data acquisition during demonstrations. Photogate timers, temperature probes, strain gauges, voltage probes, and other devices can be input directly into the computer and results can be displayed as tables or graphs. The screen can be shown to a large class by using a large monitor, a TV projection system, or an overhead projector adapter. Inexpensive software and hardware can be purchased from Vernier Software & Technology (13979 SW Millikan Way, Beaverton, OR 97005–2886; www. vernier.com). Pasco Scientific (see above for address) has data acquisition software and an extensive variety of probes for both Macintosh and Windows computers. If more sophisticated software is desired, consider the commercial package Labview (National Instruments Corporation, 11500 N. Mopac Expwy., Austin, TX 78759–3504; www.rii.com). The monograph *Photodetectors* by Jon W. McWane, J. Edward Neighbor, and Robert F. Tinker; available from the AAPT (see above for address) is a good source of technical information about photodetectors.

Laboratories. Hands-on experience with actual equipment is an extremely important element of an introductory physics course. There are many different views as to the objectives of the physics laboratory and the final decision on the types of experiments to be used has to be made by the individual instructor or department. This decision is usually based on financial and personnel considerations as well as on the pedagogical objectives of the laboratory.

Existing laboratories vary widely. Some use strictly cookbook type experiments while others allow the students to experiment freely, with practically no instructions. The equipment ranges from very simple apparatus to rather complex and sophisticated equipment. Physical phenomena may be observed directly or simulated on a computer. Data may be taken by the students or fed into a computer. The PTRA workshop manual *Role of the Laboratory in Teaching Introductory Physics* by Jim and Jane Nelson is available from the AAPT (see above for address).

The equipment described above can be used for data acquisition in a student lab. Even if data acquisition software is not used, consider having students use computers and spreadsheet programs to analyze and graph data.

Many physics departments have written their own notes or laboratory manuals and relatively few physics laboratory texts are on the market. Three such books, both available from John Wiley & Sons, are

Fundamentals of Physics Probeware Lab Manual, developed in conjunction with Pasco Scientific.

Laboratory Physics, second edition, H.F. Meiners, W. Eppenstein, R.A. Oliva, and T. Shannon. (1987).

Laboratory Experiments in College Physics, seventh edition, C.H. Bernard and C.D. Epp. (1994).

Experiments from these books are listed in the SUGGESTIONS sections of the Lecture Notes. Meiners is used to designate the Meiners, Eppenstein, Oliva, and Shannon book, Bernard is used to designate the Bernard and Epp book, and Probeware is used to designate the Pasco book. The books contain excellent experiments and activities for students. Meiners and Bernard have sections that explain laboratory procedures to students. Meiners also contains a large amount of material on the use of microprocessors in the lab.

Student supplements. Several supplements, all available from Wiley, might be recommended to the students:

A Student's Companion to Fundamentals of Physics. A study guide. The basic concepts of each chapter are reviewed in a format that helps students focus their attention on the important ideas and their relationships to each other. Hints are given for all the odd numbered end-of-chapter questions and about one-third of the odd numbered end-of chapter problems. There is also a quiz (with answers) for each chapter so students can test their understanding. A list of the problem hints in the study guide are given in Section Seven of this instructor manual.

Student Solution Manual. Contains fully worked solutions to about one-third of the end-of-chapter problems. These problems are different from those for which hints are given in the study guide. A list of the solutions in the solution manual are given in Section Seven of this Manual.

CD Physics. A CD ROM version of the text and supplements for Windows and Macintosh. This contains the complete text, the *Student Solution Manual, A Student's Companion*, interactive tutorials, interactive simulations, and a glossary. It is extensively hyperlinked.

Wiley Website. Wiley maintains a website devoted to materials for students using this text. It contains samples of worked-out solutions from the *Student Solution manual* and hints from the study guide. In addition there are self-quizzes and additional problems using graphical simulations. The site also contains links to other websites. The solutions and hints on the site are given in Section Seven of this manual.

Instructor aids. In addition to this *Instructor Manual* Wiley provides several other aids for instructors:

Instructor's Solution Manual Contains fully worked solutions to all the end-of-chapter problems.

Test Bank. Contains over 2800 multiple choice questions (with answers) for use on exams and quizzes. Both quantitative and qualitative questions are included. In each chapter, some of the questions are modeled after the checkpoints and end-of-chapter questions, as well as after the end-of-chapter problems and exercises.

A set of transparencies for overhead projectors.

eGrade Plus, WebAssign, and CAPA for homework submission and management.

Instructor's Resource CD. A CD ROM for Windows and Macintosh. It contains the *Instructor's Solution Manual* (in both Word and PDF form), reproductions of illustrations from the text (in JPEG form), and the *Test Bank*. There is a computer program that allows instructors to generate exams from the test bank questions.

SECTION THREE
LECTURE NOTES

Lecture notes for each chapter of the text are grouped under the headings BASIC TOPICS and SUGGESTIONS.

BASIC TOPICS contains the main points of the chapter in outline form. In addition, one or two demonstrations are recommended to show the main theme of the chapter. You may wish to pattern your lectures after the notes, suitably modified, or simply use them as a check on the completeness of your own notes.

The SUGGESTIONS sections recommend end-of-chapter questions and problems, video cassettes, DVDs, computer software, computer projects, alternate demonstrations, and other material that might be useful for the course. Many of the questions concentrate on points that seem to give students trouble, and it is worthwhile dealing with some of them before students tackle a problem assignment. Some questions and problems might be incorporated into the lectures while some might be assigned and used to generate discussion by students in small recitation sections. Answers to the questions appear in Section Five of this manual and answers to the problems appear in Section Six.

Chapter 1 MEASUREMENT

BASIC TOPICS

I. Base and derived units.
   A. Explain that standards are associated with base units and that measurement of a physical quantity takes place by means of comparison with a standard. Discuss qualitatively the SI standards for time, length, and mass. Show a 1 kg mass and a meter stick. Show the simple well-known procedure for measuring length with a meter stick. Many schools have atomic clocks. If yours does, here is a good place to demonstrate it.
   B. Explain that derived units are combinations of base units. Emphasize that the speed of light is now a defined unit and the meter is a derived unit. Discuss an experiment in which the time taken for light to travel a certain distance is measured. Example: the reflection of a light signal from the Moon. Use a clock and a meter stick to find your walking speed in m/s.
   C. This is a good place to review area, volume, and mass density. Use simple geometric figures (circle, rectangle, triangle, cube, sphere, cylinder, etc.) as examples.

II. Systems of units.
   A. Explain what a system of units is. Give the 1971 SI base units (Table 1–1). Stress that they will be used extensively in the course.
   B. Point out the SI prefixes (Table 1–2). The important ones for this course are mega, kilo, centi, milli, micro, nano, and pico. Discuss powers of ten arithmetic and stress the simplicity of the notation. This might be a good place to say something about significant digits.
   C. Discuss unit arithmetic and unit conversion.
   D. Most of the students' experience is with the British system. Relate the inch to the centimeter, the yard to the meter, and the slug to the kilogram. Discuss unit conversion. Use speed as an example: convert 50 mph and 3 mph to km/h and m/s. Point out the conversion tables in Appendix D.

III. Properties of standards.
  A. Discuss accessibility and invariability as desirable properties of standards.
  B. Discuss secondary standards such as the meter stick used earlier.

IV. Measurements.
  A. Stress the wide range of magnitudes measured. See Tables 1–3, 1–4, and 1–5. Explain the atomic mass unit. One atom of ^{12}C has a mass of exactly 12 u. 1 u is approximately 1.661×10^{-27} kg.
  B. Discuss indirect measurements.

SUGGESTIONS

1. Assignments
   a. To emphasize SI prefixes assign problems 3 and 10.
   b. Unit conversion is covered in many problems. Choose some, such as 2, 4, and 6 that deal with unfamiliar units. Also consider problem 9.
   c. According to the needs of the class, assign one or more problems that deal with area and volume calculations, such as 5 and 7.
   d. Assign a problem or two that deal with mass density, such as 19, 20, 21, or 23.
2. Demonstrations
   Examples of "standards" and measuring instruments: Freier and Anderson Ma1 — 3.
3. Books and Monographs
   a. *Frequency and Time Measurements*, edited by Christine Hackman and Donald B. Sullivan; available from the American Association of Physics Teachers (AAPT, One Physics Ellipse, College Park, MD 20740-3845, www.aapt.org).
   b. *SI: The International System of Units*; edited by Robert A. Nelson; available from the AAPT (see above for address).
   c. *Connecting Time and Space*; edited by Harry E. Bates; available from the AAPT (see above for address). Reprints that discuss measurements of the speed of light and the redefinition of the meter. Students will not be able to understand much of this material at this stage of the course but it is nevertheless useful for background.
   d. *Powers of Ten : A Flipbook*; by Philip Morrison and Phylis Morrison, and the Office of Charles and Ray Eames; published by W.H. Freeman and Company; available from the AAPT (see above for address).
4. Audio/Visual
   a. *Time and Place, Measuring Short Distances*; Cinema Classics DVD 1: Mechanics (I); available from Ztek Co. (PO Box 11768, Lexington, KY 40577–1768, www.ztek.com) and from the AAPT (see above for address).
   b. *Powers of Ten* from the Films of Charles and Ray Eames; produced by Pyramid Media; video tape; available from the AAPT (see above for address).
5. Laboratory
   a. Meiners Experiment 7-1: *Measurement of Length, Area, and Volume.* Gives students experience using the vernier caliper, micrometer, and polar planimeter. Good introduction to the determination of error limits (random and least count) and calculation of errors in derived quantities (volume and area).
   b. Bernard Experiments 1 and 2: *Determination of Length, Mass, and Density* and *Measurements, Measurement Errors, and Graphical Analysis.* Roughly the same as the Meiners experiment, but a laboratory balance is added to the group of instruments and the polar planimeter is not included. Graphs of mass versus radius and radius squared for a collection of disks made of the same material, with the same thickness, are used to establish the quadratic dependence of mass on radius.

c. Meiners Experiment 7-3: *The Simple Pendulum* and Bernard Experiment 3: *The Period of a Pendulum — An Application of the Experimental Method.* Students time simple pendulums of different lengths, then use the data and graphs (including a logarithmic plot) to determine the relationship between length and period. They calculate the acceleration due to gravity. This is an exercise in finding functional relationships and does not require knowledge of dynamics.

Chapter 2 MOTION ALONG A STRAIGHT LINE

BASIC TOPICS

I. Position and displacement.

A. Move a toy cart with constant velocity along a table top. Select an origin, place a meter stick and clock on the table, and demonstrate how x(t) is measured in principle. Emphasize that x is always measured from the origin; it is not the cart's displacement during any time interval.

B. Draw a graph of $x(t)$ and point out that it is a straight line. Show what the graph looks like if the cart is not moving. Point out that the line has a greater slope if the cart is going faster. Move the cart so its speed increases with time and show what the curve $x(t)$ looks like. Do with same for a cart that is slowing down.

C. Some students think of a coordinate as distance. Distinguish between these concepts. Point out that a coordinate defines a position on an axis and can be positive or negative. Demonstrate a negative velocity, both with the cart and on a graph. As another example, throw a ball into the air, pick a coordinate axis (positive in the upward direction, say), and point out when the velocity is positive and when it is negative. Draw the graph of the coordinate as a function of time. Repeat with the positive direction upward.

D. Define the displacement of an object during a time interval. Emphasize that only the initial and final coordinates enter and that an object may have many different motions between these while still having the same displacement. Point out that the displacement is zero if the initial and final coordinates are the same.

II. Velocity.

A. Define average velocity over an interval. Stress the meaning of the sign. Go over Sample Problem 2–1. Draw a graph of x versus t for an object that is accelerating. Pick an interval and draw the line between the end points on the graph. Observe that the average velocity in the interval is the slope of the line. Figs. 2–3 and 2–4 may also be used. Show how to calculate average velocity if the function $x(t)$ is given in algebraic form.

B. Define instantaneous velocity. Demonstrate the limiting process. Use a graph of x versus t for an accelerating cart to demonstrate that the line used to find the average velocity becomes tangent to the curve in the limit as Δt vanishes. Remark that the slope of the tangent line gives the instantaneous velocity. Show a plot of v versus t that corresponds to the x versus t graph used previously. Show how to calculate the instantaneous velocity if the function $x(t)$ is given in algebraic form. See Sample Problem 2–3. Stress that a value of the instantaneous velocity is associated with each instant of time. Some students think of velocity as being associated with a time interval rather than an instant of time.

C. Define instantaneous speed as the magnitude of the velocity. Compare to the average speed in an interval, which is the total path length divided by the time. Remark that the average speed is not the same as the magnitude of the average velocity if the direction of motion changes in the interval.

D. Note that many calculus texts use a prime to denote a derivative. They also define the derivative of x with respect to time by the limit of $[x(t + \Delta t) - x(t)]/\Delta t$ rather than by the limit of $\Delta x/\Delta t$. Mention the different notations in class so students can relate their physics and calculus texts.

III. Acceleration.

A. Define average and instantaneous acceleration. Show the previous v versus t graph and point out the lines used to find the average acceleration in an interval and the instantaneous acceleration at a given time. Show how to calculate the average and instantaneous acceleration if either $x(t)$ or $v(t)$ is given in algebraic form. See Sample problem 2–4.

B. Interpret the sign of the acceleration. Give examples of objects with acceleration in the same direction as the velocity (speeding up) and in the opposite direction (slowing down). Be sure to include both directions of velocity. Emphasize that a positive acceleration does not necessarily imply speeding up and a negative acceleration does not necessarily imply slowing down.

C. Use graphs of $x(t)$ and $v(t)$ to point out that an object may simultaneously have zero velocity and non-zero acceleration. Explain that if the direction of motion reverses the object must have zero velocity at some instant. Give the position as a function of time as $x(t) = At^2$, for example, and show that the velocity is 0 at $t = 0$ but the acceleration is not 0. Illustrate the function with a graph.

IV. Motion in one dimension with constant acceleration.

A. Derive the kinematic equations for $x(t)$ and $v(t)$. If students know about integration, use methods of the integral calculus (as in Section 2–8). If you use the integral calculus you might cover the graphical interpretation of an integral. See Section 2–10. In any event, show that $v(t)$ is the derivative of $x(t)$ and that a is the derivative of $v(t)$.

B. Discuss kinematics problems in terms of a set of simultaneous equations to be solved. Examples: use equations for $x(t)$ and $v(t)$ to algebraically eliminate the time and to algebraically eliminate the acceleration. The equations of constant acceleration motion are listed in Table 2–1. Some instructors teach students to use the table. Others ask students to always start with Eqs. 2–10 and 2–15, then use algebra to obtain the equations needed for a particular problem. See Sample Problem 2–5.

C. To help students see the influence of the initial conditions, sketch graphs of $v(t)$ and $x(t)$ for various initial conditions but the same acceleration. Include both positive and negative initial velocities. Draw a different set of graphs for positive and negative acceleration. Point out where the particle has zero velocity and when it returns to its initial position.

V. Free fall.

A. Give the value for g in SI units. Point out that the free-fall acceleration is essentially due to gravity and that it is directed toward the center of Earth. Say that locally Earth's surface is essentially flat and the free-fall acceleration may be taken to be in the same direction at slightly different points. Explain that $a = +g$ if down is taken to be the positive direction and $a = -g$ if up is the positive direction. Do examples using both choices. Throw a ball into the air and emphasize that its acceleration is g throughout its motion, even at the top of its trajectory.

B. Drop a small ball through two photogates, one near the top to turn on a timer and one further down to turn it off. Repeat for various distances and plot the position of the ball as a function of time. Explain that the curve is parabolic and indicates a constant acceleration.

C. Explain that all objects at the same place have the same free-fall acceleration. In reality, different objects may have different accelerations because air influences their motions differently. This can be demonstrated by placing a coin and a wad of cotton in a glass cylinder about 1 m long. Turn the cylinder over and note that the coin reaches the bottom first.

Now use a vacuum pump to partially evacuate the cylinder and repeat the experiment. Repeat again with as much air as possible pumped out.

D. Point out that free-fall problems are special cases of constant acceleration kinematics and the methods described earlier can be used. Work a few examples. For an object thrown into the air, calculate the time to reach the highest point, the height of the highest point, the time to return to the initial height, and its velocity when it returns, all in terms of the initial velocity.

SUGGESTIONS

1. Assignments
   a. To help students obtain some qualitative understanding of velocity and acceleration, ask them to discuss questions 1, 2, 3, and 6. Some aspects of motion with constant acceleration are covered in questions 4 and 7. Free fall is covered in questions 5 and 8.
   b. To make more use of the calculus assign some of problems 5, 12, and 13. Problem 13 can also be used to discuss differences between average and instantaneous velocity.
   c. To emphasize the interpretation of graphs assign a few of problems 5, 6, 13, and 18. Some of these require students to draw graphs after performing calculations.
   d. Ask students to solve a few problems dealing with motion with constant acceleration. Consider problems 21,, 22, 24, 27, 30, 33, and 35. For a little more challenge, consider problem 32.
   e. Problems 38, 42, 47, and 49 are good problems to test understanding of free-fall motion. Problem 53 is more challenging.
2. Demonstrations
   Uniform velocity and acceleration, velocity as a limiting process: Freier and Anderson Mb10 — 13, 15, 18, 21, 22.
3. Audio/Visual
   a. *Acceleration due to Gravity*; from AAPT collection 1 of single-concept films; DVD; available from Ztek Co. (PO Box 11768, Lexington, KY 40577–1768, www.ztek.com) and from the American Association of Physics Teachers (One Physics Ellipse, College Park, MD 20740-3845, www.aapt.org).
   b. *One Dimensional Motion*; *Distance, Time & Speed*; *One Dimensional Acceleration*; *Constant Velocity & Uniform Acceleration*; from the AAPT collection 2 of single-concept films; DVD; available from Ztek Co. and from the AAPT (see above for addresses).
   c. *Uniform Motion*, *Free Fall*; Cinema Classics DVD 1: Mechanics (I); available from Ztek Co. and from the AAPT (see above for addresses).
   d. *Numbers, Units, Scalars, and Vectors*; VHS video tape, DVD; Films for the Humanities and Sciences (PO Box 2053, Princeton, NJ 08543–2053; www.films.com).
4. Computer Software
   a. *Mechanics* from Exploration of Physics Volume I; Windows and Macintosh; Physics Curriculum & Instruction (22585 Woodhill Drive, Lakeville, MN 55044; www.PhysicsCurriculum.com). Simulations of physical phenomena along with graphs. Includes sections on position, velocity, acceleration, and free fall.
   b. *Forces and Motion* from Exploration of Physics Volume II; Windows and Macintosh; Physics Curriculum & Instruction (see above for address). Includes sections on velocity and acceleration graphs and on free fall, with and without air resistance.
   c. *Graphs and Tracks*; David Trowbridge; DOS, Macintosh; available from Physics Academic Software (Centennial Campus, 940 Main Campus Drive, Suite 210, Raleigh, NC 27606–5212; www.aip.org/pas). A ball rolls on a series of connected inclines. In one part the student is given graphs of the position, velocity, and acceleration and is asked to adjust the

tracks to produce the graphed motion. In a second part the student is shown the motion and asked to sketch the graphs. Complements lab experiments with a sonic ranger.

d. *Newtonian Sandbox*; Judah Schwartz; DOS, Macintosh; available from Physics Academic Software (see above for address). Generates the motion of a point particle in one and two dimensions. Plots trajectories, coordinates, velocity components, radial and angular positions, and phase space trajectories.

e. *Objects in Motion*; Peter Cramer; DOS, Macintosh; available from Physics Academic Software (see above for address). Simulates the motion of an object under various conditions and plots graphs of the position, velocity and accelerations. Situations considered are: uniform acceleration along a straight line, projectile motion, relative motion, circular motion, planetary motion, and elastic collisions.

f. *Physics Demonstrations*; Julien C. Sprott; DOS; available from Physics Academic Software (see above for address). Ten simulations of motion and sound demonstrations. Includes "the monkey and the coconut", "ballistics cat", "flame pipe", "Doppler effect".

g. *Conceptual Kinematics*; Frank Griffin and Louis Turner; DOS, Macintosh; available from Physics Academic Software (see above for address). An interactive, animated tutorial, with quiz questions for self-testing.

h. *Dynamic Analyzer*; Roger F. Sipson; DOS; available from Physics Academic Software (see above for address).

5. Computer Projects

a. Use a spreadsheet or your own computer program to demonstrate the limiting processes used to define velocity and acceleration. Given the functional form of $x(t)$, have the computer calculate and display the coordinate for some time t and a succession of later times, closer and closer to t. For each interval, have it calculate and display the average velocity. Be careful to refrain from displaying non-significant figures and be sure to stop the process before all significance is lost.

b. Have students use the root finding capability of a commercial math program or their own computer programs to solve kinematic problems for which $x(t)$ and $v(t)$ are given functions. Nearly all of them can be set up as problems that involve finding the root of either the coordinate or velocity as a function of time, followed perhaps by substitution of the root into another kinematic equation. Problems need not be limited to those involving constant acceleration. Air resistance, for example, can be taken into account. The same program can be used to solve rotational kinematic problems in Chapter 11.

6. Laboratory

a. Probeware Activity 1: *Motion in One Dimension.*

b. Motion detectors. Students use a motion detector to relate their own positions as functions of time to computer generated graphs.

c. Probeware Activity 2: *Position, Velocity, and Acceleration.* A motion detector is now used to explore one dimensional accelerated motion.

d. Several sonic rangers are reviewed in The Physics Teacher of January 1988. An extremely popular model is available from Vernier Software, 8565 SW Beaverton-Hillsdale Hwy., Portland, OR 97225-2429.

e. Meiners Experiment 7–5: *Analysis of Rectilinear Motion.* Students measure the position as a function of time for various objects rolling down an incline, then use the data to plot speeds and accelerations as functions of time. No knowledge of rotational motion is required. This experiment emphasizes the definitions of velocity and acceleration as differences over a time interval.

f. Meiners Experiment 8–1: *Motion in One Dimension* (omit the part dealing with conservation of energy). Essentially the same experiment except pucks sliding on a nearly frictionless surface are used. This experiment may be done with dry ice pucks or on an air table or air track.

g. Bernard Experiment 7: *Uniformly Accelerated Motion.* The same technique as the Meiners experiments but a variety of setups are described: the standard free fall apparatus, the free fall apparatus with an Atwood attachment, an inclined plane, an inclined air track, and a horizontal air track with a pulley attachment.

Chapter 3 VECTORS

BASIC TOPICS

I. Definition.
   A. Explain that vectors have magnitude and direction, and that they obey certain rules of addition.
   B. Example of a vector: displacement. Give the definition of displacement and point out that a displacement does not describe the path of the object. Give the definition and physical interpretation of the sum of two displacements. Demonstrate vector addition by walking along two sides of the room. Point out the two displacements and their sum. Note that the distance traveled is not the magnitude of the displacement. Go back to your original position and point out that the displacement is now zero.
   C. Compare vectors with scalars and present a list of each.
   D. Go over vector notation and insist that students use it to identify vectors clearly. In this text a vector is indicated by placing an arrow over an algebraic symbol. The italic version of the symbol, without the arrow, indicates the magnitude of the vector. Point out that many other texts use boldface type to indicate vectors.

II. Vector addition and subtraction by the graphical method.
   A. Draw two vectors tail to head, draw the resultant, and point out its direction. Explain how the magnitude of the resultant can be measured with a ruler and the orientation can be measured with a protractor. Explain how a scale is used to draw the original vectors and find the magnitude of the resultant.
   B. Define the negative of a vector and define vector subtraction as $\vec{a}-\vec{b}=\vec{a}+(-\vec{b})$. Graphically show that if $\vec{a}+\vec{b}=\vec{c}$ then $\vec{a}=\vec{c}-\vec{b}$.
   C. Show that vector addition is both commutative and associative.

III. Vector addition and subtraction by the analytic method.
   A. Derive expressions for the components of a vector, given its magnitude and the angles it makes with the coordinate axes. In preparation for the analysis of forces, find the x component of a vector in the xy plane in terms of the angles it makes with the positive and negative x axis and also in terms of the angles it makes with the positive and negative y axis.
   B. Point out that the components depend on the choice of coordinate system, and compare the behavior of vector components with the behavior of a scalar when the orientation of the coordinate system is changed. Find the components of a vector using two differently oriented coordinate systems. Point out that it is possible to orient the coordinate system so that only one component of a given vector is not zero. Remark that a pure translation of a vector (or coordinate system) does not change the components.
   C. Define the unit vectors along the coordinate axes. Give the form used to write a vector in terms of its components and the unit vectors. Explain that unit vectors are unitless so they can be used to write any vector quantity.

D. Vector addition. Give the expressions for the components of the resultant in terms of the components of the addends. Demonstrate the equivalence of the graphical and analytic methods of finding a vector sum. See the diagram to the right.

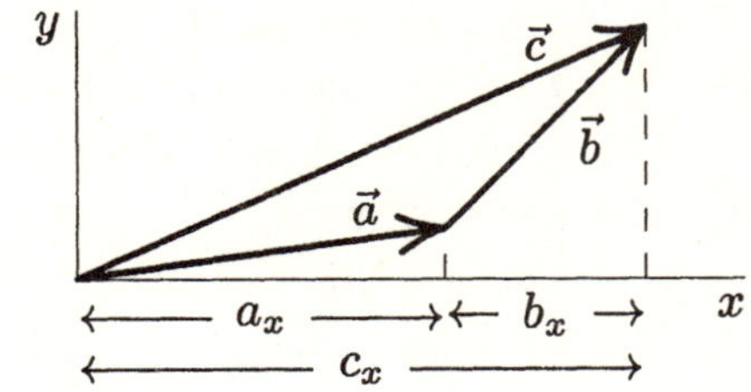

E. Give the expression for vector subtraction in terms of components. You may also wish to demonstrate the equivalence of the graphical and analytical methods of vector subtraction.

F. Show how to find the magnitude and angles with the coordinate axes, given the components. Explain that calculators give only one of the two possible values for the inverse tangent and show how to determine the correct angle for a given situation.

G. State that two vectors are equal only if their corresponding components are equal. State that many physical laws are written in terms of vectors and that many take the form of an equality between two vectors. Expressions for the laws are then independent of any coordinate system.

IV. Multiplication involving vectors.

A. Multiplication by a scalar. Give examples of both positive and negative scalars multiplying a vector. Give the components of the resulting vector as well as its magnitude and direction. Remark that division of a vector by a scalar is equivalent to multiplication by the reciprocal of the scalar.

B. Scalar product of two vectors (may be postponed until Chapter 7). Emphasize that the product is a scalar. Give the expression for the product in terms of the magnitudes of the vectors and the angle between them. To determine the angle, the vectors must be drawn with their tails at the same point. Point out that $\vec{a} \cdot \vec{b}$ is the magnitude of $\vec{a}$ multiplied by the component of $\vec{b}$ along an axis in the direction of $\vec{a}$. Explain that $\vec{a} \cdot \vec{b} = 0$ if $\vec{a}$ is perpendicular to $\vec{b}$.

C. Either derive or state the expression for a scalar product in terms of Cartesian components. See the discussion leading to Eq. 3–23. Specialize the expression to show that $\vec{a} \cdot \vec{a} = a^2$. Show how to use the scalar product to calculate the angle between two vectors if their components are known. Consider problem 31.

D. Vector product of two vectors (may be postponed until Chapter 12). Emphasize that the product is a vector. Give the expression for the magnitude of the product and the right hand rule for determining the direction. Explain that $\vec{a} \times \vec{b} = 0$ if $\vec{a}$ and $\vec{b}$ are parallel. Point out that $|\vec{a} \times \vec{b}|$ is the magnitude of $\vec{a}$ multiplied by the component of $\vec{b}$ along an axis perpendicular to $\vec{a}$ and in the plane of $\vec{a}$ and $\vec{b}$. Show that $\vec{b} \times \vec{a} = -\vec{a} \times \vec{b}$.

E. Either derive or state the expression for a vector product in terms of Cartesian components. See the discussion leading to Eq. 3–30. Give students the useful mnemonic for the vector products of the unit vectors $\hat{\mathrm{i}}$, $\hat{\mathrm{j}}$, and $\hat{\mathrm{k}}$, written in that order clockwise around a circle. One starts with the first named vector in the vector product and goes around the circle toward the second named vector. If the direction of travel is clockwise the result, is the third vector. If it is counterclockwise, the result is the negative of the third vector.

SUGGESTIONS

1. Assignments

a. Use questions 2, 3, and 4 to discuss properties of vectors. Questions 1 and 5 deal with vector addition and subtraction. Question 6 deals with the signs of components.

b. Ask students to use graphical representations of vectors to think about problems such as 8 and 10.

c. Problems 3, 4, and 8 cover the fundamentals of vector components. Problems 5 and 6 stress the physical meaning of vector components. Some good problems to test understanding of analytic vector addition and subtraction are 13, 18, 19, and 23.
d. Unit vectors are used in problems 14, 15, 16, and 20.

2. Demonstrations
   Vector addition: Freier and Anderson Mb2, 3.
3. Audio/Visual
   a. *Numbers, Units, Scalars, and Vectors*; VHS video tape, DVD; Films for the Humanities and Sciences (PO Box 2053, Princeton, NJ 08543–2053; www.films.com).
   b. *Vector Addition — Velocity of a Boat*; from AAPT collection 1 of single-concept films; DVD; available from Ztek Co. (PO Box 11768, Lexington, KY 40577–1768, www.ztek.com) and from the American Association of Physics Teachers (AAPT, One Physics Ellipse, College Park, MD 20740-3845, www.aapt.org).
   c. *Vectors*; Cinema Classics DVD 1: Mechanics (I); available from Ztek Co. and from the AAPT (see above for addresses).
   d. *Vector Addition*; Physics Demonstrations in Mechanics, Part III; VHS video tape, DVD; ≈3 min; Physics Curriculum & Instruction (22585 Woodhill Drive, Lakeville, MN 55044; www.PhysicsCurriculum.com)
4. Computer Software
   a. *Vectors*; Richard R. Silbar; Windows and Macintosh; available from Physics Academic Software (Centennial Campus, 940 Main Campus Drive, Suite 210, Raleigh, NC 27606–5212; www.aip.org/pas).
   b. *Vectors*; Windows and Macintosh; WhistleSoft, Inc.; available from Physics Academic Software (see above for address).
5. Computer Project
   Have students use a commercial math program or write their own computer programs to carry out conversions between polar and Cartesian forms of vectors, vector addition, scalar and vector products.
6. Laboratory
   Bernard Experiment 4: *Composition and Resolution of Coplanar Concurrent Forces.* Students mathematically determine a force that balances 2 or 3 given forces, then check the calculation using a commercial force table. They need not know the definition of a force, only that the forces in the experiment are vectors along the strings used, with magnitudes proportional to the weights hung on the strings. The focus is on resolving vectors into components and finding the magnitude and direction of a vector, given its components.

Chapter 4 MOTION IN TWO AND THREE DIMENSIONS

BASIC TOPICS

I. Definitions.
   A. Draw a curved particle path. Show the position vector for several times and the displacement vector for several intervals. Define average velocity over an interval. Write the definition in both vector and component form.
   B. Define velocity as $d\vec{r}/dt$. Write the definition in both vector and component form. Point out that the velocity vector is tangent to the path. Define speed of the magnitude of the velocity.
   C. Define acceleration as $d\vec{v}/dt$. Write the definition in both vector and component form. Point out that $\vec{a}$ is not zero if either the magnitude or direction of $\vec{v}$ changes with time.

D. Show that the particle is speeding up only if $\vec{a} \cdot \vec{v}$ is positive. If $\vec{a} \cdot \vec{v}$ is negative, the particle is slowing down, and if $\vec{a} \cdot \vec{v} = 0$, its speed is not changing.
E. Remark that sometimes the magnitude and direction of the acceleration are given, rather than its components. Remind students how to find the components if such is the case.
F. Go over Sample Problem 4–4 or a similar problem of your own devising. It shows how to find and use the components of the acceleration.

II. Projectile motion.
A. Demonstrate projectile motion by using a spring gun to fire a ball onto a surface at the firing height. Use various firing angles, including 45°, and point out that the maximum range occurs for a firing angle of 45°. Remark on the symmetry of the range as a function of firing angle. Mention that the maximum range occurs for a different angle when the ball is fired onto a surface at a different height and when drag is significant.
B. Draw the trajectory of a projectile, show the direction of the initial velocity, and derive its components in terms of the initial speed and firing angle.
C. Write down the kinematic equations for $x(t)$, $y(t)$, $v_x(t)$, and $v_y(t)$. At first, include both a_x and a_y but then specialize to $a_x = 0$ and $a_y = -g$ for positive y up. Stress that these form two sets of one dimensional equations, linked by the common variable t and are to be solved simultaneously. Note that a_x affects only v_x, not v_y or v_z. Make similar statements about the other components. Throw a ball vertically, then catch it. Repeat while walking with constant velocity across the room. Ask students to observe the motion of the ball relative to the chalkboard and to describe its motion relative to your hand.
D. Point out that the acceleration is the same at all points of the trajectory, even the highest point. Also point out that the horizontal component of the velocity is constant.
E. Work examples. Use punted footballs, hit baseballs, or thrown basketballs according to season.
   1. Find the time for the projectile to reach its highest point, then find the coordinates of the highest point.
   2. Find the time for the projectile to hit the ground, at the same level as the firing point. Then, find the horizontal range and the velocity components just before landing.
   3. Show that maximum range over level ground is achieved when the firing angle is 45°.
   4. Show how to work problems for which the landing point is not at the same level as the firing point.
F. Point out that all projectiles follow some piece of the full parabolic trajectory. For example, A to D could be the trajectory of a ball thrown at an upward angle from a roof to the street; B to D could be the trajectory of a ball thrown horizontally; C to D could be the trajectory of a ball thrown downward.

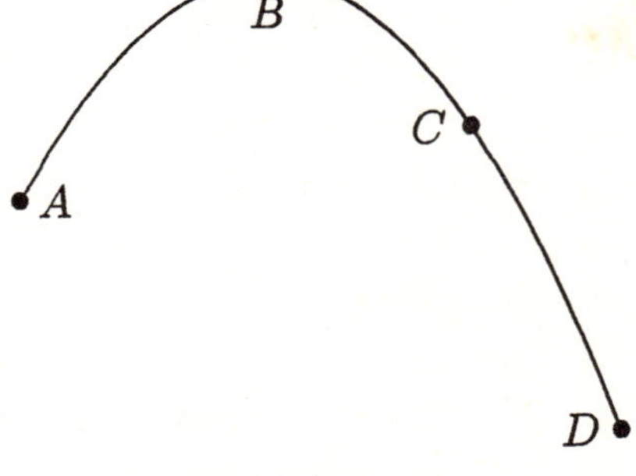

G. Explain how to find the speed and direction of travel for any time. Specialize to the time of impact on level ground and show that the speed is the same as the firing speed but that the vertical component of the velocity has changed sign. Remark that this result is true only because air resistance has been neglected.
H. Work some sample problems. Consider Sample Problems 4–6, 4–7, and 4–8 or others of your own devising.

III. Circular motion.
A. Draw the path and describe uniform circular motion, emphasizing that the speed remains constant. Remind students that the acceleration must be perpendicular to the velocity.

By drawing the velocity vector at two times, argue that the acceleration vector must be directed inward. On the diagram show the velocity and acceleration vectors for several positions of the particle.

B. Derive $a = v^2/r$. As an alternative to the derivation given in the text, write the equations for the particle coordinates as functions of time, then differentiate twice.

C. Example: calculate the speed of an Earth satellite, given the orbit radius and the acceleration to due to gravity at the orbit. Emphasize that the acceleration is toward Earth.

IV. Relative motion.

A. Material in this section is used in Chapter 5 to discuss inertial frames and in Chapter 11 to discuss rolling without slipping. It is also useful as a prelude to relativity.

B. Relate the position of a particle as given in coordinate system A to the position as given in coordinate system B by $\vec{r}_{\rm PA} = \vec{r}_{\rm PB} + \vec{r}_{\rm BA}$, where $\vec{r}_{\rm BA}$ is the position of the origin of B relative to the origin of A. Differentiate to show that $\vec{v}_{\rm PA} = \vec{v}_{\rm PB} + \vec{v}_{\rm BA}$ and $\vec{a}_{\rm PA} = \vec{a}_{\rm PB} + \vec{a}_{\rm BA}$, where $\vec{v}_{\rm BA}$ and $\vec{a}_{\rm BA}$ are the velocity and acceleration, respectively, of B relative to A.

C. Discuss examples of a ball thrown or rolled in accelerating and non-accelerating trains. The discussion may be carried out for either one- or two-dimensional motion.

D. Remark that $\vec{a}_{\rm PB} = \vec{a}_{\rm PA}$ if the two coordinate systems are not accelerating with respect to each other. This is an important point for the discussion of inertial reference frames in Chapter 5.

E. Work several problems dealing with airplanes flying in the wind and boats sailing in moving water. Emphasize that relative motion problems are chiefly exercises in vector addition. To help students understand some of the problems explain that a boat's "heading" is its direction of motion in a frame attached to the water, while its direction of travel is its direction of motion in a frame attached to the ground.

SUGGESTIONS

1. Assignments
   a. Assign some of problems 3, 5, 6, 9, 12, and 14 to have students think about the analysis of motion in two dimensions.
   b. Use questions 3 through 10 to generate discussions of ideal projectile motion.
   c. Ask questions 11, 12, and 13 in connection with centripetal acceleration.
   d. Have students work several of the projectile motion problems (17 through 43). Some of these deal with sports. See, for example, problems 18, 26, 28, 32, 34, 37, 39, and 43.
   e. Assign two or three of problems 45, 47, 49, and 51 in connection with uniform circular motion.
   f. Assign one or two problems that deal with relative motion. Good examples are 56, 57, 58, 60, and 66.
2. Demonstrations
   Projectile motion: Freier and Anderson Mb14, 16, 17, 19, 20, 23, 24, 28.
3. Audio/Visual
   a. *A Matter of Relative Motion*, *Galilean Relativity — Ball Dropped from Mast of Ship*; *Object Dropped from Aircraft*, *Projectile Fired Vertically*; from the AAPT collection 1 of single-concept films; DVD; available from Ztek Co. (PO Box 11768, Lexington, KY 40577–1768, www.ztek.com) and from the American Association of Physics Teachers (AAPT, One Physics Ellipse, College Park, MD 20740-3845, www.aapt.org).
   b. *Projectile Motion*, *Circular Motion*; Cinema Classics DVD 2: Mechanics (II) and Heat; available from Ztek Co. and the AAPT (see above for addresses).
   c. *Projectile Motion*; VHS video tape, DVD (20 min); Films for the Humanities and Sciences (PO Box 2053, Princeton, NJ 08543–2053; www.films.com).

d. *Circular and Rotational Motion*; VHS video tape, DVD (21 min); Films for the Humanities & Sciences (see above for address).
e. *Reference Frames* from Skylab Physics Videodisc; video disk; available from Ztek Co. (see above for address).
f. *Projectile Motion*; from Physics Demonstrations in Mechanics, Part I; VHS video tape, DVD; ≈3 min; Physics Curriculum & Instruction (22585 Woodhill Drive, Lakeville, MN 55044; www.PhysicsCurriculum.com).
g. *Circular Motion*; from Physics Demonstrations in Mechanics, Part I; VHS video tape, DVD; ≈3 min; Physics Curriculum & Instruction (see above for address).
h. *Velocity and Acceleration Vectors*; *Frame of Reference*; from Physics Demonstrations in Mechanics, Part III; VHS video tape, DVD; ≈3 min each; Physics Curriculum & Instruction (see above for address).
i. *Projectile Motion*; VHS video tape, DVD (part of a collection); Films for the Humanities and Sciences (PO Box 2053, Princeton, NJ 08543–2053; www.films.com).
j. *Circular and Rotational Motion*; VHS video tape, DVD (part of a collection); Films for the Humanities and Sciences (see above for address).

4. Computer Software
   a. *Mechanics* from Exploration of Physics Volume I; Windows and Macintosh; Physics Curriculum & Instruction (see above for address). Includes sections on projectile motion and centripetal force.
   b. *Forces and Motion* from Exploration of Physics Volume II; Windows and Macintosh; Physics Curriculum & Instruction (see above for address). Includes sections on velocity and acceleration graphs and on projectile motion, with and without air resistance.
   c. *Mechanics in Motion*; Stephen Saxon; Windows; available from Physics Academic Software (Centennial Campus, 940 Main Campus Drive, Suite 210, Raleigh, NC 27606–5212; www.aip.org/pas). Contains projectile, pendulum, and collision simulators. Can also be used to demonstrate conservation of energy and rotational motion.
   d. *Newtonian Sandbox*. See Chapter 2 SUGGESTIONS.
   e. *Objects in Motion*. See Chapter 2 SUGGESTIONS.
   f. *Physics Demonstrations*. See Chapter 2 SUGGESTIONS.
   g. *Dynamic Analyzer*. See Chapter 2 SUGGESTIONS.

5. Computer Projects
   a. Have students use a commercial math program or their own root finding programs to solve projectile motion problems.
   b. Have students use a spreadsheet or write a computer program to tabulate the coordinates and velocity components of a projectile as functions of time. Have them change the initial velocity and observe changes in the coordinates of the highest point and in the range. Ask them to find the firing angle for the greatest horizontal coordinate when the landing point is above or below the firing point.

6. Laboratory
   a. Probeware Activity 3A: *Projectile Motion Part 1 — Change Initial Speed* and Probeware Activity 3B: *Projectile Motion Part 2 — Change Launch Angle*. Photogates and a time-of-flight detector are used to investigate some of the basic ideas of projectile motion.
   b. Meiners Experiment 7–9: *Ballistic Pendulum — Projectile Motion* (use only the first method in connection with this chapter). Students find the initial velocity of a ball shot from a spring gun by measuring its range. Emphasizes the use of kinematic equations.
   c. *Inelastic Impact and the Velocity of a Projectile* (use only Procedure B with this chapter). In addition to using range data to find the initial velocity, students plot the range as a

function of firing angle.

Chapter 5 FORCE AND MOTION — I

BASIC TOPICS

I. Overview

A. Explain that objects may interact with each other and, as a result, their velocities change. State that the strength of an interaction depends on properties of the objects and their relative positions. Gravitational mass is responsible for gravitational interactions, electric charge is responsible for electric and magnetic interactions.

B. Explain that we split the problem into two parts and say that each body exerts a force on the other and that the net force on a body changes its velocity. Remark that an equation that gives the force in terms of the properties of the objects and their positions is called a force law. Force laws are discussed throughout the course. The dominant theme of this chapter, however, is the relationship between the net force and the acceleration it produces.

II. Newton's first law.

A. State the law: if an object does not interact with any other objects, its acceleration is zero.

B. Point out that the acceleration depends on the reference frame used to measure it and that the first law can be true for only a select set of frames. Cover the essential parts of the relative motion section of Chapter 4, if they were not covered earlier. Define an inertial frame. Tell students that an inertial frame can be constructed, in principle, by finding an object that is not interacting with other objects and then attaching a reference frame to it. Any frame that moves with constant velocity relative to an inertial frame is also an inertial frame, but one that is accelerating relative to an inertial frame is not.

C. Explain that we may take a reference frame attached to Earth as an inertial frame for the description of most laboratory phenomena but we cannot for the description of ocean and wind currents, space probes, and astronomical phenomena.

III. Newton's second law.

A. Explain that the environment influences the motion of an object and that force measures the extent of the interaction. The result of the interaction is an *acceleration.* Place a cart at rest on the air track. Push it to start it moving and note that it continues at constant velocity. After it is moving, push it to increase its speed, then push it to decrease its speed. In each case note the direction of the force and the direction of the acceleration. Also give an eraser a shove across a table and note that it stops. Point out that the table top exerts a force of friction while the eraser is moving. Push the eraser at constant velocity and explain that the force of your hand and the force of friction sum to zero.

B. Define force in terms of the acceleration imparted to the standard 1 kg mass. Explain how this definition can be used to calibrate a spring, for example. Point out that force is a vector, in the same direction as the acceleration. If two or more forces act on the standard mass, its acceleration is the same as when a force equal to the resultant acts. Unit: newton. Explain that $1\,\mathrm{N}$ is $1\,\mathrm{kg{\cdot}m/s^2}$.

C. Have three students pull on a rope, knotted together as shown. Ask one to increase his or her pull and ask the others to report what they had to do to remain stationary.

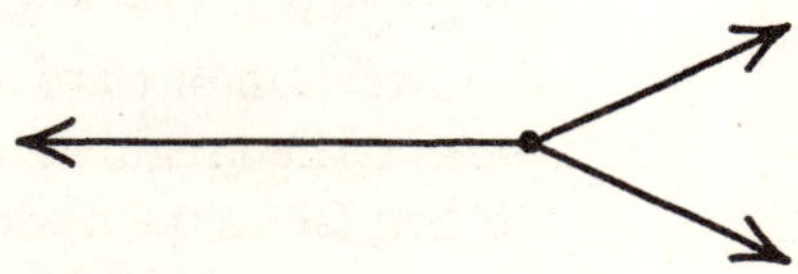

D. Define mass in terms of the ratio of the acceleration imparted to the standard mass and to the unknown mass, with the same force acting. Attach identical springs to two identical carts, one empty and the other containing a lead brick. Pull with the same force (same elongation of the springs) and observe the difference in acceleration. Unit: kilogram.

E. State the second law. Stress that the force that appears is the net or resultant force. Explain that the law holds in inertial frames. Point out that this is an experimentally established law and does not follow as an identity from the definitions of force and mass. Emphasize that $m\vec{a}$ is not a force.

F. Discuss examples: calculate the constant force required to stop an object in a given time, given the mass and initial velocity; calculate the force required to keep an object in uniform circular motion, given its speed and the radius of its orbit. Calculate the acceleration of an object being pushed by two forces in opposite directions and note that the acceleration vanishes if the forces have equal magnitudes. Emphasize that the forces continue to act but their sum vanishes. Some students believe that the forces literally cancel each other and no longer act.

IV. Newton's third law.

A. State the law. Stress that the two forces in question act on different bodies and each helps to determine the acceleration of the body on which it acts. Explain that the third law describes a characteristic of force laws. State that the two forces in an action-reaction pair are of the same type: gravitational, for example.

B. Discuss examples. Hold a book stationary in your hand, identify action-reaction pairs (hand-book, book-Earth). Now allow your hand and the book to accelerate downward with an acceleration less than g and again identify action-reaction pairs. Note that you can control the acceleration of the book by means of the force you exert but once you exert a given force you cannot control the force that the book exerts on you.

C. Attach a force probe to each of two air-track carts. Use a computer to plot the force that each exerts on the other as the carts collide. Point out that at each instant the forces have the same magnitude and are in opposite directions.

V. Applications of Newton's laws involving a single object.

A. Go over the steps used to solve a one-body problem: identify the body and all forces acting on it, draw a free-body diagram, choose a coordinate system, write the second law in component form, and finally solve for the unknown.

B. Some special forces should be explained. They are important for many of the problems but are rarely mentioned explicitly. Warn students they must take these forces into account if they act.

1. Point out that the magnitude of the gravitational force is mg, where g is the local acceleration due to gravity and m is the mass of the object. It is directed toward the center of Earth. Explain that the magnitude of this force is the weight of the object. Explain that weight varies with altitude and slightly from place to place on the surface of Earth, but mass does not vary. Emphasize that the appearance of g in the formula for the gravitational force does not imply that the acceleration of the body is g.

2. Point out that a massless rope transmits force unaltered in magnitude and that the magnitude of the force it exerts on objects at each end is called the tension force. If a person pulls an object by exerting a force on a string attached to the object, the motion is as if the person pulled directly on the object. The string serves to define the direction of the force. A frictionless, massless pulley serves to change the direction but not the magnitude of the tension force of the rope passing over it.

3. Explain that the normal force of a surface on an object originates in elastic and ultimately electric forces. It prevents the object from moving through the surface. State that it is perpendicular to the surface. If the surface is at rest, the normal force adjusts so the acceleration component perpendicular to the surface vanishes. More generally,

the object and the surface have the same perpendicular acceleration component. Place a book on the table and press on it. State that the normal force is greater than when you were not pressing. Hold the book against the wall by pressing on it and mention that the normal force is horizontal.

C. Set up the situation described in Sample Problem 5–7 using an inclined air track but attach a calibrated spring scale to the support at the top of the incline and tie the other end of the scale to the block. Calculate the tension force of the string and compare the result to the reading on the scale. Cut the string, then calculate the acceleration.

D. Consider a person standing on a scale in an elevator. State that the scale measures the normal force and calculate its value for an elevator at rest, one accelerating upward, one accelerating downward with $a < g$, and one in free fall. See Sample Problem 5–8.

VI. Applications of Newton's laws involving more than one object.

A. Explain that when two or more objects are involved, a free-body diagram must be drawn for each. A Newton's second law equation, in component form, is also written for each object. Point out that differently oriented coordinate systems may be used for different bodies. Show how to invoke the third law when necessary. Explain that the same symbol should be used for the magnitude of the two forces of an action-reaction pair and that their opposing directions are taken into account when drawing the free-body diagram and in writing the second law equations.

B. Explain that in some cases both objects can be considered as a single object. Say that the objects must have the same acceleration and that the forces they exert on each other must not be requested. The mass of the single object is then the sum of the masses of the constituent objects and internal forces are not included in the analysis.

C. Use examples to show how rods, strings, and pulleys relate the motions of bodies in various cases. Explain that, in addition to the second law equations, there will be equations relating the accelerations of the objects. Show that these equations depend on the choice of coordinate systems.

D. Consider several examples, carefully explaining each step. If you have not developed an application of your own, work Sample Problem 5–9 in the text. If possible, give a demonstration.

SUGGESTIONS

1. Assignments

   a. Use questions 1 through 7 to help students think about the influence of forces on bodies. some of these emphasize that the net force is a vector sum and others exercise Newton's first law. Assign one or two of problems 1, 2, and 3.

   b. Use questions 9 and 10 to help students think about normal forces.

   c. Use question 8 and problem 9 to help students with tensions in ropes.

   d. Assign problem 2 to emphasize the definition of force and problem 4 or 5 to demonstrate Newton's second law.

   e. Use problems 21 and 43 to discuss Newton's third law.

   f. Assign a few applications problems from the group 13 through 56, according to the needs and interests of the class.

   g. As a prelude to Chapter 9 (where the center of mass and conservation of momentum are discussed) assign problem 27.

2. Demonstrations

   a. Inertia: Freier and Anderson Mc1 — 5, Me1.

   b. $\vec{F} = m\vec{a}$: Freier and Anderson Md2, Ml1.

   c. Third-law pairs: Freier and Anderson Md1, 3, 4.

d. Mass and weight: Freier and Anderson Mf1, 2.
e. Tension in a string: Freier and Anderson Ml1.

3. Books and Monographs

Resource Letters, Book Four; American Association of Physics Teachers (AAPT, One Physics Ellipse, College Park MD 20740–3845; www.aapt.org). Contains a resource letter on mechanics.

4. Audio/Visual

a. *Dynamics*; VHS video tape, DVD; Films for the Humanities and Sciences (PO Box 2053, Princeton, NJ 08543–2053; www.films.com).
b. *Frames of Reference*; DVD; available from Ztek Co. (PO Box 11768, Lexington, KY 40577–1768, www.ztek.com) and from the American Association of Physics Teachers (see above for address).
c. *Human Mass Measurement* from Skylab Physics; video disk; available from Ztek Co. (see above for address).
d. *Newton's First and Second Laws*; *Newton's Third Law*; *Inertial Forces*; *Translational Acceleration*; from the AAPT collection 2 of single-concept films; DVD; available from Ztek Co. (see above for address).
e. *Inertial Forces - Translational Acceleration*; from the AAPT Miller collection of single-concept films; DVD; available from Ztek Co. and from the AAPT (see above for addresses).
f. *Forces*, *Newton's Laws*; Cinema Classics DVD 1: Mechanics (I); available from Ztek Co and from the AAPT (see above for addresses).
g. *Newton's 1st Law*; *Newton's 2nd Law*; *Newton's 3rd Law*; Physics Demonstrations in Mechanics, Part II; VHS video tape, DVD; ≈3 min each; Physics Curriculum & Instruction (22585 Woodhill Drive, Lakeville, MN 55044; www.PhysicsCurriculum.com).
h. *Newton's 1st Law*; Physics Demonstrations in Mechanics, Part III; VHS video tape, DVD; ≈3 min; Physics Curriculum & Instruction (see above for address).

5. Computer Software

a. *Freebody*; Graham Oberum; Macintosh; available from Physics Academic Software (Centennial Campus, 940 Main Campus Drive, Suite 210, Raleigh, NC 27606–5212; www.aip.org/pas) and from the AAPT (see above for address) . Students draw force vectors and can change the length and orientation of the vectors in response to questions. The screen gives the components.
b. *Force and Motion Microworld*; Ping-Kee L. Tao and Ming-Wai Tse; available from Physics Academic Software (see above for address). Uses velocity graphs to display the effects of force on the motion of an object. Includes drag forces.
c. *Forces and Motion* from Exploration of Physics Volume II; Windows and Macintosh; Physics Curriculum & Instruction Physics Curriculum & Instruction (see above for address). Includes sections on Newton's laws of motion.
d. *Forces*; Windows; available from Physics Academic Software and from the AAPT (see above for addresses). Covers most of the principles, with examples of electrostatic and electromagnetic forces.
e. *Force and Motion*; interactive CD-ROM; Films for the Humanities and Sciences (see above for address).
f. *Dynamic Analyzer*. See Chapter 2 SUGGESTIONS.
g. *Newtonian Sandbox*. See Chapter 2 SUGGESTIONS.

6. Laboratory

a. Probeware Activity 4A: *Newton's Second Law Part 1 — Constant Mass* and Probeware Activity 4B: *Newton's Second Law Part 2 — Constant New Force.* A motion detector is used to relate net force and acceleration. A small cart on a track is accelerated by means of a weight attached to a string and hung over a pulley.

Meiners Experiment 8–2: *Concept of Mass: Newton's Second Law of Motion.* Students measure the accelerations of two pucks that interact via a spring on a nearly frictionless surface and compare the ratio to the ratio of their masses. This experiment may be done with dry ice pucks or on an air table or air track.

b. Probeware Activity 5A: *Newtons's Third Law Part 1 — Collisions* and Probeware Activity 5B: *Newton's Third Law Part 2 — Tug-of-War.* A force sensor is used to generate a computer plot of the forces of two carts on each other during a collision and to compare the forces on the ends of a rope during a tug-of-war.

Chapter 6 FORCE AND MOTION — II

BASIC TOPICS

I. Frictional forces.

A. Place a large massive wooden block on the lecture table. Attach a spring scale, large enough to be read easily. If necessary, tape sandpaper to the table under the block. Pull weakly on the scale and note that the reading is not zero although the block does not move. Pull slightly harder and note that the reading increases but the block still does not move. Remark that there must be a force of friction opposing the pull and that the force of friction increases as the pull increases. Now increase your pull until the block moves and note the reading just before it starts to move. Pull the block at constant speed and note the reading. Have the students repeat the experiment in a qualitative manner, using books resting on their chair arms. To show that the phenomenon depends on the nature of the surface, the demonstration can be repeated after waxing the wooden block and table top.

B. Give a brief qualitative discussion about the source of frictional forces. Stress that the force of static friction has whatever magnitude and direction are required to hold the two bodies in contact at rest relative to each other, up to a certain limit in magnitude. Define the coefficient of static friction and explain the use of $f_s < \mu_s F_N$. In particular, explain that if the surface is stationary the force of static friction is determined by the condition that the object on it has zero acceleration. To test if an object remains at rest, the frictional force required to produce zero relative acceleration is calculated and compared with $\mu_s F_N$.

C. Define the coefficient of kinetic friction and explain that $f_k = \mu_k F_N$ gives the frictional force as long as the object is sliding on the surface. Also explain that if the surface is stationary the force of kinetic friction is directed opposite to the velocity of the object sliding on it.

D. Work some examples:

1. Find the angle of an inclined plane for which sliding starts; find the angle for which the body slides at constant speed. These examples can be analyzed in association with a demonstration and the students can use the data to find the coefficients of friction.
2. Analyze an object resting on the floor, with a person applying a force that is directed at some angle above the horizontal. Find the minimum applied horizontal force that will start the object moving and point out that it is a function of the angle between the applied force and the horizontal.
3. Consider the same situation but with the object moving. Find its acceleration. This and the previous example demonstrate the dependence of the normal force and the force of friction on the externally applied force.

4. To give an illuminating variant, consider a book being held against the wall by a horizontal force. Calculate the minimum applied force that will keep the book from falling.

II. Drag forces and terminal speed.
   A. Make or buy a small toy parachute. Drop two weights side by side and note they reach the floor at the same time. Attach the parachute to one and repeat. Explain that the force of the air reduces the acceleration.
   B. State that for turbulent flow of air around an object the magnitude of the drag force is given by $D = \frac{1}{2}C\rho Av^2$, where A is the effective cross-sectional area, ρ is the density of air, and v is the speed of the object relative to the air. C is a drag coefficient, usually determined by experiment. Remark that a parachute increases the cross-sectional area. State that the drag force is directed opposite to the velocity in still air.
   C. Explain that as an object falls its speed approaches terminal speed as a limit. Write down Newton's second law for a falling object and point out that the drag and gravitational forces are in opposite directions. Suppose the object is dropped from rest and point out that the acceleration is g at first but as the object gains speed its acceleration decreases in magnitude. At terminal speed the acceleration is zero and remains zero, so the velocity no longer changes. Show that zero acceleration leads to $v_t = \sqrt{2mg/C\rho A}$. Point out Table 6–1, which gives some terminal speeds.
   D. Remark that if an object is thrown downward with a speed that is greater than terminal speed it slows down until terminal speed is reached.
   E. Qualitatively discuss projectile motion with drag. The horizontal component of the velocity tends to zero while the vertical component tends to the terminal speed. Contrast the trajectory with one in the absence of air resistance.

III. Uniform circular motion.
   A. Point out that for uniform circular motion to occur there must be a radially inward force of constant magnitude and that something in the environment of the body supplies the force. Whirl a mass tied to a string around your head and explain that the string supplies the force. Set up a loop-the-loop with a ball or toy cart on a track and explain that the combination of the normal force of the track and the force of gravity supplies the centripetal force. Have students identify the source of the force in examples and problems as they are discussed.
   B. Point out that $F = mv^2/r$ is just $F = ma$ with the expression for centripetal acceleration substituted for a.
   C. Discuss problem solving strategy. After identifying the forces, find the radial component of the resultant and equate it to mv^2/r.
   D. Examples:
      1. Find the speed and period of a conical pendulum.
      2. Find the speed with which a car can round an unbanked curve, given the coefficient of static friction.
      3. Find the angle of banking required to hold a car on a curve without aid of friction.
      4. Analyze the loop-the-loop and point out that the ball leaves the track when the normal force vanishes. Show that the critical speed at the top is given by $v^2/r = g$.

SUGGESTIONS

1. Assignments
   a. Discuss some or all of questions 1 through 7 in connection with the force of static friction and the onset of sliding. Kinetic friction is the subject of questions 6, 7, and 9. Consider asking question 9 in connection with problem 30.

b. Ways in which the coefficient of static friction is used are emphasized in problems 1, 3, and 15. Problem 29 is more challenging. To help students understand the role played by the normal force in the onset of sliding, assign problem 11. Problem 7 deals with the role of the normal force in kinetic friction.
c. Problems 14 and 30 provide some interesting applications of the laws of friction.
d. Use questions 10 and 11 in your discussion of centripetal acceleration and force. Problems 36 and 37 cover the role of friction in rounding a level curve. Also consider problems 46 and 47.

2. Demonstrations
   a. Friction: Freier and Anderson Mk.
   b. Inclined plane: Freier and Anderson Mj2.
   c. Centripetal acceleration: Freier and Anderson Mb29, 31, Mm1, 2, 4 — 8, Ms5.
3. Audio/Visual
   a. *Trajectories*; from AAPT collection 2 of single-concept films; DVD; available from Ztek Co. (PO Box 11768, Lexington, KY 40577–1768, www.ztek.com).
   b. *Inertial Forces — Centripetal Acceleration*; from the AAPT Miller collection of single-concept films; video DVD; available from Ztek Co. (see above for address) and from the American Association of Physics Teachers (AAPT, One Physics Ellipse, College Park MD 20740–3845; www.aapt.org).
4. Computer Software
   a. *Forces and Motion* from Exploration of Physics Volume II; Windows and Macintosh; Physics Curriculum & Instruction Physics Curriculum & Instruction (22585 Woodhill Drive, Lakeville, MN 55044; www.PhysicsCurriculum.com). Includes sections on velocity and acceleration graphs and on friction and centripetal acceleration.
   b. *Dynamic Analyzer*. See Chapter 2 SUGGESTIONS.
5. Computer Projects
   a. Have students use a computer program to investigate objects that are subjected to time dependent forces. To check the program first have them consider a constant force and compare machine generated functions with the known kinematic equations.
   b. Have students modify the program to integrate Newton's second law for velocity dependent forces, then have them investigate the motion of an object subjected to a force that is proportional to v or v^2. It is instructive to have them plot the velocity components as functions of time for a projectile fired straight up or down, subject to air resistance. Consider initial velocities that are both greater and less than the terminal velocity. Also have them study the maximum height and range of projectiles with various coefficients of air resistance.
6. Laboratory
   a. Meiners Experiment 7–6: *Coefficient of Friction — The Inclined Plane.* Students determine the coefficients of static and sliding friction for three blocks on an inclined plane. They devise their own experimental procedures.
   b. Meiners Experiment 7–7: *Radial Acceleration* (Problem I only). The centripetal force and the speed of a ball on a string, executing uniform circular motion, are measured for various orbit radii. Essentially a verification of $F = mv^2/r$.
   c. Meiners Experiment 7–8: *Investigation of Uniform Circular Motion*, or Bernard Experiment 13: *Centripetal Force.* Students measure the force acting on a body undergoing uniform circular motion, with the centripetal force provided by a spring.
   d. Meiners Experiment 8–3: *Centripetal Force.* Students measure the speed of a puck undergoing uniform circular motion on a nearly frictionless surface. The data is used to calculate

the centripetal force.

Chapter 7 KINETIC ENERGY AND WORK

BASIC TOPICS

I. Kinetic energy and the work-kinetic energy theorem.

A. Define kinetic energy for a particle. Remind students that kinetic energy is a scalar and depends on the speed but not on the direction of the velocity. Point out that $v^2 = v_x^2 + v_y^2$ for two-dimensional motion and remark that the appearance of velocity components in the expression does *not* mean K has components.

B. Consider a ball thrown into the air. Neglect air resistance and point out that during the upward part of the motion the force of gravity slows the ball and the kinetic energy decreases. As the ball falls, the force of gravity speeds the ball and the kinetic energy increases. Remind students that for a constant force (and acceleration) $v^2 = v_0^2 + 2a\,\Delta x$ (which was derived in the study of kinematics). Multiply by $m/2$ to obtain $K = K_0 + F\,\Delta x$. Say that for a constant force acting on a particle that moves in one dimension $W = F\,\Delta x$ is the work done by the force F as the particle travels through the displacement Δx. State that $K = K_0 + W$ is an example of the work-kinetic energy theorem: the change in the kinetic energy of a particle during a given interval equals the work done on the particle by the total force during the interval.

C. Point out that only the component of a force parallel or antiparallel to the velocity changes the speed. Other components change the direction of motion. Positive total work results in an increase in kinetic energy and speed, negative total work results in a decrease. Remind students of previous examples in which the object moves with constant speed (including uniform circular motion). The total work is zero and the kinetic energy does not change. Avoid quantitative calculations involving frictional forces.

D. Explain that the work-kinetic energy theorem can be applied only to particles and objects that can be treated as particles. To give an example in which it cannot be applied directly, consider a car crashing into a rigid barrier: the barrier does no work but the kinetic energy of the car decreases.

E. Explain that observers in different inertial frames will measure different values of the net work done and for the change in the kinetic energy but both will find $W_{\text{net}} = \Delta K$.

F. Use Newton's second law to prove the theorem for motion in one dimension. If the students are mathematically sophisticated, extend the theorem to the general case. Stress that it is the total work (done by the resultant force) that enters the theorem.

II. Work done by a constant force.

A. Write down $W = \vec{F} \cdot \vec{d} = Fd\cos\phi$ and point out ϕ on a diagram. Explain that this is the work done *on* a particle *by* the constant force $\vec{F}$ as the particle undergoes a displacement $\vec{d}$. Explain that work can be calculated for each individual force and that the total work done on the particle is the work done by the resultant force. Point out that work is a scalar quantity. Also point out that work is zero for a force that is perpendicular to the displacement and that, in general, only the component of $\vec{F}$ tangent to the path contributes to the work. The force does no work if the displacement is zero. Emphasize that work can be positive or negative, depending on the relative orientation of $\vec{F}$ and $\vec{d}$. For a constant force, the work depends only on the displacement, not on details of the path. Unit: joule.

B. Calculate the work done by the force of gravity as a mass falls a distance h and as it rises a distance h. Emphasize the sign. Calculate the work done by a non-horizontal force used to pull a box across a horizontal floor. Point out that the work done by the normal force and the work done by the force of gravity are zero. Consider both an accelerating box and one moving with constant velocity. Repeat the calculation for a crate being pulled up an incline by a force applied parallel to the incline. Show the work done by gravity is $-mgh$, where h is the change in the height of the crate.

III. Work done by a variable force.

A. For motion in one dimension, discuss the integral form for work as the limit of a sum over infinitesimal path segments. Explain that the sum can be carried out by a computer even if the integral cannot be evaluated analytically.

B. Examples: derive expressions for the work done by an ideal spring and a force of the form k/x^2. If you have not yet discussed the force of an ideal spring, do so now as a preface to the calculation of work. Explain how the spring constant can be found by hanging a mass from the spring and measuring the extension. Demonstrate changes in the spring length during which the spring does positive work and during which the spring does negative work.

C. As an example consider a stone dropped onto a vertical spring and calculate the maximum compression of the spring, given the mass of the stone, the height from which it is dropped, and the spring constant of the spring.

D. For motion in more than one dimension, write down the expression for the work in the form $\int_i^f \vec{F} \cdot d\vec{r}$ and explain its interpretation as the limit of a sum over infinitesimal path segments. Explain that this is the general definition of work. Calculate the work done by the applied force, the force of gravity, and the tension in the string as a simple pendulum is pulled along its arc until it is displaced vertically through a height h by a horizontal applied force $\vec{F}$.

IV. Power.

A. Define power as $P = dW/dt$. Unit: watt.

B. Show that $P = \vec{F} \cdot \vec{v}$. Explain that the work done over a time interval is given by $\int P\,dt$.

SUGGESTIONS

1. Assignments

a. The idea of kinetic energy is covered in question 1. Assign one or more of problems 2, 3, and 5 to exercise the concept.

b. The idea of work is covered in questions 2 and 6. Problems 7, 8, and 9 are good examples of the quantitative aspects. The work done by the net force, as opposed to individual forces, is emphasized in problems 13 and 14.

c. To discuss the work done by gravity, ask question 5, then assign problems 16, 17, and 18. To discuss the force exerted by an ideal spring and the work done by it, ask question 8, then assign problems 24 and 26.

d. Question 10 and problem 29 are a good introduction to the work-kinetic energy theorem. Also assign problem 30 or 35.

e. Assign problem 32 in connection with the work done by a variable force. Also consider problems 33, 34, and 39.

f. Assign one or more of problems 40, 42, 44, and 46 in connection with power.

2. Demonstrations

Work: Freier and Anderson Mv1.

3. Audio/Visual

a. *Work and Energy*; Cinema Classics DVD 5: Conservation Laws; available from Ztek Co. (PO Box 11768, Lexington, KY 40577–1768; www.ztek.com) and from the American Association of Physics Teachers (One Physics Ellipse, College Park, MD 20740-3845, www.aapt.org).

b. *Work and Energy*; Physics Demonstrations in Mechanics, Part VI; VHS video tape, DVD; ≈3 min; Physics Curriculum & Instruction, 22585 Woodhill Drive, Lakeville, MN 55044.
c. *Friction, Work, and Energy*; VHS video tape, DVD (part of a collection); Films for the Humanities and Sciences (PO Box 2053, Princeton, NJ 08543–2053; www.films.com).
d. *Motion of Bodies and Mechanical Energy*; VHS video tape, DVD; Films for the Humanities and Sciences (PO Box 2053, Princeton, NJ 08543–2053; www.films.com).
e. *Energy and Force: Part 1*; VHS video tape; Films for the Humanities and Sciences (see above for address).
f. *Energy and Force: Part 2*; VHS video tape; Films for the Humanities and Sciences (see above for address).

4. Computer Project
   Have students use a commercial math program or write a program to numerically evaluate the integral for work, then use the program to calculate the work done by various forces, given as functions of position. Include a nonconservative force and use the program to show the work done on a round trip does not vanish.
5. Laboratory
   a. Probeware Activity 6: *Work and Energy.* Net work and change in kinetic energy are compared. A force probe is used to measure the force and photogates are used to measure the speed of a cart that is pulled by a hanging weight.
   b. Meiners Experiment 7–16: *Elongation of an Elastomer.* Students measure the elongation of an elastomer for a succession of applied forces and use a polar planimeter to calculate the work done by the force. The experiment may also be done in connection with Chapter 13.
   c. Bernard Experiment 10: *Mechanical Advantage and Efficiency of Simple Machines.* This experiment can be used to broaden the course to include these topics. A lever, an inclined plane, a pulley system, and a wheel and axle are studied. In each case the force output is measured for a given force input and the work input is compared to the work output.
   d. Probeware Activity 7A: *Hooke's Law Part 1: Find the Spring Constant* and Probeware Activity 7B: *Hooke's Law part 2: Work Done By a Spring.* A force probe is used to determine the force of a spring as a function of its elongation and compression and a motion detector is used to measure the speed of a cart being pushed by a spring. The change in kinetic energy of the cart is then related to the work done by the spring.

Chapter 8 POTENTIAL ENERGY AND CONSERVATION OF ENERGY

BASIC TOPICS

I. Potential energy, conservative and nonconservative forces
   A. Explain that potential energy is an energy of configuration. The potential energy of a system of objects depends on the relative positions of the objects. A system consisting of an object and Earth has a potential energy that depends on the separation of Earth and the object, for example.
   B. State that a potential energy can be associated with a force only if that force is conservative and explain that a force is conservative if the work done by the force when the system starts and ends with the same configuration is zero, no matter what the configurations and no matter what motions occur between the beginning and end of the interval. Show that this implies that the work done by the force between any given starting and ending configurations is the same no matter what intervening configurations are assumed by the system.

C. Discuss the force of gravity and the force of an ideal spring as examples. For either or both of these, show that the work done depends only on the end points and not on the path between, then argue that the work vanishes for a round trip. Point out that on some parts of the path the force does positive work while on other parts it does negative work. Demonstrate that the work done by a spring is independent of the path by considering two different motions with the same end points. For the first motion, have the mass go directly from the initial point to the final point; for the second, have it first go away from the final point before going there.

D. Use a force of friction with constant magnitude as an example of a nonconservative force. Consider a block on a horizontal table top and argue that the work done by the force cannot vanish over a round trip since it is negative for each segment. Suppose the block moves around a circular path and friction is the only force that does work. Argue that the object returns to its initial position with less kinetic energy than it had when it started. State that a potential energy cannot be associated with a frictional force.

E. Use a cart on a linear air track to demonstrate these ideas. Couple each end of the cart, via a spring, to a support at the corresponding end of the air track. Give the cart an initial velocity and tell students to observe its speed each time it returns to its initial position. Point out that the kinetic energy returns to nearly the same value and that the springs do zero work during a round trip. Reduce or eliminate air flow to show the influence of a nonconservative force. If this is done rapidly and skillfully, you can cause the cart to stop far from the starting point.

II. Potential energy.

A. Give the definition of potential energy in terms of work for motion in one dimension. See Eq. 8–6 and emphasize that the change in potential energy is the negative of the work done *by* the force responsible for the potential energy.

B. Discuss the following properties:

1. The zero is arbitrary. Only potential energy differences have physical meaning.
2. The potential energy is a *scalar* function of position.
3. The force is given by $F = -dU/dx$ in one dimension.
4. Unit: joule.

C. Derive expressions for the potential energy functions associated with the force of gravity (uniform gravitational field) and the force of an ideal spring. Stress that the potential energy is a property of the object-Earth or spring-mass system and depends on the *configuration* of the system.

D. Use the work-kinetic energy theorem to show that $W = \Delta U$ is the work that must be applied by an external agent to increase the potential energy by ΔU if the kinetic energy does not change. Show that ΔU is recovered as kinetic energy when the external agent is removed. Example: raising an object in a gravitational field.

III. Conservation of energy.

A. Explain that if all the forces acting between the objects of a system are conservative and the net work done by external forces on objects of the system is zero then $K + U =$ constant. This follows from the work-kinetic energy theorem with the work of the conservative forces represented by the negative of the change in potential energy. The negative sign in Eq. 8–6 is essential to obtain this result. Emphasize that U is the sum of the individual potential energies if more than one conservative force acts. Define the total mechanical energy as $E_{\text{mec}} = K + U$.

B. Discuss the conversion of kinetic to potential energy and vice versa. Drop a superball on a rigid table top and point out when the potential and kinetic energies are maximum and when they are minimum. The question of elasticity can be glossed over by saying that to a good approximation the ball rebounds with unchanged speed. Also discuss the energy in a spring-mass system. Return to the cart on the air track and discuss its motion in terms of $K + U =$ constant. To avoid later confusion in the students' minds, start the motion with neither K nor U equal to zero. Emphasize that the energy remains in the system but changes its form during the motion. The agent of the change is the work done by the forces of the springs.

C. Show how to calculate the total energy for a spring-mass system from the initial conditions. Write the conservation principle in the form $\frac{1}{2}mv^2 + U(x) = \frac{1}{2}mv_0^2 + U(x_0)$. Use conservation of energy to find expressions for the maximum speed, maximum extension, and maximum compression, given the total energy.

D. Use the example of a ball thrown upward to demonstrate that conservation of energy must be applied to a system rather than to a single particle. Remark that Earth does work on the ball, the ball does work on Earth, and the change in potential energy is the negative of the sum. Show it is mgh, where h is the change in their separation. Remark that both kinetic energies change and the total change is the negative of the change in the potential energy. Explain that because Earth is so massive the change in its kinetic energy is small and may be neglected.

E. Discuss potential energy curves. Use the curve for a spring-mass system, then a more general one, and show how to calculate the kinetic energy and speed from the coordinate and total energy. Point out the turning points on the curves and discuss their physical significance. Use $F = -dU/dx$ to argue that the particle turns around at a turning point. For an object on a frictionless roller coaster track, find the speed at various points and identify the turning points.

F. Define stable, unstable, and neutral equilibrium. Use a potential energy curve (a frictionless roller coaster, say) to illustrate. Emphasize that $dU/dx = 0$ at an equilibrium point.

IV. Potential energy in two and three dimensions.

A. Define potential energy as a line integral and explain that it is the limit of a sum over infinitesimal path segments. Remark that conservation of energy leads to $\frac{1}{2}mv^2 + U(x, y, z) =$ constant. Explain that $v^2 = v_x^2 + v_y^2 + v_z^2$ and that v^2 is a scalar.

B. Example: simple pendulum. Since the gravitational potential energy depends on height, in the absence of nonconservative forces the pendulum has the same swing on either side of the equilibrium point and always returns to the same turning points. Demonstrate with a pendulum hung near a blackboard and mark the end points of the swing on the board. For a more adventurous demonstration, suspend a bowling ball pendulum from the ceiling and release the ball from rest in contact with your nose. Stand very still while it completes its swing and returns to your nose.

V. External work and thermal energy.

A. Explain that when forces due to objects external to the system do work W on the system the energy equation becomes $\Delta K + \Delta U = W$ if the internal forces are conservative. K is the total kinetic energy of all objects in the system and U is the total potential energy of their interactions with each other.

B. Show that if the external force is conservative the system can be enlarged to include the (previously) external agent. Then, $W = 0$ and U must be augmented to include the new interactions. Give the example of a ball thrown upward in Earth's gravitational field.

C. Explain that if some or all of the internal forces are nonconservative then the total energy must include an thermal energy term to take account of energy that enters or leaves some or all of the objects and contributes to the energy (kinetic and potential) of the particles that make up the objects. Distinguish between thermal energy and the energy associated with the motion and interactions of the object as a whole. Write $\Delta K + \Delta U + \Delta E_{\text{th}} = W$.

D. Refer back to the block sliding on the horizontal table top, discussed earlier. Explain that when the block stops all the original kinetic energy has been converted to thermal energy. As an example assume the table top exerts a constant frictional force of magnitude f. Explain that $K_i = fd$, where K_i is the initial kinetic energy and d is the distance the block slides before stopping. Say that if the system consists of only the block, then $-K_i + \Delta E_{\text{th}} = W$, where W is the work done by friction on the block and ΔE_{th} is the change in the thermal energy of the block. Note that $-fd$ is not the work done by friction. Now take the system to consist of the both the block and table top. Then $-K_i + \Delta E_{\text{th}} = 0$, where ΔE_{th} is the total change in the thermal energies of the block and table top. Argue that $\Delta E_{\text{th}} = fd$. Explain that the division of thermal energy between the block and the table top cannot be calculated without a detailed model of friction.

E. Explain that the quantity $\int \vec{f} \cdot d\vec{r}$ along the path of an object does NOT give the work done by friction but it does contribute to the change in kinetic energy of the object, along with similar contributions from other forces, if any. For a block sliding on a table top, the work done by friction is algebraically greater than the value of the integral and the difference is the increase in the thermal energies of the block and table top.

F. Explain that there may be other forms of internal energy. For example, the chemical energy stored in the fuel of a car and the kinetic energy of the moving pistons are forms of internal energy if the car is taken to be an object of a system. Write $E = K + U + E_{\text{th}} + E_{\text{int}}$, where E_{int} is the total internal energy of the system, exclusive of the thermal energy. E_{int} is not considered in detail in this chapter but is considered in Chapter 9 and other chapters.

SUGGESTIONS

1. Assignments
   a. Use question 1 to discuss the idea of a conservative force.
   b. Question 3 and problems 2, 3, 5, and 6 test basic understanding of gravitational potential energy. Use problem 1 to test basic understanding of elastic potential energy.
   c. Test for understanding of the conservation of mechanical energy by asking questions 2 and 3 and assigning some of problems 9, 10, 12, 17, 26, 29, 31, and 34. Some of these are related to previous problems. Some combine gravitational and elastic potential energies.
   d. Draw several potential energy curves and have the class analyze the particle motion for various values of the total energy. This can provide particularly useful feedback as to how well the students have mastered the idea of energy conservation. Also ask question 5 and assign problems 37 and 39.
   e. Assign a few questions and problems dealing with applied and dissipative forces. Consider questions 4, 6, 7, 8, and 9 and problems 41, 43, 44, 45, 51, 52, and 58.

2. Demonstrations
   a. Conservation of energy: Freier and Anderson Mn1 — 3, 6.
   b. Nonconservative forces: Freier and Anderson Mw1.

3. Books and Monographs
   The Bicycle by Phillip DiLavore; available from the American Association of Physics Teachers (AAPT, One Physics Ellipse, College Park MD 20740–3845; www.aapt.org). Deals chiefly with energy. Students do not need to know about rotational motion.

4. Audio/Visual
   a. *Gravitational Potential Energy*; *Conservation of Energy — Pole Vault and Aircraft Take-off*; from the AAPT collection 1 of single-concept films; DVD; available from Ztek Co. (PO Box 11768, Lexington, KY 40577–1768, www.ztek.com).
   b. *Energy Conservation*; Cinema Classics DVD 5: Conservation Laws; available from Ztek Co. (see above for address) and the AAPT (see above for address).

c. *Conservation of Energy*; *Work and Conservation of Energy*; from Physics Demonstrations in Mechanics, Part I; VHS video tape, DVD; ≈3 min each; Physics Curriculum & Instruction (22585 Woodhill Drive, Lakeville, MN 55044; www.PhysicsCurriculum.com).

5. Computer Software

   Momentum and Energy from Exploration of Physics Volume II; Windows and Macintosh; Physics Curriculum & Instruction (see above for address). Includes sections on conservation of energy.

6. Laboratory

   a. Probeware Activity 8: *Conservation of Energy.* Students use a motion detector to compare the potential energy of a cart at the top of an incline with its kinetic energy at the bottom. Bernard Experiment 9: *Work, Energy, and Friction.* A string is attached to a car on an incline and passes over a pulley at the top of the incline. Weights on the free end of the string are adjusted so the car rolls down the incline at constant speed. The work done by gravity on the weights and on the car is calculated and used to find the change in mechanical energy due to friction. The coefficient of friction is computed. The experiment is repeated for the car rolling up the incline and for various angles of incline. It is also repeated with the car sliding on its top and the coefficients of static and kinetic friction are found.

Chapter 9 CENTER OF MASS AND LINEAR MOMENTUM

BASIC TOPICS

I. Center of mass.

   A. Spin a chalkboard eraser as you toss it. Point out that, if the influence of air can be neglected, one point (the center of mass) follows the parabolic trajectory of a projectile although the motions of other points are more complicated.

   B. Define the center of mass by giving its coordinates in terms of the coordinates of the individual particles in the system. As an example, consider a system consisting of three discrete particles and calculate the coordinates of the center of mass, given the masses and coordinates of the particles. Point out that no particle need be at the center of mass.

   C. Extend the definition to include a continuous mass distribution. Note that if the object has a point, line, or plane of symmetry, the center of mass must be at that point, on that line, or in that plane. Examples: a uniform sphere or spherical shell, a uniform cylinder, a uniform square, a rectangular plate, or a triangular plate. Show how to compute the coordinates of the center of mass of a complex object comprised of a several simple parts, a table for example. Each part is replaced by a particle with mass equal to the mass of the part, positioned at the center of mass of the part. The center of mass of the particles is then found. Explain how to find the center of mass of a simple shape, such as a rectangular or circular plate, with a hole cut in it.

   D. Explain that the general motion of a rigid body may be described by giving the motion of the center of mass and the motion of the object around the center of mass.

   E. Derive expressions for the velocity and acceleration of the center of mass in terms of the velocities and accelerations of the particles in the system.

   F. Derive $\vec{F}_{\text{net}} = M\vec{a}_{\text{com}}$ and emphasize that $\vec{F}_{\text{net}}$ is the net *external* force on all objects of the system. As an example, consider a two-particle system with external forces acting on both particles and each particle interacting with the other. Invoke Newton's third law to show that the internal forces cancel when all forces are summed.

   G. State that if $\vec{F}_{\text{net}} = 0$ and the center of mass is initially at rest, then it remains at the same point no matter how individual parts of the system move. Refer to the two carts of

C above.Work a sample problem. For example, consider a person running from one end to the other of a slab that is free to slide on a horizontal frictionless surface or two skaters who pull themselves toward each other with a rope or pole. Ask how far the cart and running person or the two skaters each move.

II. Momentum.
   A. Define momentum for a single particle. If you are including modern physics topics in the course, also give the relativistic definition of momentum.
   B. Show that Newton's second law can be written $\vec{F}_{\text{net}} = d\vec{p}/dt$ for a particle. Emphasize that the mass of the particle is constant and that this form of the law does not imply that a new term $\vec{v}\,dm/dt$ has been added to $\vec{F}_{\text{net}} = m\vec{a}$.
   C. State that the total momentum of a system of particles is the *vector* sum of the individual momenta and show that $\vec{P} = M\vec{v}_{\text{com}}$.
   D. Show that Newton's second law for the center of mass can be written $\vec{F}_{\text{net}} = d\vec{P}/dt$, where $\vec{P}$ is the total momentum of the system. Stress that $\vec{F}_{\text{net}}$ is the net external force and that this equation is valid only if the mass of the system is constant.

III. Impulse.
   A. Define the impulse of a force as the integral over time of the force. Note that it is a vector. Clearly distinguish between impulse (integral over time) and work (integral over path). Draw a force versus time graph for the force of one body on the other during a one-dimensional collision and point out the impulse is the area under the curve.
   B. Define the time averaged force and show that the impulse is the product $\vec{F}_{\text{avg}}\Delta t$, where Δt is the time of interaction. Remark that we can use the average force to estimate the strength of the interaction during the collision.
   C. Use Newton's second law to show that the impulse on a body equals the change in its momentum. Remark that the change in momentum depends not only on the force but also on the duration of the interaction.

IV. Conservation of linear momentum.
   A. Point out that $\vec{P}$ = constant if $\vec{F}_{\text{net}} = 0$. Stress that one examines the *external* forces to see if momentum is conserved in any particular situation. Point out that one component of $\vec{P}$ may be conserved when others are not.
   B. Put two carts, connected by a spring, on a horizontal air track and set them in oscillation by pulling them apart and releasing them from rest. Explain that the center of mass does not accelerate and the total momentum of the system is constant. Use the conservation of momentum principle to derive an expression for the velocity of one cart in terms of the velocity of the other. Push one cart and explain that the center of mass is now accelerating and the total momentum is changing.
   C. Consider a projectile that splits in two and find the velocity of one part, given the velocity of the other. Point out that mechanical energy is conserved for the cart-spring system but is not for the fragmenting projectile. The exploding projectile idea can be demonstrated with an air track and two carts, one more massive than the other. Attach a brass tube to one cart and a tapered rubber stopper to the other. Arrange so that the tube is horizontal and the stopper fits in its end. The tube has a small hole in its side, through which a firecracker fuse fits. Start the carts at rest and light a firecracker in the tube. The carts rapidly separate, strike the ends of the track, come back together again, and stop. Arrange the initial placement so the carts strike the ends of the track simultaneously. Explain that $\vec{P} = 0$ throughout the motion. For a less dramatic demonstration, tie two carts together with a compressed spring between them, then cut the string.
   D. Explain that observers in two different inertial frames will measure different values of the momentum for a system but they will agree on the conservation of momentum. That is,

if the net external force is found to vanish in one inertial frame, it vanishes in all inertial frames.

E. Illustrate the use of conservation of momentum to solve problems by considering the firing of a cannon initially resting on a frictionless surface. Assume the barrel is horizontal and calculate the recoil velocity of the cannon. Explain that muzzle velocity is measured relative to the cannon and that we must use the velocity of the cannonball relative to Earth.

V. Properties of collisions

A. Set up a collision between two carts on an air track. Point out the interaction interval and the intervals before and after the interaction.

B. Explain that for the collisions considered two bodies interact with each other over a short period of time and that the times before and after the collision are well defined. The force of interaction is great enough that external forces can be ignored during the interaction time.

C. Refer to the air track collision and point out that it is the impulse of one body on the other that changes the momentum of the second body. Repeat the air track collision. Measure the velocity of one cart before and after the collision and calculate the change in its momentum. Equate this to the impulse the other cart exerts. Estimate the collision time and calculate the average force exerted on the cart.

D. Use the third law to show that two bodies in a collision exert equal and opposite impulses on each other and show that, if external impulses can be ignored, then total momentum is conserved. Refer to the air track collision. Again stress that external forces are neglected during the collisions considered here.

E. Force probes, with input to a computer, can be used to plot the forces acting during a collision as functions of time. The curves can be integrated to find the impulse.

VI. Two-body collisions in one dimension.

A. Define the terms "elastic", "inelastic", and "completely inelastic". Distinguish between the transfer of kinetic energy from one colliding object to the another and the loss of kinetic energy to internal energy.

B. Two-body completely inelastic collisions.

1. Derive an expression for the velocity of the bodies after the collision in terms of their masses and initial velocities.
2. Demonstrate the collision on an air track, using carts with velcro bumpers. Point out that the kinetic energy of the bodies is not conserved and calculate the energy loss. Remark that the energy is dissipated by the mechanism that binds the objects to each other. Some goes to internal energy, some to deformation energy. Note that $\frac{1}{2}(m_1 + m_2)v_{\text{com}}^2$ is retained. If we use a reference frame attached to the center of mass to describe the collision, we would find the combined bodies at rest after the collision and all kinetic energy lost.

C. Two-body elastic collisions.

1. Derive expressions for the final velocities in terms of the masses and initial velocities.
2. Specialize the general result to the case of equal masses and one body initially at rest. Demonstrate this collision on the air track using carts with spring bumpers. Point out that the carts exchange velocities.
3. Specialize the general result to the case of a light body, initially at rest, struck by a heavy body. Demonstrate this collision on the air track. Point out that the velocity of the heavy body is reduced only slightly and that the light body shoots off at high speed. Relate to a bowling ball hitting a pin.
4. Specialize the general result to the case of a heavy body, initially at rest, struck by a light body. Demonstrate this collision on the air track. Point out the low speed

acquired by the heavy body and the rebound of the light body. Relate to a ball rebounding from a wall. A nearly elastic collision can be obtained with a superball.

5. Point out that, although the total kinetic energy does not change, kinetic energy is usually transferred from one body to the other. Consider a collision in which one body is initially at rest and calculate the fraction that is transferred. Show that the fraction is small if either mass is much greater than the other and that the greatest fraction is transferred if the two masses are the same. This is important, for example, in deciding what moderator to use to thermalize neutrons from a fission reactor.

D. Point out that while the greatest energy loss occurs when the interaction is completely inelastic, there are many other inelastic collisions in which less than the maximum energy loss occurs. Note that it is possible to have a collision in which kinetic energy increases (an explosive impact, for example).

VII. Two-body collisions in two dimensions.

A. Write down the equations for the conservation of momentum, in component form, for a collision with one body initially at rest. Mention that these can be solved for two unknowns.

B. Consider an elastic collision for which one body is at rest initially and the initial velocity of the second is given. Write the conservation of kinetic energy and conservation of momentum equations. Point out that the outcome is not determined by the initial velocities but that the impulse of one body on the other must be known to determine the velocities of the two bodies after the collision. State that the impulse is usually not known and in practice physicists observe one of the outgoing particles to determine its direction of motion. Carry out the calculation: assume the final direction of motion of one body is known and calculate the final direction of motion of the other and both final speeds.

C. State, or perhaps prove, that if the particles have the same mass then their directions of motion after the collision are perpendicular to each other.

D. Consider a completely inelastic collision for which the two bodies do not move along the same line initially. Mention that the outcome of this type collision *is* determined by the initial velocities. Calculate the final velocity. Calculate the fraction of energy dissipated.

VIII. Variable mass systems.

A. Derive the rocket equations, Eqs. 9–87 and 9–88. Emphasize that we must consider the rocket and fuel to be a single constant mass system. Derive expressions for the momentum before and after a small amount of fuel is expelled, in terms of the mass of the rocket and fuel together, the mass of the fuel expelled, the initial velocity of the rocket, the change in its velocity, and the relative velocity of the expelled fuel. Assume no external forces act on the rocket-fuel system and equate the two expressions for the total momentum.

B. To demonstrate, screw several hook eyes into a toy CO_2 propelled rocket, run a line through the eyes, and string the line across the lecture hall. Start the rocket from rest and have the students observe its acceleration as it crosses the hall.

C. As a second example, consider the loading of sand on a conveyor belt and calculate the force required to keep the belt moving at constant velocity.

SUGGESTIONS

1. Assignments

a. Use problems 1, 3, and 6 to generate discussion about the position of the center of mass. To present a challenge, assign problem 8.

b. Questions 1 and 2 are good tests of understanding of the motion of the center of mass. Discuss them as an introduction to the problems. Assign problems 11, and 12. Assign some problems in which the center of mass does not move: 16 and 17, for example.

c. To emphasize the vector nature of momentum, assign problems 19 and 20.

d. To help students with the concept of impulse and the impulse-momentum theorem ask question 5. Also consider problems 23 and 32.
e. To discuss conservation of momentum use questions 6 and 7 and assign problems such as 35 and 39. To emphasize the difference between conservation of energy and conservation of momentum assign problem 44.
f. Use question 9 to discuss the motion of the center of mass during a collision.
g. Assign some problems dealing with inelastic collisions, such as 51. Assign and discuss a problem for which the mechanism of kinetic energy loss is given explicitly. See problem 53.
h. Assign a problem, such as 56, that ask students to determine if a collision is elastic.
i. Ask question 8 in support of the discussion of elastic collisions. Assign problem 55. For some fun, carry out the demonstration described in problem 63.
j. Demonstrate the ballistic pendulum and show how it can be used to measure the speed of a bullet. Assign problem 46.
k. Assign some problems that deal with two-dimensional collisions: the group 65 through 73. Include both elastic and inelastic collisions.
l. Assign problems such as 70, 72, and 73, which are concerned with variable mass systems.

2. Demonstrations
   a. Center of mass, center of gravity: Freier and Anderson Mp7, 12, 13.
   b. Motion of center of mass: Freier and Anderson Mp1, 2, 16 — 19.
   c. Conservation of momentum: Freier and Anderson Mg4, 5, Mi2.
   d. Collisions: Freier and Anderson Mg1 — 3, Mi1, 3, 4, Mw3, 4.
   e. Rockets: Freier and Anderson Mh.

3. Books and Monographs

   Rockets by David Keeports; available from the American Association of Physics Teachers (AAPT, One Physics Ellipse, College Park MD 20740–3845; www.aapt.org).

4. Audio/Visual
   a. *Finding the Speed of a Rifle Bullet*; from the AAPT collection 1 of single-concept films; DVD; from Ztek Co. (PO Box 11768, Lexington, KY 40577–1768, www.ztek.com).
   b. *Center of Mass*, *Conservation of Linear/Angular Momentum*, *Conservation of Momentum*; from the AAPT collection 2 of single-concept films; DVD; available from Ztek Co. (see above for address).
   c. *Linear Momentum*; Cinema Classics DVD 5: Conservation Laws; available from Ztek Co. (see above for address) and the AAPT (see above for address).
   d. *Motion of Center of Mass*; *Conservation of Momentum*; from Physics Demonstrations in Mechanics, Part II; VHS video tape, DVD; ≈3 min; Physics Curriculum & Instruction (22585 Woodhill Drive, Lakeville, MN 55044; www.PhysicsCurriculum.com).
   e. *Dynamics*; VHS video tape, DVD; Films for the Humanities and Sciences (see above for address).
   f. *Impulse and Momentum*;*Conservation of Momentum*; from Physics Demonstrations in Mechanics, Part V; VHS video tape, DVD; ≈3 min; Physics Curriculum & Instruction (see above for address).
   g. *Motion of Center of Mass*; from Physics Demonstrations in Mechanics, Part V; VHS video tape, DVD; ≈3 min; Physics Curriculum & Instruction (see above for address).
   h. *Linear Momentum and Newtons' Laws of Motion*; VHS video tape, DVD (part of a collection); Films for the Humanities and Sciences (PO Box 2053, Princeton, NJ 08543–2053; www.films.com).
   i. *One-Dimensional Collisions*, *Two-Dimensional Collisions*, *Scattering of a Cluster of Objects*, *Dynamics of a Billiard Ball*, *Inelastic One-Dimensional Collisions*, *Inelastic Two-*

Dimensional Collisions, and *Colliding Freight Cars*; from the AAPT collection 1 of single-concept films; DVD; available from Ztek Co. (see above for address).

j. *Drops and Splashes*, *Collisions in Two Dimensions*, *Inelastic Collisions*; from the AAPT collection 2 of single-concept films; DVD; available from Ztek Co. (see above for address).

k. *Elastic Collisions*, *Inelastic Collisions*, *Collisions*; Cinema Classics DVD 5: Conservation Laws; available from Ztek Co. and from the AAPT (see above for addresses).

l. *Collisions* from Skylab Physics; DVD; available from Ztek Co. (see above for address).

m. *Characteristics of Collisions*; *Elastic Collision*; from Physics Demonstrations in Mechanics, Part V; VHS video tape, DVD; ≈3 min; Physics Curriculum & Instruction (see above for address).

5. Computer Software
   a. *Objects in Motion.* See Chapter 2 SUGGESTIONS.
   b. *Mechanics* from Exploration of Physics Volume I; Windows and Macintosh; Physics Curriculum & Instruction (see above for address). Includes sections on center of mass and 2D collisions.
   c. *Forces and Motion* from Exploration of Physics Volume II; Windows and Macintosh; Physics Curriculum & Instruction (see above for address). Includes a section on the center of mass.
   d. *Momentum and Energy* from Exploration of Physics Volume II; Windows and Macintosh; Physics Curriculum & Instruction (see above for address). Includes sections on elastic and inelastic collisions and on conservation of momentum.
6. Computer Projects
   a. Have students use a spreadsheet or write a computer program to follow individual particles in a two or three particle system, given the force law for the forces they exert on each other. The program should integrate Newton's second law for each particle. Have the students use their data to verify the conservation of momentum.
   b. Have students use a commercial math program or write a program to graph the total final kinetic energy as a function of the final velocity of one object in a two-body, one-dimensional collision, given the initial velocities and masses of the two objects. Ask them to run the program for specific initial conditions and identify elastic, inelastic, completely inelastic, and explosive collisions on their graphs.
7. Laboratory
   a. Meiners Experiment 8–7: *Linear Momentum.* Essentially the same as 8–6 but data is analyzed to give the individual momenta and total momentum as functions of time. Kinetic energy is also analyzed.
   b. Meiners Experiment 7–10: *Impulse and Momentum.* Students use a microprocessor to measure the force as a function of time as a toy truck hits a force transducer. They numerically integrate the force to find the impulse, then compare the result with the change in momentum, found by measuring the velocity before and after the collision.
   c. Probeware Activity 9: *Impulse v Change in Momentum.* Much like Meiners Experiment 7–10. A force probe is used to measure the force on a cart as it collides with a magnetic bumper. The force is integrated by computer to find the impulse. A motion sensor is used to find the change in the velocity of the cart.
   d. Bernard Experiment 8: *Impulse, Momentum, and Energy.* Part A deal with a mass that is hung on a string passing over a pulley and attached to an air track glider. The glider accelerates from rest for a known time and a spark timer is used to find its velocity at the end of the time. The impulse is calculated and compared with the momentum. In part B a glider is launched by a stretched rubber band and a spark record of its position as a function of time is made while it is in contact with the rubber band. A static technique is

used to measure the force of the rubber band for each of the recorded glider positions and the impulse is approximated. The result is again compared with the final momentum of the glider.

e. Meiners Experiment 7–9: *Ballistic Pendulum — Projectile Motion.* A ball is shot into a trapping mechanism at the end of a pendulum. The initial speed of the ball is found by applying conservation of momentum to the collision and conservation of energy to the subsequent swing of the pendulum. Also see Bernard Experiment 11: *Inelastic Impact and the Velocity of a Projectile.*

f. Meiners Experiment 8–6: *Center of Mass Motion.* Two pucks are connected by a rubber band or spring and move toward each other on a nearly frictionless surface. A spark timer is used to record their positions as functions of time. Students calculate and study the position of the center of mass as a function of time. They also find the center of mass velocity. Can be performed with dry ice pucks or on an air table or air track.

g. Meiners Experiment 8–8: *Two-Dimensional Collisions.* Same as Meiners 8–5 but the pucks are allowed to scatter out of the original line of motion. Students must measure angles and calculate components of the momenta. The experiment may be performed with dry ice pucks or on an air table.

h. Bernard Experiment 12: *Elastic Collision — Momentum and Energy Relations in Two Dimensions.* A ball rolls down an incline on a table top and strikes a target ball initially at rest at the edge of the table. The landing points of the balls on the floor are used to find their velocities just after the collision. The experiment is run without a target ball to find the velocity of the incident ball just before the collision. Data is used to check for conservation of momentum and energy. Both head-on and grazing collisions are investigated. A second experiment, similar to Meiners 8–8, is also described.

i. Probeware Activity 10A: *Conservation of Linear Momentum Part 1 — Inelastic Collision* and Probeware Activity 10B: *Conservation of Linear Momentum Part 2 — Elastic Collision.* Motion detectors are used to find the velocities of two carts before and after they collide. The momentum of each of the carts and the net momentum before the collision is compared with the same quantity after the collision.

j. Meiners Experiment 7–11: *Scattering* (for advanced groups). The deflection of pellets from a stationary disk is used to investigate the scattering angle as a function of impact parameter and to find the radius of the disk.

k. Meiners Experiment 8–5: *One-Dimensional Collisions.* A puck moving on a nearly frictionless surface collides with a stationary puck. A spark timer is used to record the positions of the pucks as functions of time. Students calculate the velocities, momenta, and energies before and after the collision. May be performed with dry ice pucks or on an air table or track.

Chapter 10 ROTATION

BASIC TOPICS

I. Rotation about a fixed axis.

A. Spin an irregular object on a fixed axis. A bicycle wheel or spinning platform with the object attached can be used. Draw a rough diagram, looking along the rotation axis. Explain that each point in the body has a circular orbit and that, for any selected point, the radius of the orbit is the perpendicular distance from the point to the rotation axis. Contrast to a body that is simultaneously rotating and translating.

B. Define angular position θ (in radians and revolutions), angular displacement $\Delta\theta$, angular velocity ω (in rad/s, deg/s, and rev/s), and angular acceleration α (in rad/s^2, deg/s^2, and rev/s^2). Treat both average and instantaneous quantities but emphasize that the instantaneous quantities are most important for us. Remind students of radian measure.

C. Use Fig. 10–4 to show how an angular displacement is measured. By convention in this text position angles are positive in the counterclockwise direction. Remark that as the body rotates counterclockwise, say, θ continues increase beyond 2π rad.

D. Interpret the signs of ω and α. Give examples of spinning objects for which ω and α have the same sign and for which they have opposite signs.

E. Point out the analogy to one-dimensional linear motion. θ corresponds to x, ω to v, and α to a.

F. Point out that ω and α can be thought of as the components of vectors $\vec{\omega}$ and $\vec{\alpha}$, respectively. For fixed axis rotation, the vectors lie along the rotation axis, with the direction of $\vec{\omega}$ determined by a right hand rule: if the fingers curl in the direction of rotation, then the thumb points in the direction of $\vec{\omega}$. If $d\omega/dt > 0$, then $\vec{\alpha}$ is in the same direction; if $d\omega/dt < 0$, then it is in the opposite direction. Use Fig. 10–7 to explain that a vector cannot be associated with a finite angular displacement because displacements do not add as vectors.

II. Rotation with constant angular acceleration.

A. Emphasize that the discussion here is restricted to rotation about a fixed axis but that the same equations can be used when the rotation axis is in linear translation. This type motion will be discussed in the next chapter.

B. Write down the kinematic equations for $\theta(t)$ and $\omega(t)$. Make a comparison with the analogous equations for linear motion (see Table 10–1).

C. Point out that the problems of rotational kinematics are similar to those for one-dimensional linear kinematics and that the same strategies are used for their solution.

D. Go over examples. Calculate the time and number of revolutions for an object to go from some initial angular velocity to some final angular velocity, given the angular acceleration. If time permits, consider both a body that is speeding up and one that is slowing down. For the latter, calculate the time to stop and the number of revolutions made while stopping. Calculate the time to rotate a given number of revolutions and the final angular velocity, again given the angular acceleration.

III. Linear speed and acceleration of a point rotating about a fixed axis.

A. Write down $s = \theta r$ for the arc length. Explain that it is a rearrangement of the defining equation for the radian and that θ must be in radians for it to be valid.

B. Wrap a string on a large spool that is free to rotate about a fixed axis. Mark the spool so the angle of rotation can be measured. Slowly pull out the string and explain that the length of string pulled out is equal to the arc length through which a point on the rim moves. Compare the string length to θr for $\theta = \pi/2$, π, $3\pi/2$, and 2π rad. Show that $s = \theta r$ reduces to the familiar result for $\theta = 2\pi$ rad.

C. Differentiate $s = \theta r$ to obtain $v = r\omega$ and $a_t = \alpha r$. Emphasize that radian measure *must* be used. Point out that v gives the speed and a_t gives the acceleration of the string as it is pulled provided it does not slip on the spool. Point out that all points in a rotating rigid body have the same value of ω and the same value of α but points that are different distances from the rotation axis have different values of v and different values of a_t.

D. Point out that the velocity is tangent to the circular orbit but that the total acceleration is not. a_t gives the tangential component while $a_r = v^2/r = \omega^2 r$ gives the radial component. The tangential component is not zero only when the point on the rim speeds up or slows down in its rotational motion while the radial component is not zero as long as the object is turning. For students who have forgotten, reference the derivation of $a_r = v^2/r$, given in Chapter 4.

E. Explain how to find the magnitude and direction of the total acceleration in terms of ω, α, and r.

IV. Kinetic energy of rotation and rotational inertia.

A. By substituting $v = r\omega$ into $K = \frac{1}{2}mv^2$, show that $K = \frac{1}{2}mr^2\omega^2$ for a particle moving around a circle with angular velocity ω and, by summing over all particles in a rigid body, show that $K = \frac{1}{2}I\omega^2$, where $I = \sum m_i r_i^2$ if the body is rotating about a fixed axis. Explain that I is called the rotational inertia of the body. Mention that many texts call it the moment of inertia.

B. Point out that rotational inertia depends on the distribution of mass and on the position and orientation of the rotation axis. Explain that two bodies may have the same mass but quite different rotational inertias. State that Table 10–2 gives the rotational inertia for various objects and axes. Particularly point out the rotational inertia of a hoop rotating about the axis through its center and perpendicular to its plane. Note that all its mass is the same distance from the rotation axis. Also point out the rotational inertias of a cylinder rotating about its axis and a sphere rotating about a diameter. Note that the mass is now distributed through a range of distances from the rotation axis and the rotational inertia is less than that for a hoop with the same mass and radius.

C. Optional: show how to convert the sum for I to an integral. Use the integral to find the rotational inertia for a thin rod rotating about an axis through its center and perpendicular to its length. If the students have experience with volume integrals using spherical coordinates, derive the expression for the rotational inertia of a sphere.

D. Prove the parallel axis theorem. The proof can be carried out using a sum for I rather than an integral. Explain its usefulness for finding the rotational inertia when the rotation axis is not through the center of mass. Emphasize that the actual axis and the axis through the center of mass must be parallel for the theorem to be valid. Use the parallel axis theorem to obtain the rotational inertia for the rotation of a thin rod about one end from the rotational inertia for rotation about the center, given in Table 10–2.

V. Torque.

A. Define torque for a force acting on a single particle. Consider forces that lie in planes perpendicular to the axis of rotation and take $\tau = rF\sin\phi$, where $\vec{r}$ is a vector that is perpendicular to the rotation axis and points from the axis to the point of application of the force. ϕ is the angle between $\vec{r}$ and $\vec{F}$ when they are drawn with their tails at the same point. The definition will be generalized in the next chapter. Explain that $\tau = rF_t = r_\perp F$, where F_t is the tangential component of $\vec{F}$ and $r_\perp$ is the moment arm.

B. Explain that the torque vanishes if $\vec{F}$ is along the same line as $\vec{r}$ and that only the component of $\vec{F}$ that is perpendicular to $\vec{r}$ produces a torque. This is a mechanism for picking out the part of the force that produces angular acceleration, as opposed to the part that is associated with centripetal acceleration. Also explain that the same force can produce a larger torque if it is applied at a point farther from the rotation axis.

C. Use a wrench tightening a bolt as an example. The force is applied perpendicular to the wrench arm and long moment arms are used to obtain large torques.

D. Explain the sign convention for torques applied to a body rotating about a fixed axis. For example, torques tending to give the body a counterclockwise (positive) angular acceleration are positive while those tending to give the body a clockwise angular acceleration are negative. Remark that the convention is arbitrary and the opposite convention may be convenient for some problems.

VI. Newton's second law for rotation.

A. Use a single particle on a circular orbit to introduce the topic. Start with $F_t = ma_t$ and show that $\tau = I\alpha$, where $I = mr^2$. Explain that this equation also holds for extended bodies, although I is then the sum given above.

B. Remark that problems are solved similarly to linear second law problems. Tell students to identify torques, draw a force diagram, choose the direction of positive rotation, and substitute the total torque into $\tau_{net} = I\alpha$. Remark that the point of application of a force is important for rotation, so the object cannot be represented by a dot on a force diagram. Tell students to sketch the object and place the tails of force vectors at the application points.

C. Wrap a string around a cylinder, free to rotate on a fixed horizontal axis. Attach the free end of the string to a mass and allow the mass to fall from rest. Note that its acceleration is less than g, perhaps by dropping a free mass beside it. See Sample Problem 10–8.

VII. Work-kinetic energy theorem for rotation.

A. Use $dW = \vec{F} \cdot d\vec{s}$ to show that the work done by a torque is given by $W = \int \tau\, d\theta$ and that the power delivered is given by $P = \tau\omega$.

B. Use $\tau\, d\theta = I\alpha\, d\theta = \frac{1}{2} I\, d(\omega^2)$ to show that $W = \frac{1}{2} I(\omega_f^2 - \omega_i^2)$.

C. For the situation of Sample Problem 10–8 use conservation of energy to find the angular velocity of the cylinder after the mass has fallen a distance h. Use rotational kinematics and the value for the angular acceleration found in the text to check the answer.

SUGGESTIONS

1. Assignments
   a. Use questions 1, 2, and 3 to discuss graphical interpretations of angular position and velocity.
   b. Use techniques of the calculus to derive the kinematic equations for constant angular acceleration. That is, integrate $\alpha =$ constant twice with respect to time. Assign problem 4 or 6.
   c. Assign some problems that deal with rotation with constant angular acceleration: 11, 13, and 17, for example.
   d. To discuss the relationship between angular and linear variables, assign some of problems 21, 22, 23, and 28.
   e. Use question 4 to guide students through a qualitative discussion of rotational inertia. Assign problems 37, 39, and 41.
   f. Use problem 47 or 48 to discuss the calculation of torque.
   g. To help students think about torque and $\tau_{net} = I\alpha$, discuss questions 5, 6, 7,8, and 9. Assign some of problems 53, 54, and 57. To deal with a situation in which the dynamics of more than one object is important, demonstrate the Atwood machine and discuss problem 55.
   h. Use question 10 to discuss the work done by a torque and changes in rotational kinetic energy. Discuss conservation of mechanical energy and assign problems 65 and 67.

2. Demonstrations
   a. Rotational dynamics: Freier and Anderson Ms7, Mt 5, 6, Mo5.
   b. Rotational work and energy: Freier and Anderson Mv2, Mr5, Ms2.

3. Computer Software

   Mechanics from Exploration of Physics Volume I; Windows and Macintosh; Physics Curriculum & Instruction (22585 Woodhill Drive, Lakeville, MN 55044; www.PhysicsCurriculum.com). Includes a section on torque.

4. Computer Project

   Ask students to use a commercial math program or their own root finding programs to solve rotational kinematic problems.

5. Laboratory
   a. Probeware Activity 11: *Rotational Motion.* A motion detector is used to plot the angular position and angular velocity of an accelerating disk. The graphs are compared to the analogous graphs for the linear motion of a cart.
   b. Meiners Experiment 7–14: *Rotational Inertia.* The rotational inertia of a disk is measured dynamically by applying a torque (a falling mass on a string wrapped around a flange on the disk). A microprocessor is used to measure the angular acceleration. Small masses are attached to the disk and their influence on the rotational inertia is studied. The acceleration of the mass can also be found by timing its fall through a measured distance. Then, $a_t = \alpha r$ is used to find the angular acceleration of the disk. Also see Bernard Experiment 14: *Moment of Inertia.*

Chapter 11 ROLLING, TORQUE, AND ANGULAR MOMENTUM

BASIC TOPICS

I. Rolling.
   A. Remark that a rolling object can be considered to be rotating about an axis through the center of mass while the center of mass moves. The text considers the special case for which the axis of rotation does not change direction. Point out that the rotational motion obeys $\tau_{\text{net}} = I_{\text{com}}\alpha$ and the translational motion of the center of mass obeys $\vec{F}_{\text{net}} = m\vec{a}_{\text{com}}$, where τ_{net} is the sum of external torques and $\vec{F}_{\text{net}}$ is the sum of external forces. Emphasize that one of the forces acting may be the force of friction produced by the surface on which the object rolls.
   B. Explain that the speed of a point at the top of a rolling object is $v_{\text{com}} + \omega R$ and the speed of a point at the bottom is $v_{\text{com}} - \omega R$. Specialize to the case of rolling without slipping. Point out that the point in contact with the ground has zero velocity, so $v_{\text{com}} = \omega R$. Use Fig. 11–5 as evidence. Also point out that tire tracks in the snow are clean (not smudged) if the tires do not slip.
   C. Explain that a wheel rolling without slipping can be viewed as rotating about an axis through the point of contact with the ground. Use this and the parallel axis theorem to show that the kinetic energy is $\frac{1}{2}Mv_{\text{com}}^2 + \frac{1}{2}I_{\text{com}}\omega^2$.
   D. Consider objects rolling down an inclined plane and show how to calculate the speed at the bottom using energy considerations. If time permits, carry out an analysis using the equations of motion and show how to find the frictional force that prevents slipping.
   E. Roll a sphere, a hoop, and a cylinder, all with the same radius and mass, down an incline. Start the objects simultaneously at the same height and ask students to pick the winner. Point out that the speed at the bottom is determined by the dimensionless parameter $\beta = I/MR^2$ and not by I, M, and R alone. All uniform cylinders started from rest reach the bottom in the same time and have the same speed when they get there.
   F. Consider a ball striking a bat. Show how to find the point at which the ball should hit so the instantaneous center of rotation is at the place where the bat is held. The striking point is called the center of percussion. When the ball hits there the batter feels no sting.

II. Torque and angular momentum.
   A. Define torque as $\vec{\tau} = \vec{r} \times \vec{F}$ and explain that this is the general definition. Review the vector product, give the expression for the magnitude ($\tau = rF\sin\phi$), and give the right hand rule for finding the direction. Explain that $\vec{\tau} = 0$ if $\vec{r} = 0$, $\vec{F} = 0$, or $\vec{r}$ is parallel (or antiparallel) to $\vec{F}$.

B. Consider an object going around a circle and suppose a force is applied tangentially. Take the origin to be on the rotation axis but not at the circle center and show the general definition reduces to the expression used in Chapter 11 for the component of the torque along the rotation axis: $\tau = F_t R$, where R is the radius of the circle (not the distance from the origin to the point of application).

C. Use vector notation to define angular momentum for a single particle ($\vec{\ell} = m\vec{r} \times \vec{v}$). Give the expression for the magnitude and the right-hand rule for the direction.

D. Derive the relationship $\ell = mr^2\omega$ between the magnitude of the angular momentum and the angular velocity for a particle moving on a circle centered at the origin. Also find the angular momentum if the origin is on a line through the circle center, perpendicular to the circle, but not at the center. Explain that the component along the rotation axis is $mr^2\omega$ and is independent of the position of the origin along the line and that the component perpendicular to the axis rotates with the particle.

E. To show that a particle may have angular momentum even if it is not moving in a circle, calculate the angular momentum of a particle moving with constant velocity along a line not through the origin. Point out that the angular momentum depends on the choice of origin. In preparation for G below you might want to find the time rate of change of $\vec{\ell}$.

F. Use Newton's second law to derive $\vec{\tau} = d\vec{\ell}/dt$ for a particle. Consider a particle moving in a circle, subjected to both centripetal and tangential forces. Take the origin to be at the center of the circle and show that $\vec{\tau} = d\vec{\ell}/dt$ reduces to $F_t = ma_t$, as expected. Take the origin to be on the line through the center, perpendicular to the circle, but not at the center. Show that the torque associated with the centripetal force produces the change in $\vec{\ell}$ expected from the discussion of D.

G. Show that the magnitude of the torque about the origin exerted by gravity on a falling mass is mgd, where d is the perpendicular distance from the line of fall to the origin. Write down the velocity as a function of time and show that the angular momentum is $mgtd$. Remark that $\tau_{\text{net}} = d\ell/dt$ by inspection. See Sample Problem 11–5.

III. Systems of particles.

A. Explain that the total angular momentum for a system of particles is the vector sum of the individual momenta.

B. Show that $\vec{\tau}_{\text{net}} = d\vec{L}/dt$ for a system of particles for which internal torques cancel. Emphasize that $\vec{\tau}_{\text{net}}$ is the result of summing all torques on all particles in the system and that $\vec{L}$ is the sum of all individual angular momenta. Demonstrate in detail the cancellation of internal torques for two particles that interact via central forces. Point out that this equation is the starting point for investigations of the rotational motion of bodies.

C. Show that the component along the rotation axis of the total angular momentum of a rigid body rotating about a fixed axis is $I\omega$. Use the example of a single particle to point out that the angular momentum vector is along the rotation axis if the body is symmetric about the axis but that otherwise it is not. Emphasize that for fixed axis rotation we are chiefly interested in the components of angular momentum and torque along the rotation axis.

D. Make a connection to material of the last chapter by showing that $L = I\omega$ and $\tau_{\text{net}} = dL/dt$ lead to $\tau_{\text{net}} = I\alpha$ for a rigid body rotating about a fixed axis. Here τ_{net} is the component of the total external torque along the rotation axis.

IV. Conservation of angular momentum.

A. Point out that $\vec{L}$ =constant if $\vec{\tau}_{\text{net}} = 0$. State that different objects in a system may change each other's angular momentum but the changes sum vectorially to zero. Also explain that the rotational inertia of an object may change while it is spinning. Then, $I_i\omega_i = I_f\omega_f$ if the net external torque vanishes.

B. As examples consider a mass dropped onto the rim of a freely spinning platform, a person running tangent to the rim of a merry-go-round and jumping on, and a spinning skater whose rotational inertia is changed by dropping her arms.

C. The third example can be demonstrated easily if you have a rotating platform that can hold a person. Have a student hold weights in each hand to increase the rotational inertia. Start him spinning with arms extended, then have him bring his arms in toward the sides of his body. See Fig. 11–17. Also carry out the spinning bicycle wheel demonstration described in the text. See Fig. 11–20.

SUGGESTIONS

1. Assignments
   a. In connection with rolling without sliding assign problems 4, 6, and 7. For a little greater challenge, assign problem 14.
   b. The definition of torque is covered in questions 1, 2, and 3. The definition of angular momentum is covered in questions 5 (particle) and 6 (system of particles). Assign problems 22 and 26 to stress the importance of the origin in calculations of torque and angular momentum. Problem 27 asks students to calculate the angular momentum if the cartesian components of the position and momentum vectors are given. Problem 24 deals with both angular momentum and torque. Be sure students can calculate the angular momentum of an object moving along a straight line and the angular momentum of a rigid body rotating about a fixed axis. See problems 25 and 37. Also consider discussing the angular momentum of a projectile.
   c. Newton's second law in angular form is covered in questions 7 and 8. Assign problem 32 or 33.
   d. Assign questions and problems dealing the conservation of angular momentum. Consider questions 9 and 10. Problem 47 includes motion along a straight line. Problems 41 and 46 deal with changes in rotational inertia. Problems 42 and 43 deal with inelastic rotational collisions. Assign one or more from each of these groups.
2. Demonstrations
   a. Rolling: Freier and Anderson Mb4, 7, 30, Mo3, Mp3, Mr1, 4, Ms1, 3, 4, 6.
   b. Conservation of angular momentum: Freier and Anderson Mt1 — 4, 7, 8, Mu1.
   c. Gyroscopes: Freier and Anderson Mu2 — 18.
3. Audio/Visual
   a. *Human Momenta, Initial Translation and Rotation* from Skylab Physics; DVD; available from Ztek Co. (PO Box 11768, Lexington, KY 40577–1768, www.ztek.com).
   b. *Conservation of Linear/Angular Momentum*; from the AAPT collection 2 of single-concept films; DVD; available from Ztek Co. (see above for address).
   c. *Angular Momentum*; Cinema Classics DVD 6: Angular Momentum and Modern Physics; available from Ztek Co. (see above for address) and the American Association of Physics Teachers (AAPT, One Physics Ellipse, College Park MD 20740–3845; www.aapt.org).
   d. *Rotational Dynamics*; from Physics Demonstrations in Mechanics, Part VI; VHS video tape, DVD; ≈3 min; Physics Curriculum & Instruction (22585 Woodhill Drive, Lakeville, MN 55044; www.PhysicsCurriculum.com).
   e. *Conservation of Angular Momentum*; *Center of Percussion*; from Physics Demonstrations in Mechanics, Part II; VHS video tape, DVD; ≈3 min; Physics Curriculum & Instruction (see above for address).
4. Computer Software

   Forces and Motion from Exploration of Physics Volume II; Windows and Macintosh;

Physics Curriculum & Instruction (see above for address). Includes a section on angular momentum.

5. Computer Project
   Given the law for the torque between two rotating rigid bodies a computer program or spreadsheet can be used to integrate Newton's second law for rotation and tabulate the angular positions and angular velocities as functions of time. The data can be used to verify the conservation of angular momentum.
6. Laboratory
   a. Meiners Experiment 7–12: *Rotational and Translational Motion.* Students measure the center of mass acceleration of various bodies rolling down an incline and calculate the center of mass velocities at the bottom. Results are compared to measured velocities. It is also instructive to use energy methods to find the final speeds.
   b. Meiners Experiment 7–13: *Rotational Kinematics and Dynamics.* Students find the velocity and acceleration of a ball rolling around a loop-the-loop and analyze the forces acting on it.
   c. Meiners Experiment 8–9: *Conservation of Angular Momentum.* Uses the Pasco rotational dynamics apparatus. A ball rolls down a ramp and becomes coupled to the rim of a disk that is free to rotate on a vertical axis. Students measure the velocity of the ball before impact and the angular velocity of the disk-ball system after impact, then check for conservation of angular momentum.

Chapter 12 EQUILIBRIUM AND ELASTICITY

BASIC TOPICS

I. Conditions for equilibrium.
   A. Write down the equilibrium conditions for a rigid body: $\vec{F}_{\text{net}} = 0$, $\vec{\tau}_{\text{net}} = 0$ (about any point). Remind students that only external forces and torques enter. Explain that these conditions mean that the acceleration of the center of mass and the angular acceleration about the center of mass both vanish. The body may be at rest or its center of mass may be moving with constant velocity or the body may be rotating with constant angular momentum. Point out that the equilibrium conditions form six equations that are to be solved for unknowns, usually the magnitudes of some of the forces or the angles made by some of the forces with fixed lines. Explain that we will be concerned chiefly with static equilibrium for which $\vec{P} = 0$ and $\vec{L} = 0$. Remark that the subscript "ext" is usually omitted.
   B. Show that, for a body in equilibrium, $\vec{\tau}_{\text{net}} = 0$ about *every* point.
   C. Explain that the gravitational forces and torques, acting on individual particles of the body, can be replaced by a single force acting at a point called the center of gravity. If the gravitational field is uniform over the body, the center of gravity coincides with the center of mass and the magnitude of the replacement force is Mg, where M is the total mass. It points downward.

II. Solution of problems.
   A. Give the problem solving steps: isolate the body, identify the forces acting on it, draw a force diagram, choose a reference frame for the resolution of the forces, choose a reference frame for the resolution of the torques, write down the equilibrium conditions in component form, and solve these simultaneously for the unknowns. Point out that the two reference frames may be different and that the reference frame for the resolution of torques can often be chosen so that one or more of the torques vanish.

B. Work examples. Consider a ladder leaning against a wall (Sample Problem 12–2) or an object hanging from a boom (Sample Problem 12–3). In each case show how the situation can be analyzed qualitatively to find the directions of the forces, then solve quantitatively.

III. Elasticity.

A. Point out that you have been considering mainly rigid bodies until now. Real objects deform when external forces are applied. Explain that deformations are often important for determining the equilibrium configuration of a system.

B. Consider a rod of unstrained length L subjected to equal and opposite forces F applied uniformly at each end, perpendicular to the end. Define the stress as F/A, where A is the area of an end. Define strain as the fractional change in length $\Delta L/L$ caused by the stress. Explain that stress and strain are proportional if the stress is sufficiently small. Define Young's modulus E as the ratio of stress to strain and show that $\Delta L = FL/EA$. Explain that Young's modulus is a property of the object and point out Table 12–1.

C. Explain that if the stress is small, the object returns to its original shape when the stress is removed and it is said to be *elastic*. Explain what happens if the stress is large and define *yield strength* and *ultimate strength.*

D. Calculate the fractional change in length for compressional forces acting on rods made of various materials. Use data from Table 12–1.

E. Explain that shearing occurs when the forces are parallel to the ends. Define the stress as F/A and the strain as $\Delta x/L$ where Δx is the displacement of one end relative to the other. Define the shear modulus G as the ratio of stress to strain and show that $\Delta x = FL/GA$.

F. Explain hydraulic compression. Define pressure as the force per unit area exerted by the fluid on the object. Explain that the pressure is now the stress and the fractional volume change $\Delta V/V$ is the strain. Define the bulk modulus B by $p = B\Delta V/V$.

G. To show how elastic properties are instrumental in determining equilibrium go over Sample Problem 12–6 or a similar problem.

SUGGESTIONS

1. Assignments

   a. Use questions 1, 2, 3, and 4 to help students gain understanding of the equilibrium conditions in specific situations. Assign a few problems, such as 3 and 10, for which only the total force is important. Assign others, such as 7, 11, and 11, for which torque is also important. To provide a greater challenge assign a few of problems 21, 28, and 31.

   b. The fundamentals of elasticity are covered in problems 36 (Young's modulus), and 37 (shear). Also assign one or both of problems 39 and 40, in which the laws of elasticity are used in conjunction with the equilibrium conditions to solve for forces and their points of application.

2. Demonstrations

   Freier and Anderson Mo1, 2, 4, 6 — 9, Mp4 — 6, 9, 11, 14, 15, Mq1, 2.

3. Audio/Video

   Linear Momentum and Newtons' Laws of Motion; VHS video tape, DVD (part of a collection); Films for the Humanities and Sciences (PO Box 2053, Princeton, NJ 08543–2053; www.films.com).

4. Laboratory

   a. Bernard Experiment 5: *Balanced Torques and Center of Gravity.* A non-uniform rod is pivoted on a fulcrum. A single weight is hung from one end and the pivot point moved until equilibrium is obtained. The data is used to find the center of gravity and mass of the rod. Additional weights are hung and equilibrium is again attained. The data is used to check that the net force and net torque vanish.

b. Bernard Experiment 6: *Equilibrium of a Crane.* Students study a model crane: a rod attached to a wall pivot at one end and held in place by a string from the other end to the wall. Weights are attached to the crane and the equilibrium conditions are used to calculate the tension in the rod and in the string. The latter is measured with a spring balance.
c. Meiners Experiment 7–16: *Elongation of an Elastomer* (see Chapter 7 notes).
d. Meiners Experiment 7–17: *Investigation of the Elongation of an Elastomer with a Microcomputer.* Same as Meiners 7–16 but a microprocessor is used to plot the elongation as a function of applied force. A polar planimeter is used to calculate the work done.

Chapter 13 GRAVITATION

BASIC TOPICS

I. Newton's law of gravity.
   A. This is an important chapter. It is the first chapter devoted to a force law and its ramifications. Students get a glimpse of how a force law and the laws of motion are used together. It reviews the concepts of potential energy, angular momentum, and centripetal acceleration in the context of some important applications. In addition, the discussion of the gravitational fields of continuous mass distributions is a precursor to Gauss' law.
   B. Write down the equation for the magnitude of the force of one point mass on another. Explain that the force is one of mutual attraction and is along the line joining the masses. Give the value of G ($6.67 \times 10^{-11}\,\text{N}\cdot\text{m}^2/\text{kg}^2$) and explain that it is a universal constant determined by experiment. If you have a Cavendish balance, show it but do not take the time to demonstrate it. As a thought experiment dealing with the magnitude of G, consider a pair of 100-kg spheres falling from a height of 100 m, initially separated by a bit more than their radii. As they fall, their mutual attraction pulls them only slightly closer together. Air resistance has more influence.
   C. Explain that the same mathematical form holds for bodies with spherically symmetric mass distributions (this was tacitly assumed in B) if r is now the separation of their centers. Explain that the force on a point mass anywhere inside a uniform spherical shell is zero. (Optional: use integration to prove that this follows from Newton's law for point masses.) Use this to derive an expression for the force on a point mass inside a spherically symmetric mass distribution. See Sample Problem 13–4.
   D. Point out the assumed equivalence of gravitational and inertial mass.
   E. Use Newton's law of gravity to calculate the acceleration a_g due to gravity for objects near the surface of Earth and justify the use of a constant acceleration due to gravity in previous chapters. Remark that the acceleration due to gravity is independent of the mass of the body.
   F. Optional: Discuss factors that influence a_g and apparent weight. Explain the difference between a_g and the free-fall acceleration g arising from earth's rotation.

II. Gravitational potential energy.
   A. Use integration to show that the gravitational potential energy of two point masses is given by $U = -GMm/r$ if the zero is chosen at $r \to \infty$. Demonstrate that this result obeys $F = -dU/dr$.
   B. Argue that the work needed to bring two masses to positions a distance r apart is independent of the path. Divide an arbitrary path into segments, some along lines of gravitational force and others perpendicular to the gravitational force.
   C. Consider a body initially at rest far from Earth and calculate its speed when it gets to Earth's surface. Calculate the escape velocity for Earth and for the Moon.

D. Show how to calculate the gravitational potential energy of a collection of discrete masses. Warn the students about double counting the interactions — a term of the sum is associated with each *pair* of masses. Relate this energy to the binding energy of the system.

III. Planetary motion and Kepler's laws.

A. Consider a single planet in orbit about a massive sun. The center of mass for the system is essentially at the sun and it remains stationary.

B. Explain that the orbit is elliptical with the sun at one focus. This is so because the force is proportional to $1/r^2$ and the planet is bound. Draw a planetary orbit and point out the semimajor axis, the perihelion point, and the aphelion point. Define eccentricity. Show that $R_p = a(1-e)$ and $R_a = a(1+e)$, where a is the semimajor axis, R_p is the perihelion distance, and R_a is the aphelion distance.

C. Explain that the displacement vector from the sun to the planet sweeps out equal areas in equal time intervals. Sketch an orbit to illustrate. Show that the torque acting on the planet is zero because the force is along the displacement vector; then show that conservation of angular momentum leads to the equal area law. Note that the result is true for any central force.

D. For circular orbits, show that the square of the period is proportional to the cube of the orbit radius and that the constant of proportionality is independent of the planet's mass. State that the result is also true for elliptical orbits if the radius is replaced by the semimajor axis. Verify the result for planets in nearly circular orbits. The data can be found in Table 14–3.

E. For a body held by gravitational force in circular orbit about another, much more massive body, show that the kinetic energy is proportional to $1/r$ and that the total mechanical energy is $-GMm/2r$. Explain that the energy is zero for infinite separation with the bodies at rest, that a negative energy indicates a bound system, and that a positive energy indicates an unbound system. Describe the orbits of recurring and non-recurring comets. Explain that the expression for the energy is valid for elliptical orbits if r is replaced by the semimajor axis. Remark that the energy of a satellite cannot be altered without changing the semimajor axis of its orbit.

F. Remark that the laws of planetary motion hold for moons (including artificial satellites) traveling around planets, binary star systems, stars traveling around the center of a galaxy, and for galaxies in clusters. Explain that when the masses of the two objects are comparable, both objects travel around the center of mass and it is the relative displacement that obeys Kepler's laws. When discussing stars in galaxies you might show how the law of periods has been used to argue for the existence of dark matter.

SUGGESTIONS

1. Assignments

a. To stress Newton's force law, ask question 1 and assign problem 1. Also assign problem 15 to test if students know the source of the value for a_g. To discuss symmetry, ask question 3.

b. Use problems 4 through 6 to test for understanding of the superposition principle.

c. Discuss problem 21 in connection with calculations of the gravitational force of a spherically symmetric mass distribution on a point mass. Problem 20 is fundamental to the shell theorem.

d. The essentials of gravitational potential energy are covered in problems 24 and 31. Conservation of mechanical energy is important for the solution to problem 32. Some of these can be used later as models for electrostatic potential energy. Question 10 covers some important qualitative aspects of gravitational work and potential energy. Escape velocity and energy are covered in several problems. Consider problems 25, 27, and 33.

e. To discuss planetary orbits assign some of problems 39, 44, 45, and 56.

2. Audio/Visual
   a. *Gravitation*; VHS video tape, DVD (part of a collection); Films for the Humanities and Sciences (PO Box 2053, Princeton, NJ 08543–2053; www.films.com).
   b. *The Determination of the Newtonian Constant of Gravitation* ; VHS video tape, DVD; Films for the Humanities and Sciences (see above for address).
   c. *Retrograde Motion — Heliocentric Model and Geocentric Model*; *Kepler's Laws*; *Jupiter Satellite Orbits*; from the AAPT collection 1 of single-concept films; DVD; available Ztek Co. (PO Box 11768, Lexington, KY 40577–1768, www.ztek.com).
   d. *Measurement of "G" — The Cavendish Experiment*; from the AAPT collection 2 of single-concept films; DVD; available from Ztek Co. (see above for address).
   e. *Planetary Motion*; Cinema Classics DVD 2: Mechanics (II) and Heat; available from Ztek Co. (see above for address) and the American Association of Physics Teachers (AAPT, One Physics Ellipse, College Park MD 20740–3845; www.aapt.org).
   f. *Newton's Law of Universal Gravitation*; from Physics Demonstrations in Mechanics, Part IV; VHS video tape, DVD; ≈3 min; Physics Curriculum & Instruction (22585 Woodhill Drive, Lakeville, MN 55044; www.PhysicsCurriculum.com).
3. Computer Software
   a. *Orbits*; James B. Harold, Kenneth Hennacy, and Edward Redish; Windows; available from Physics Academic Software (Centennial Campus, 940 Main Campus Drive, Suite 210, Raleigh, NC 27606–5212; www.aip.org/pas). Calculates and plots the trajectories of up to seven bodies. Two can be massive and influence the motions of the others. The rest are light. The user can change the value of the gravitational constant and can shift the view to various reference frames.
   b. *Planets and Satellites*; Windows; available from Physics Academic Software and from the AAPT (see above for addresses).
   c. *Mechanics* from Exploration of Physics Volume I; Windows and Macintosh; Physics Curriculum & Instruction (see above for address). Includes a section or planetary orbits.
   d. *Astronomy* from Exploration of Physics Volume II; Windows and Macintosh; Physics Curriculum & Instruction (see above for address). Includes sections on a comparison of geocentric and heliocentric planetary systems and Kepler's laws.
   e. *Objects in Motion*. See Chapter 2 SUGGESTIONS.
4. Computer Project

   Have students use a spreadsheet or write a computer program to integrate Newton's second law for a $1/r^2$ central force and use it to investigate satellite motion. Try some projects in which the orbit is changed and have students compute the energy required for the change.
5. Laboratory

   Meiners Experiment 7–21: *Analysis of Gravitation.* Students use the Leybold-Heraeus Cavendish torsional balance to determine G. Requires extremely careful work and a solid vibration free wall to mount the apparatus.

Chapter 14 FLUIDS

BASIC TOPICS

I. Pressure and density.

A. Introduce the subject by giving a few examples of fluids, including both liquids and gases. Remark that fluids cannot support shear.

B. Define density as the mass per unit volume in a region of the fluid. Point out that the limit is a macroscopic limit: the limiting volume still contains many atoms. The density is a scalar and is a function of position in the fluid.

C. Explain that fluid in any selected volume exerts a force on the material across the boundary of the volume. The boundary may be a mathematical construct and the material on the other side may be more of the same fluid. The boundary may also be a container wall or an interface with another fluid. Explain that, for a small segment of surface area, the force exerted by the fluid is normal to the surface and is proportional to the area. The pressure is the force per unit area and $\vec{F} = p\vec{A}$, where the magnitude of $\vec{A}$ is the area and the direction of $\vec{A}$ is outward, normal to the surface. Units: Pa (= N/m^2), atmosphere, bar, torr, mm of Hg. Give the conversions or point out Appendix D in the text.

D. Show that in equilibrium with y measured positive above some reference height $dp/dy = -\rho g$, where ρ is the fluid density. Then, note that $p_2 - p_1 = -\int \rho g\, dy$, where the integral limits are y_1 and y_2. Point out that the difference in pressure arises because a fluid surface is supporting the fluid above it. Finally, point out that if the fluid is incompressible and homogeneous, then ρ is a constant. If $y_2 - y_1$ is sufficiently small that g is also constant, then $p_2 - p_1 = -\rho g(y_2 - y_1)$. Point out that if p_0 is the surface pressure, then the pressure a distance h below the surface is $p = p_0 + \rho gh$. Note that the pressure is the same at all points at the same depth in a homogeneous fluid. Explain that p_0 is atmospheric pressure if the surface is open to the air and is zero if the fluid is in a tube with the region above the surface evacuated.

E. Connect a length of rubber tubing to one arm of a U-tube partially filled with colored water. Blow into the tube, then suck on it. In each case note the change in water level. Insert the U-tube into a deep beaker of water, with the free end of the tubing out of the water. As the open end is lowered, the change in the level of the colored water will indicate the increase in pressure. Go over Sample Problem 14–3 to show the equilibrium positions of two immiscible liquids of different densities. Show how to obtain the pressure at the top of one arm in terms of the pressure at the top of the other arm, the densities, and the quantities of fluids. Point out that the pressures are the same and are the atmospheric pressure if the U-tube is open. Explain that the pressure is always the same at two points that are at the same height *and* can be joined by a line along which neither ρ nor g vary. Use the diagram associated with the problem to point out two places at the same height where the pressure is the same and two places at the same height where the pressures are different.

II. Measurement of pressure.

A. This section not only describes some pressure measuring instruments but also provides some applications of previous material, especially the variation of pressure with depth in a fluid.

B. Show a mercury barometer. A lens system or an overhead projector suitably propped on its side can be used to project an image of the mercury column on a screen for viewing by a large class. Use $p = p_0 + \rho gh$ to show why the height of the column is proportional to the pressure at the mercury pool. Emphasize that the pressure at the top of the column is nearly zero and that this is important for the operation of the barometer.

C. Show a commercial open-tube manometer or explain that such an instrument is similar to the U-tube demonstration done earlier. Explain gauge pressure and emphasize that the instrument measures gauge pressure.

III. Pascal's and Archimedes' principles.

A. State Pascal's principle. Start with $p = p_0 + \rho gh$, consider a change in p_0, and show $\Delta p = \Delta p_0$ if the fluid is incompressible. You can demonstrate the transmission of pressure with a soda bottle full of water, fitted with a tight rubber stopper. Wrap a towel around the neck of the bottle and hit the stopper sharply. With some practice you can blow the bottom out of the bottle cleanly.

B. Apply the principle to a hydraulic jack. Show that $F_1/A_1 = F_2/A_2$. Also explain that if the fluid is incompressible, F_1 and F_2 do the same work. The point of application of the smaller force moves the greater distance. A hydraulic jack can be made from a hot water bottle, fitted with a narrow rubber tube. Put the bottle on the floor and fasten the tube to a tall ringstand so it is vertical. Place a thin wooden board on the bottle to distribute the weight and have a student stand on it. To change the pressure, use a plunger or rubber squeeze ball from an atomizer or blow into the tube.

C. State Archimedes' principle. Stress that the force is due to the surrounding fluid. Contrast the case of an immersed body surrounded by fluid with one placed on the bottom of the container. Consider a flat board floating on the surface of a liquid, compute the net upward force in terms of the difference in pressure and use $p = p_0 + \rho gh$ to show that this is the weight of the displaced liquid.

D. Explain why some objects sink while others float.

E. Fill a large mouthed plastic vessel with water precisely up to an overflow pipe. Immerse a dense object tied by string to a spring balance. Weigh the object while it is immersed and weigh the displaced water. Observe that the buoyant force is the same as the weight of the displaced water.

IV. Fluids in motion.

A. Describe:

1. Steady and non-steady flow. Emphasize that the velocity and density fields are independent of time if the flow is steady. They may depend on position, however.
2. Compressible and incompressible flow. Emphasize that the density is independent of both position and time if the flow is incompressible.
3. Rotational and irrotational flow.
4. Viscous and nonviscous flow.

B. Describe streamlines for steady flow and point out that streamlines are tangent to the fluid velocity and that no two streamlines cross. Remark that the velocity is not necessarily constant along a streamline. Describe a tube of flow as a bundle of streamlines. Sketch a tube of flow with streamlines far apart at one end and close together at the other. Explain that since streamlines do not cross the boundaries of a tube of flow they are close together where the tube is narrow and far apart where the tube is wide. Remark that particles do not cross the boundaries of a tube of flow.

V. Equation of continuity

A. Define volume flow rate (volume flux) and mass flow rate (mass flux). Consider a tube of flow with cross-sectional area A at one point and give the physical significance of $A\rho v$ and Av. Remark that the first can be measured in kg/s and the second in m^3/s. Show how to convert m^3/s to gal/s and L/s.

B. State the equation of continuity: $A\rho v =$ constant along a streamline if there are no sources or sinks of fluid and if the flow is steady. Argue that if the equation were not true there would be a build up or depletion of fluid in some regions and the flow would not be steady.

C. Discuss the special case of an incompressible fluid and explain that the fluid speed is great where the tube of flow is narrow and vice versa. Point out that the fluid velocity is great where the streamlines are close together and small where they are far apart. Use the diagram of section IVB above as an example.

VI. Bernoulli's equation.

A. Apply the work-kinetic energy theorem to a tube of flow to show that for steady, nonviscous, incompressible flow $p + \frac{1}{2}\rho v^2 + \rho g y =$ constant along a streamline. Point out that this equation also gives the pressure variation in a static fluid ($v = 0$ everywhere).

B. Remark that a typical fluid dynamics problem gives the conditions v, p, y at one point on a streamline and asks for conditions at another. The equation of continuity and Bernoulli's equation can be solved simultaneously for two quantities.

C. Work a sample problem. Consider horizontal flow ($y =$ constant) through a pipe that narrows. Give the fluid velocity where the pipe is wide and use the equation of continuity to calculate the velocity where it is narrow. Then, use Bernoulli's equation to calculate the pressure difference. Emphasize that the pressure must decrease to provide the force that accelerates the fluid as it passes into the narrow region.

D. Now work the same problem but suppose the height of the pipe increases along the direction of flow. Point out the difference in the answers for the pressure.

SUGGESTIONS

1. Assignments
   a. Use question 1 to discuss pressure. Problems 1, 2, and 4 cover the definition of pressure. Problem 14 deals with the variation of pressure with depth. Problem 19 includes torque.
   b. Use problem 22 in connection with Pascal's principle.
   c. Questions 4, 5, and 6 all provide good examples of Archimedes' principle. Pick several to illustrate applications of the principle. Also assign problems 24 and 25 and some of problems 31, 33, 35, 36, and 38.
   d. Use problems 42 and 47 as part of the discussion of the equation of continuity.
   e. The fundamentals of Bernoulli's equation are covered in problems 45, 47, and 48. Also consider problems 55 and 59. Some of these require students to combine the equation of continuity and Bernoulli's equation. Work one or two of these as examples in lecture and assign others.

2. Demonstrations
   a. Force and pressure: Freier and Anderson Fa, Fb, Fc, Fd, Fe, Ff, Fh.
   b. Archimedes' principle: Freier and Anderson Fg.
   c. Bernoulli's principle: Freier and Anderson Fj, Fl1.

3. Books and Monographs
   Hydraulic Devices; by Malcolm Goldber, John P. Ouderkirk, and Bruce B. Marsh ; available from the American Association of Physics Teachers (AAPT, One Physics Ellipse, College Park MD 20740–3845; www.aapt.org).

4. Audio/Visual
   Pressure; VHS video tape, DVD; Films for the Humanities and Sciences (PO Box 2053, Princeton, NJ 08543–2053; www.films.com).

5. Computer Software
   a. *Fluids* from Exploration of Physics Volume I; Windows and Macintosh; Physics Curriculum & Instruction (22585 Woodhill Drive, Lakeville, MN 55044; www.PhysicsCurriculum.com). Simulated experiments with analysis. Includes exercises dealing with density, pressure, buoyancy and, the Bernoulli equation.

b. *Fluids* from Exploration of Physics Volume II; Windows and Macintosh; Physics Curriculum & Instruction (see above for address). Includes sections on measuring pressure, volume, density, and buoyancy.

6. Laboratory
   a. Probeware Activity 12: Buoyant Force. a force sensor is used to measure the force on an object as it is lowered into water. A force versus depth is generated. Data is used to compute the density of the water.
   b. Meiners Experiment 7–7: *Radial Acceleration* (Problem II only). Students measure the orbit radii of various samples floating on the surface of water in a spinning globe and analyze the forces on the samples. This experiment is an application of buoyancy forces to rotational motion.
   c. Bernard Experiment 16: *Buoyancy of Liquids and Specific Gravity.* Archimedes' principle is checked by weighing the water displaced by various cylinders. Buoyant forces are measured by weighing the cylinders in and out of water. The same cylinder is immersed in various liquids and the results are used to find the specific gravities of the liquids.

Chapter 15 OSCILLATIONS

BASIC TOPICS

I. Oscillatory motion.
   A. Set up an air track and a cart with two springs, one attached to each end. Mark the equilibrium point, then pull the cart aside and release it. Point out the regularity of the motion and show where the speed is the greatest and where it is the least. By reference to the cart define the terms periodic motion, equilibrium point, period, frequency, cycle, and amplitude.
   B. Explain that $x(t) = x_m \cos(\omega t + \phi)$ describes the coordinate of the cart as a function of time if $x = 0$ is taken to be the equilibrium point, where the force of the springs on the cart vanishes. State that this type motion is called simple harmonic. Show where $x = 0$ is on the air track, then show what is meant by positive and negative x. Sketch a mass on the end of a single spring and explain that the mass also moves in simple harmonic motion if dissipative forces are negligible.
   C. Discuss the equation for $x(t)$.
      1. Explain that x_m is the maximum excursion of the mass from the equilibrium point and that the spring is compressed by x_m at one point in a cycle. x_m is called the *amplitude* of the oscillation. Explain that the amplitude depends on initial conditions. Draw several $x(t)$ curves, identical except for amplitude. Illustrate with the air track apparatus.
      2. Note that ω is called the angular frequency of the oscillation and is given in radians/s. Define the *frequency* by $f = \omega/2\pi$ and the *period* by $T = 1/f$. Show that $T = 2\pi/\omega$ is in fact the period by direct substitution into $x(t)$; that is, show $x(t) = x(t + T)$. Explain that the angular frequency does not depend on the initial conditions. For the cart on the track, use a timer to show that the period, and hence ω, is independent of initial conditions. Draw several $x(t)$ curves, for oscillations with different periods. Replace the original springs with stiffer springs and note the change in period. Also replace the cart with a more massive cart and note the change in period.
      3. Define the phase of the motion and explain that the phase constant ϕ is determined by initial conditions. Draw several $x(t)$ curves, identical except for ϕ, and point out the different conditions at $t = 0$. Remark that the curves are shifted copies of each other. Illustrate various initial conditions with the air track apparatus.

D. Derive expressions for the velocity and acceleration as functions of time for simple harmonic motion. Show that the speed is a maximum at the equilibrium point and is zero when $x = \pm x_m$. Also show that the magnitude of the acceleration is a maximum when $x = \pm x_m$ and is zero at the equilibrium point. Relate these results to $F(x)$.

E. Show that the initial conditions are given by $x_0 = x_m \cos\phi$ and $v_0 = -x_m\omega\sin\phi$. Solve for x_m and ϕ: $x_m^2 = x_0^2 + v_0^2/\omega^2$ and $\tan\phi = -v_0/\omega x_0$. Calculate x_m and ϕ for a few special cases: $x_0 = 0$ and v_0 positive, $x_0 = 0$ and v_0 negative, x_0 positive and $v_0 = 0$, x_0 negative and $v_0 = 0$. Tell students how to test the result given by a calculator for ϕ to see if π must be added to it.

II. The force law.

A. State the force law for an ideal spring: $F = -kx$. Point out that the negative sign is necessary for the force to be a restoring force. Hang identical masses on springs with different spring constants, measure the elongations, and calculate the spring constants. Remark that stiff springs have larger spring constants than weak springs. Remark that the expression for the force is an idealization. It is somewhat different for real springs.

B. Start with Newton's second law and derive the differential equation for $x(t)$. Show that $x = x_m \cos(\omega t + \phi)$ satisfies the equation if $\omega = \sqrt{k/m}$ and explain that this is the most general solution for a given spring constant and mass.

C. Show a vertical spring-mass system. Point out that the equilibrium point is determined by the mass, force of gravity, and the spring constant. Show, both analytically and with the apparatus, that the force of gravity does not influence the period, phase, or amplitude of the oscillation.

III. Energy considerations.

A. Derive expressions for the kinetic and potential energies as functions of time. Show that the total mechanical energy is constant by adding the two expressions and using the trigonometric identity $\sin^2\alpha + \cos^2\alpha = 1$. Remark that the energy is wholly kinetic at the equilibrium point and wholly potential at a turning point. It changes from one form to the other as the mass moves between these points.

B. Show how to use the conservation of energy to find the amplitude, given the initial position and velocity, to find the maximum speed, and to find the speed as a function of position.

IV. Applications.

A. Demonstrate a torsional pendulum and discuss it analytically. Derive the differential equation for the angle as a function of time and compare with the differential equation for a spring to obtain the angular frequency and period in terms of the spring constant and the rotational inertia.

B. Demonstrate a simple pendulum and discuss it analytically in the small amplitude approximation. Derive the differential equation for the angle as a function of time and obtain expressions for the angular frequency and period from the equation. Emphasize that the angular displacement must be measured in radians for the small amplitude approximation to be valid. Have students use their calculators to find the sines of some angles, in radians, starting with large angles and progressing to small angles.

C. Demonstrate a physical pendulum. Use Newton's second law for rotation to obtain the differential equation for the angular displacement. Obtain expressions for its angular frequency and period in the small amplitude approximation. Remind students that the rotational inertia depends on the position of the pivot and show them how to use the parallel axis theorem to find its value.

V. Simple harmonic motion and uniform circular motion.

A. This section is particularly important if you intend to include wave interference and diffraction in the course.

B. Mount a bicycle wheel vertically and arrange for it to be driven slowly with uniform angular speed. Attach a tennis ball to the rim and project the shadow of the ball on the wall. Note that the shadow moves up and down in simple harmonic motion. Point out that the period of the wheel and the period of the shadow are the same. It is possible to suspend a mass on a spring near the wall and adjust the angular speed and initial conditions so the mass and shadow move together for several cycles. A period of about 1 s works well.

C. Analytically show that the projection of the position vector of a particle in uniform circular motion undergoes simple harmonic motion. Mention the converse: if an object simultaneously undergoes simple harmonic motion in two orthogonal directions, with the same amplitude and frequency, but a $\pi/2$ phase difference, the result is a circular orbit.

VI. Damped and forced harmonic motion.

A. Write the differential equation for a spring-mass oscillator with a damping term proportional to the velocity. Treat the case $(b/2m)^2 < k/m$ and write the solution, including the expression for the angular frequency in terms of k, m, and b. If there is time, prove it is the solution by direct substitution into the differential equation or leave the proof as an exercise for the students. Remark that the natural angular frequency is nearly $\sqrt{k/m}$ if damping is small.

B. Show a graph of the displacement as a function of time. See Fig. 15–16. Point out the exponential decay of the amplitude. Mention that the oscillator loses mechanical energy to dissipative forces.

C. Explain that if $(b/2m)^2 > k/m$ then the mass does not oscillate but rather moves directly back to the equilibrium point. The displacement is a decreasing exponential function of time. To demonstrate under- and over-damping, attach a vane to a pendulum. Experiment with the size so the pendulum oscillates in air but does not when the vane is in water.

D. Write the differential equation for a forced spring-mass system, including a damping term. Assume an applied driving force of the form $F_m \cos(\omega_d t)$ and point out that ω_d is not necessarily the same as the natural angular frequency of the oscillator.

E. Mention that when the system is first started transients are present and the motion is somewhat complicated. However, it settles down to a sinusoidal motion with an angular frequency that is the same as that of the driving force.

F. Also point out that in steady state the amplitude is constant in time but that it depends on the frequency of the driving force. Illustrate with Fig. 15–17, which shows the amplitude as a function of the driving frequency for various values of the damping coefficient. Mention that the amplitude is the greatest when the driving frequency nearly matches the natural frequency and say this is the resonance condition. Also mention that at resonance the amplitude is greater for smaller damping and that small damping produces a sharper resonance than large damping.

G. Resonance can be demonstrated with three identical springs and two equal masses, as shown. Fasten the bottom spring to a heavy weight on the floor and drive the upper spring by hand (perhaps standing on a table). Obtain resonance at each of the normal modes (masses moving in the same and opposite directions). After showing the two resonances, drive the system at a low frequency to show a small response, then drive it at a high frequency to again show a small response. Repeat at a resonance frequency to show the larger response. To show pronounced damping effects, attach a large stiff piece of aluminum plate to each mass.

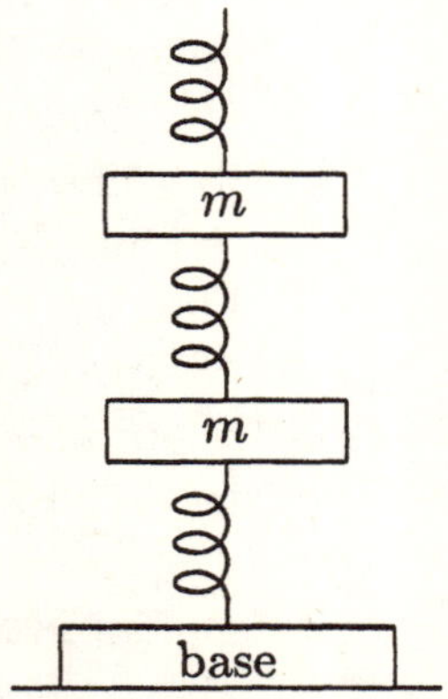

SUGGESTIONS

1. Assignments
   a. Ask some of questions 1 through 6 as part of the discussion of the conditions for simple harmonic motion and of the parameters of that motion. Question 5 is a good test of understanding of the phase constant. Also assign problems 15 and 28.
   b. Assign question 10 and problem 5 in support of the spring-mass demonstration and discussion. Also assign problem 25. Problems 20 and 23 deal with vertical oscillators. Assign one of these for variety.
   c. Springs in parallel and series test understanding of the spring force law. Consider problems 11, 24, and 26. Also consider problem 22.
   d. Assign problems 31 and 33 in support of the discussion of energy. When assigning problem 33 also ask for the maximum speed of the mass.
   e. If oscillators other than a spring-mass system are considered, assign problem 39 (angular simple harmonic motion), 46 (simple pendulum), and 51 (physical pendulum).
   f. Use problems 59 and 60 to test for understanding of damped harmonic motion and problems 62 and 63 to test for understanding of forced harmonic motion.
2. Demonstrations
   a. Simple harmonic motion: Freier and Anderson Mx1, 2, 3, 4, 7.
   b. Pendulums: Freier and Anderson Mx6, 9, 10, 11, 12, My1, 2, 3, 8, Mz1, 2, 3, 6, 7, 9.
3. Audio/Visual
   a. *Simple Harmonic Motion*; *The Stringless Pendulum*; *Sand Pendulum*; from the AAPT collection 2 of single-concept films; DVD; available from Ztek Co. (PO Box 11768, Lexington, KY 40577–1768, www.ztek.com).
   b. *Oscillations* from Skylab Physics; DVD; available from Ztek Co. (see above for address).
   c. *Tacoma Narrows Bridge Collapse*; from the AAPT Miller collection of single-concept films; DVD; available from Ztek Co. (see above for address) and from the American Association of Physics Teachers (AAPT, One Physics Ellipse, College Park MD 20740–3845; www.aapt.org).
   d. *Coupled Oscillators — Equal Masses*; *Coupled Oscillators — Unequal Masses*; from the AAPT Miller collection of single-concept films; DVD; available from Ztek Co. and from the AAPT (see above for addresses).
   e. *Periodic Motion*; from Cinema Classics DVD 2: Mechanics (II) and Heat; available from Ztek Co. and the AAPT (see above for addresses).
   f. *Twin Views of the Tacoma Narrows Bridge Collapse*; VHS video tape; available from the AAPT (see above for address).
4. Computer Software
   a. *Physics of Oscillation*; Eugene L. Butikov; Windows; available from Physics Academic Software (Centennial Campus, 940 Main Campus Drive, Suite 210, Raleigh, NC 27606–5212; www.aip.org/pas).
   b. *Mechanics* from Exploration of Physics Volume I; Windows and Macintosh; Physics Curriculum & Instruction (22585 Woodhill Drive, Lakeville, MN 55044; www.PhysicsCurriculum.com). Simulated experiments with analysis. Includes sections on springs, pendulums, damped oscillators, and 2D oscillators.
   c. *Vibrations, Waves, and Sound* from Exploration of Physics Volume II; Windows and Macintosh; Physics Curriculum & Instruction (see above for address). Includes sections on simple harmonic motion and resonance of a damped mass on a spring.
   d. *Dynamic Analyzer*. See Chapter 2 SUGGESTIONS.
   e. *Oscillations and Waves*; interactive CD-ROM; Films for the Humanities and Sciences (PO Box 2053, Princeton, NJ 08543–2053; www.films.com).

5. Laboratory
   a. Probeware Activity 13: *Harmonic Motion — Mass on a Spring.* A motion detector is used to plot the position of a weight hanging from a spring. The period of oscillation is then calculated and compared with the theoretical value.
   b. Probeware Activity 14: *Simple Harmonic Motion — Simple Pendulum.* A motion detector is used to plot the position of the bob of a simple pendulum. The period of oscillation is computed and compared with the theoretical value.
   c. Meiners Experiment 7–2: *The Vibrating Spring.* Students time a vertical vibrating spring with various masses attached, then use the data and a logarithmic plot to determine the relationship between the period and mass.
   d. Bernard Experiment 15: *Elasticity and Vibratory Motion.* The experiment is much the same as Meiners 7–2, in that a graph is used to determine the relationship between the mass on a spring and the period of oscillation. This measurement is preceded by a static determination of the spring constant.
   e. Meiners Experiment 7–4: *The Vibrating Ring.* Students time the oscillations of various diameter rings, hung on a knife edge, then use the data and a logarithmic plot to determine the relationship between the period and ring diameter. A good example of a physical pendulum.
   f. Meiners Experiment 7–15: *Investigation of Variable Acceleration.* A pendulum swings above a track and a spark timer is used to record its position as a function of time. Its velocity and acceleration are investigated.
   g. Meiners Experiment 7–19: *Harmonic Motion Analyzer.* This apparatus allows students to vary the spring constant, mass, driving frequency, driving amplitude, and damping coefficient of a spring-mass system. They can measure the amplitude, period, and relative phase of the oscillating mass. A variety of experiments can be performed.
   h. Meiners Experiment 8–4: *Linear Oscillator.* A spark timer is used to record the position of an oscillating mass on a spring, moving horizontally on a nearly frictionless surface. The period as a function of mass can be investigated and the conservation of energy can be checked.
   i. Meiners Experiment 7–18: *Damped Driven Linear Oscillator.* The amplitude and relative phase of a driven damped spring-mass system are measured as functions of the driving frequency and are used to plot a resonance curve.
   j. Meiners Experiment 7–20: *Analysis of Resonance with a Driven Torsional Pendulum.* The driving frequency and driving amplitude of a driven damped torsional pendulum are varied and the frequency, amplitude, and relative phase are measured. Damping is electromagnetic and can be varied or turned off. A variety of experiments can be performed.

Chapter 16 WAVES — I

BASIC TOPICS

I. General properties of waves.
   A. Explain that wave motion is the mechanism by which a disturbance created at one place travels to another. Use the example of a pulse on a taut string and point out that the displaced string causes neighboring portions of the string to be displaced. Stress that the individual particles have limited motion (perhaps perpendicular to the direction of wave travel), whereas the pulse travels the length of the string. Demonstrate by striking a taut string stretched across the room. Point out that energy is transported by the wave from one place to another. Ask the students to read the introductory section of the chapter for other examples of waves.

B. Point out that a wave on a string travels in one dimension, water waves produced by dropping a pebble travel in two, and sound waves emitted by a point source travel in three.
C. Explain the terms longitudinal and transverse. Demonstrate longitudinal waves with a slinky.
D. State that waves on a taut string of uniform density travel with constant speed and that this course deals chiefly with idealized waves that do not change shape. Take the string to lie along the x axis and draw a distortion in the shape of a pulse, perhaps a sketch of $\exp[-\alpha(x - x_0)^2]$. Remark that the initial displacement of the string can be described by giving a function $f(x)$. Now suppose the pulse moves in the positive x direction and draw the string at a later time. Point out that the maximum has moved from x_0 to $x_0 + vt$, where v is the wave speed. Remark that the displacement can be calculated by substituting $x - vt$ for x in the function $f(x)$. Substantiate the remark by showing that $x - vt = x_0$ if x is the coordinate of the pulse maximum at time t. Explain that $x + vt$ is substituted if the pulse travels in the negative x direction. Emphasize the relative signs of kx and ωt.

II. Sinusoidal traveling waves.

A. Write $f(x) = y_m \sin(kx)$ for the initial displacement of the string and sketch the function. Identify the amplitude as giving the limits of the displacement and point it out on the sketch. Also point out the periodicity of the function and identify the wavelength on the sketch. Show that k must be $2\pi/\lambda$ for $f(x)$ to equal $f(x + n\lambda)$ for all integers n. Remark that k is called the angular wave number of the wave.
B. Substitute $x - vt$ for x in $f(x)$ and explain you will assume the wave travels in the positive x direction. Show that the result is $y(x, t) = y_m \sin(kx - \omega t)$, where $\omega = kv$.
C. State that the motion of the string at any point is simple harmonic and that ω is the angular frequency. Show that at a given place on the string the motion repeats in a time equal to $2\pi/\omega$. This is the period T. Remind students that the frequency is $f = 1/T = \omega/2\pi$.
D. Remark that any given point on the string reaches its maximum displacement whenever a maximum on the wave passes that point. Since the time interval is one period a sinusoidal wave travels one wavelength in one period and $v = \lambda/T = \lambda f = \omega/k$, in agreement with the derivation of $y(x, t)$.
E. Explain that $y(x, t) = y_m \sin(kx + \omega t)$ represents a sinusoidal wave traveling in the negative x direction.
F. Show that the string velocity is $u(x, t) = \partial y/\partial t = -\omega y_m \cos(kx - \omega t)$. Point out that x is held constant in taking the derivative since the string velocity is proportional to the difference in the displacement of the *same* piece of string at two slightly different times. Remark that different points on the string may have different velocities at the same time and the same point may have a different velocity at different times. Contrast this behavior with that of the wave velocity. Point out that for a transverse wave u is transverse.
G. Explain that the wave speed for an elastic medium depends on the inertia and elasticity of the medium. State that, for a taut string, $v = \sqrt{\tau/\mu}$, where τ is the tension in the string and μ is the linear density of the string. Show how to measure μ for a homogeneous, constant radius string. The expression for v may be derived as in Section 17–6 of the text.
H. Point out that the frequency is usually determined by the source and that doubling the frequency for the same string with the same tension halves the wavelength. The product λf remains the same. Remark that if a wave goes from one medium to another the speed and wavelength change but the frequency remains the same. Work an example: given the two densities and the frequency, calculate the wave speed and wavelength in each segment. Draw a diagram of the wave.

III. Energy considerations.

A. Point out that the energy in the wave is the sum of the kinetic energy of the moving string and the potential energy the string has because it is stretched in the region of the disturbance. Energy moves with the disturbance.

B. Show that the kinetic energy of an infinitesimal segment of string is given by $dK = \frac{1}{2}\,dm\,v^2 = \frac{1}{2}(\mu\,dx)(\omega^2 y_m^2)\cos^2(kx - \omega t)$. State that this energy is transported to a neighboring portion in time $dt = v\,dx$, so $dK/dt = \frac{1}{2}\mu v\omega^2 y_m^2 \cos^2(kx - \omega t)$ gives the rate at which kinetic energy is transported past the point with coordinate x, at time t. Explain that when this is averaged over a cycle the result is $(dK/dt)_{\text{avg}} = \frac{1}{4}\mu v\omega^2 y_m^2$. Remark that this is not zero. Although kinetic energy moves back and forth as the string oscillates, there is a net flow.

C. State without proof that the average rate at which potential energy is transported is exactly the same as the rate for kinetic energy, so the average rate of energy flow is $P_{\text{avg}} = \frac{1}{2}\mu v\omega^2 y_m^2$. Note that this depends on the square of the amplitude and on the square of the frequency.

IV. Superposition and interference.

A. Stress that displacements, not intensities, add. State that if y_1 and y_2 are waves that are simultaneously present, then $y = y_1 + y_2$ is the resultant wave. Using diagrams of two similar sinusoidal waves, show that the resultant amplitude can be twice the amplitude of one of them, can vanish, or can have any value in between. Mention that the medium must be linear.

B. Start with the waves $y_1 = y_m \sin(kx - \omega t + \phi)$ and $y_2 = y_m \sin(kx - \omega t)$ and show that $y = 2y_m \cos(\phi/2)\sin(kx - \omega t + \phi/2)$. Show that maximum constructive interference occurs if $\phi = 2n\pi$, where n is an integer and maximum destructive interference occurs if $\phi = (2n+1)\pi$, where n is again an integer. Remark that the maximum amplitude is $2y_m$ and the minimum is zero. The derivation depends heavily on the trigonometric identity given as Eq. 16–50. You may wish to verify this identity for the class. Use the expressions for the sine and cosine of the sum of two angles to expand the right side of Eq. 16–50.

C. Interference can easily be demonstrated with a monaural amplifier, a signal generator, a microphone, an oscilloscope, and a pair of speakers. Fix the position of speaker S_1 and, with S_2 disconnected, show the wave form on the oscilloscope. Then, connect S_2 and show the wave form as S_2 is moved. Because both speakers are driven by the same amplifier, the only phase difference is due to the path difference.

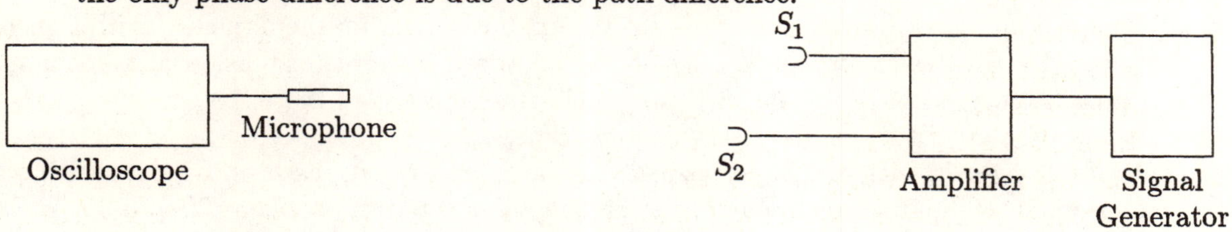

D. Explain that a phasor is an arrow that rotates around the location of its tail. Its length, to some scale, is taken to be the amplitude of a sinusoidal traveling wave and its angular velocity is taken to the angular frequency of the wave. Show that its projection on an axis through the tail behaves like the displacement in a wave. Point out the significance of the phase constant ϕ for the phasor rotation. Show how to use phasors to add two sinusoidal waves with the same frequency and wavelength but with different amplitudes and phase constants. Develop the expression $y_m^2 = y_{1m}^2 + y_{2m}^2 + 2y_{1m}y_{2m}\cos\phi$ for the amplitude of the resultant wave. Show how to use the law of sines to obtain the phase constant of the resultant wave.

V. Standing waves.

A. Use a mechanical oscillator to set up a standing wave pattern on a string. Otherwise, draw the pattern. Point out nodes and antinodes. Explain that all parts of the string vibrate either in phase or 180° out of phase and that the amplitude depends on position along the string. The disturbance does not travel. If possible, use a stroboscope to show the standing wave pattern. CAUTION: students with epilepsy should not watch this demonstration.

B. Explain that a standing wave can be constructed from two sinusoidal traveling waves of the same frequency and amplitude, traveling with the same speed in opposite directions. Use the trigonometric identity of Eq. 16–50 to show that $y_1 + y_2 = 2y_m \sin(kx)\cos(\omega t)$ if the phase constant for each wave is zero. Find the coordinates of the nodes and show they are half a wavelength apart. Also find the coordinates of the antinodes and show they lie halfway between nodes.

C. Point out that standing waves can be created by a wave and its reflection from a boundary. By means of a diagram show how the incident and reflected waves cancel at the fixed end of a string.

D. Remark that for a string fixed at both ends, each end must be a node. Derive the expression for the standing wave frequencies of such a string. Draw diagrams showing the string at maximum displacement for the lowest three or four frequencies.

E. Place two speakers, driven by the same signal generator and amplifier, well apart on the lecture table, facing the class. Standing waves are created throughout the room. Have each student place a finger in one ear and move his head slowly from side to side in an attempt to find the nodes and antinodes. Use a frequency of about 1 kHz.

F. Consider a driven string and describe resonance. Explain that the amplitude becomes large when the driving frequency matches a standing wave frequency. Explain that at resonance the energy supplied by the driving force is dissipated and that off resonance the string does work on the driving mechanism.

G. You may wish to explain that when the string is driven at a non-resonant frequency, each traveling wave and its reflection from an end produce a standing wave, just as at resonance. The standing waves produced by successive reflections, however, do not coincide and a jumble results.

SUGGESTIONS

1. Assignments
   a. Use question 4 to discuss wave speed. To emphasize the mathematical description of a traveling wave, assign problems 6 and 8. Wave speed in terms of tension and linear mass density is covered in problems 13 through 21. Assign a few of these.
   b. Assign problem 29 when discussing energy transport.
   c. Question 4 deals with the superposition of waves and questions 5 and 7 deal with wave interference. The fundamentals of interference are covered in problems 29 and 30. Include problem 34 if you discuss phasors.
   d. Assign questions 9, 10, and 11 and problems 38, 43, and 44 in connection with standing waves. The superposition of traveling waves to form a standing wave is covered in problems 47, 50, and 53. For a challenge assign problem 55.
2. Demonstrations
   a. Traveling waves: Freier and Anderson Sa3, 4, 5, 6, 12, 13.
   b. Reflection: Freier and Anderson Sa7, 12, 14.
   c. Standing waves: Freier and Anderson Sa8, 9.
3. Audio/Visual
   a. *Superposition*; *Vibrations of a Wire*; *Vibrations of a Drum*; from the AAPT collection 1 of single-concept films; DVD; available from Ztek Co. (PO Box 11768, Lexington, KY 40577–1768, www.ztek.com).

b. *Nonrecurrent Wavefronts*; from the AAPT Miller collection of single-concept films; DVD; available from Ztek Co. (see above for address) and from the American Association of Physics Teachers (AAPT, One Physics Ellipse, College Park MD 20740–3845; www.aapt-.org).
c. *Wave Propagation*, *Periodic Waves*, *Superposition*, *Standing Waves*; Cinema Classics DVD 3: Waves (I); available from Ztek Co. and from the AAPT (see above for addresses).
d. *Mechanical Resonance*; *Velocity/Wavelength & Frequency*; *Standing Waves*; *Change in Medium/Interference*; from Physics Demonstrations in Sound & Waves, Part I; VHS video tape, DVD; ≈3 min each; Physics Curriculum & Instruction (22585 Woodhill Drive, Lakeville, MN 55044; www.PhysicsCurriculum.com).

4. Computer Software
   a. *Waves* from Exploration of Physics Volume I; Windows and Macintosh; Physics Curriculum & Instruction (see above for address). Simulated experiments with analysis. Deals with waves on a taut string, interference, and standing waves.
   b. *Vibrations, Waves, and Sound* from Exploration of Physics Volume II; Windows and Macintosh; Physics Curriculum & Instruction (see above for address). Includes sections on frequency, phase, amplitude, and superposition.
   c. *Physics Simulation Programs*; Robert H. Good; DOS; available from Physics Academic Software (Centennial Campus, 940 Main Campus Drive, Suite 210, Raleigh, NC 27606–5212; www.aip.org/pas). Contains simulations of traveling and standing waves.
   d. *WaveMaker*; Freeman Deutsch, Philip Sadler, Charles Whitney, Stephen Engquist, and Linda Shore; Macintosh; available from Physics Academic Software (see above for address). Beads are attached to elastic, massless strings and oscillate transversely. The user can control the masses and the spring constants. The program sill plot the position, velocity, and acceleration of any bead. Demonstrates beats, reflection at fixed and free ends, normal oscillations, wave superposition, and transmission through a boundary between two different media.
   e. *Wave Motion*; VHS video tape, DVD; Films for the Humanities and Sciences (PO Box 2053, Princeton, NJ 08543–2053; www.films.com).
   f. *Oscillations and Waves*; interactive CD-ROM; Films for the Humanities and Sciences (see above for address).
   g. *Harmonic Motion and Waves*; VHS video tape, DVD (part of a collection); Films for the Humanities and Sciences (see above for address).

5. Computer Project

   Have students use a commercial math program, a spreadsheet, or their own programs to investigate energy in a string carrying a wave. The program should calculate the kinetic, potential, and total energies at a given point and time, given the string displacement as a function of position and time. Use the program to plot the energies as functions of time for a given position. Consider a pulse, a sinusoidal wave, and a standing wave. Demonstrate that energy passes the point in the first two cases but not in the third. For sinusoidal and standing waves, the program should also calculate averages over a cycle.

6. Laboratory
   a. Meiners Experiment 12–1: *Transverse Standing Waves* (Part A). Several harmonics are generated in a string by varying the driving frequency. Frequency ratios are computed and compared with theoretical values. Values of the wave speed found using λf and using $\sqrt{\tau/\mu}$ are compared. The experiment can be repeated for various tensions and various linear densities.
   b. Bernard Experiment 22: *A Study of Vibrating Strings.* A horizontal string is attached to a driven tuning fork vibrator. It passes over a pulley and weights are hung on the end.

The weights are adjusted so standing wave patterns are obtained and the wavelength of each is found from the measured distance between nodes. Graphical analysis is used to find the relationship between the wave velocity and the tension in the string and to find the frequency. Several strings are used to show the relationship between the wave velocity and the linear density.

Chapter 17 WAVES — II

BASIC TOPICS

I. Qualitative description of sound waves.
   A. Explain that the disturbance that is propagated is a deviation from the ambient density and pressure of the material in which the wave exists. This comes about through the motion of particles. If Chapter 14 was not covered, you should digress to discuss density and pressure briefly. Point out that sound waves in solids can be longitudinal or transverse but sound waves in fluids are longitudinal: the particles move along the line of wave propagation. Waves in crystalline solids moving in low symmetry directions are examples that are neither transverse nor longitudinal. Use a slinky to show a longitudinal wave and point out the direction of motion of the particles. State that sound can be propagated in all materials.
   B. Draw a diagram, similar to Fig. 17–3, to show a compressional pulse. Point out regions of high, low, and ambient density. Also show the pulse at a later time.
   C. Similarly, diagram a sinusoidal sound wave in one dimension and draw a rough graph of the pressure as a function of position for a given time. Give the rough frequency limits of audible sound and mention ultrasonic and infrasonic waves.
   D. Discuss the idea that the wave velocity depends on an elastic property of the medium (bulk modulus) and on an inertia property (ambient density). Recall the definition of bulk modulus (or introduce it) and show by dimensional analysis that v is proportional to $\sqrt{B/\rho}$. Assert that the constant of proportionality is 1. Point out the wide range of speeds reported in Table 17–1.

II. Interference.
   A. Remind students of the conditions for interference. Consider two sinusoidal sound waves with the same amplitude and frequency, traveling in the same direction. Explain that constructive interference occurs if they are in phase and complete destructive interference occurs if they are π rad out of phase.
   B. Explain that a phase difference can occur at a detector if two waves from the same source travel different distances. Show that the phase difference is given by $k\Delta x$ $(= 2\pi\Delta x/\lambda)$.
   C. Interference of sound waves can be demonstrated by wiring two speakers to an audio oscillator and putting the apparatus on a slowly rotating platform. Students will hear the changes in intensity.

III. Mathematical description of one-dimensional sound waves.
   A. If desired, derive $v = \sqrt{B/\rho}$ as it is done in the text.
   B. Write $s = s_m \cos(kx - \omega t)$ for the displacement of the material at x. Show how to calculate the pressure as a function of position and time. Relate the pressure amplitude to the displacement amplitude. Explain that a sinusoidal pressure wave traveling in the positive x direction is written $\Delta p(x,t) = \Delta p_m \sin(kx - \omega t)$, where $\Delta p_m = v\rho\omega s_m$. State that Δp is the deviation of the pressure from its ambient value. Remind students that $k = 2\pi/\lambda$, $f = \omega/2\pi$, and $\lambda f = v$.

C. Remark that power is transmitted by a sound wave because each element of fluid does work on neighboring elements. Show that the kinetic energy in an infinitesimal length dx of a sinusoidal sound wave traveling along the x axis is $dK = \frac{1}{2}A\rho\omega^2 s_m^2 \sin^2(kx - \omega t)\,dx$, where A is the cross-sectional area. Show that its average over a cycle is $(dK)_{\text{avg}} = \frac{1}{4}A\rho\omega^2 s_m^2$. Argue that this energy moves to a neighboring segment in time $dt = dx/v$ and show that the rate of kinetic energy flow is, on average, $(dK/dt)_{\text{avg}} = \frac{1}{4}A\rho v\omega^2 s_m^2$, where v is the speed of sound. Tell students that the rate of flow of potential energy is exactly the same, so the rate of energy flow is $P_{\text{avg}} = \frac{1}{2}A\rho v\omega^2 s_m^2$.

D. Define intensity as the average rate of energy flow per unit area and show that it is given by $I = \frac{1}{2}\rho v\omega^2 s_m^2$. Show that conservation of energy implies that the intensity decreases as the reciprocal of the square of the distance as a spherical wave moves outward from an isotropic point source.

E. Show a scale of the range of human hearing in terms of intensity. Introduce the idea of sound level and define the bel and decibel. Discuss both absolute (relative to $10^{-12}\,\text{W/m}^2$) and relative intensities. Remark that an increase in intensity by a factor of 10 means an increase in sound level by 10 db. If you have a sound level meter, use an oscillator, amplifier, and speaker to demonstrate the change of a few db in sound level.

IV. Standing longitudinal waves and sources of sound.

A. Use a stringed instrument or a simple taut string to demonstrate a source of sound. Point out that the wave pattern on the string is very nearly a standing wave, produced by a combination of waves reflected from the ends. If the string is vibrating in a single standing wave pattern, then sound waves of the same frequency are produced in the surrounding medium. Demonstrate the same idea by striking a partially filled bottle, then blowing across its mouth. Also blow across the open end of a ball point pen case. If you have them, demonstrate Chladni plates.

B. Derive expressions for the natural frequencies and wavelengths of air pipes open at both ends and closed at one end. Stress that pressure nodes occur near open ends and that pressure antinodes occur at closed ends. Define the terms fundamental and harmonic.

C. Optional: Discuss the quality of sound for various instruments in terms of harmonic content. If possible, demonstrate the instruments.

D. Demonstrate voice patterns by connecting a microphone to an oscilloscope and keeping the setup running through part or all of the lecture. This is particularly instructive in connection with part C.

V. Beats.

A. Demonstrate beats using two separate oscillators, amplifiers, and speakers, operating at nearly, but not exactly, the same frequency. If possible, show the time dependence of the wave on an oscilloscope. Remark that the sound is like that of a pure note but the intensity varies periodically. Explain that this technique is used to tune instruments in an orchestra.

B. Remark that you will consider displacement oscillations at a point in space when two sound waves of the same amplitude and nearly the same frequency are present. Write the expression for the sum of $s_1 = s_m \cos(\omega_1 t)$ and $s_2 = s_m \cos(\omega_2 t)$, where $\omega_1 \approx \omega_2$, but the two frequencies are not exactly equal. Show that $s_1 + s_2 = 2s_m \cos(\omega' t)\cos(\omega t)$, where $\omega' t = (\omega_1 - \omega_2)/2$ and $\omega = (\omega_1 + \omega_2)/2$. Remark that because the difference in frequencies is much smaller than either constituent frequency we can think of the oscillation as having an angular frequency of $\omega = (\omega_1 + \omega_2)/2$ and a time dependent amplitude. Note that the angular frequency of the amplitude is $\omega' = |\omega_1 - \omega_2|/2$ but the angular frequency of the intensity is $\omega_{\text{beat}} = |\omega_1 - \omega_2|$. The latter is the beat angular frequency.

VI. Doppler effect.

A. Explain that the frequency increases when the source is moving toward the listener, decreases when the source is moving away, and that similar effects occur when the listener is moving toward or away from the source. Use Fig. 17–19 to illustrate the physical basis of the phenomenon.

B. Derive expressions for the frequency when the source is moving and for the frequency when the listener is moving. Point out that the velocities are measured relative to the medium carrying the sound.

C. The effect can be demonstrated by placing an auto speaker and small audio oscillator (or sonalert type oscillator) on a rotating table. The sonalert can also be secured to a cable and swung in a circle. Show the effect of a passive reflector by moving a hand-held sonalert rapidly toward and away from the blackboard.

SUGGESTIONS

1. Assignments
   a. The speed of sound is emphasized in problem 3.
   b. Ask question 3 and assign problems 17 and 19 in connection with interference.
   c. Use problems 22 and 23 to discuss sound intensity and problems 24 and 25 to discuss sound level. They will help students with the concepts of bel and decibel. Also consider problem 24.
   d. Ask questions 4 through 7 and assign problems 36, 42, and 44 when discussing standing waves.
   e. Assign problems 46 and 46 in connection with beats.
   f. Use question 10 and problem 59 in a discussion of the Doppler effect. Assign problems 57 and 61. Assign problem 63 in connection with sonic booms.
2. Demonstrations
   a. Wavelength and speed of sound in air: Freier and Anderson Sa16, 17, 18, Sh1.
   b. Sound not transmitted in a vacuum: Freier and Anderson Sh2.
   c. Sources of sound, acoustical resonators: Freier and Anderson Sd3, Se, Sf, Sj6.
   d. Harmonics: Freier and Anderson Sj2 — 5
   e. Beats: Freier and Anderson Si4 — 6.
   f. Doppler shift: Freier and Anderson Si1 — 3.
3. Books and Monographs
   a. *Resource Letters, Book Four* and *Resource Letters, Book Five*; available from the American Association of Physics Teachers (AAPT, One Physics Ellipse, College Park MD 20740–3845; www.aapt.org). Contains resource letters on sound and acoustics.
   b. *Musical Acoustics*; edited by Thomas D. Rossing; available from the AAPT (see above for address). Reprints.
4. Audio/Visual
   a. *Waves and Sound*; VHS video tape, DVD; Films for the Humanities and Sciences (PO Box 2053, Princeton, NJ 08543–2053; www.films.com).
   b. *Experiments on the Doppler Effect*; VHS video tape, DVD; Films for the Humanities and Sciences (see above for address).
   c. *Longitudinal Waves*; *Longitudinal Standing Waves*; from Physics Demonstrations in Sound & Waves, Part I; VHS video tape, DVD; ≈3 min each; Physics Curriculum & Instruction (22585 Woodhill Drive, Lakeville, MN 55044; www.PhysicsCurriculum.com).
   d. *Nature of Sound Waves*; *Propagation of Sound*; *Transmission of Sound*; *Refraction of Sound*; *Interference of Sound*; *diffraction of Sound*; *Doppler Effect*; from Physics Demonstrations in Sound & Waves, Part II; VHS video tape, DVD; ≈3 min each; Physics Curriculum & Instruction (see above for address).

e. *Standing Sound Waves*; *Standing Sound Waves in Two Dimensions*; *Resonance/Real Time*; *Superposition Principle*; from Physics Demonstrations in Sound & Waves, Part III; VHS video tape, DVD; ≈3 min each; Physics Curriculum & Instruction (see above for address).

5. Computer Software
   a. *Waves* from Exploration of Physics Volume I; Windows and Macintosh; Physics Curriculum & Instruction (22585 Woodhill Drive, Lakeville, MN 55044; www.PhysicsCurriculum.com). Simulated experiments with analysis. Some sections deal with beats and the Doppler shift.
   b. *Vibrations, Waves, and Sound* from Exploration of Physics Volume II; Windows and Macintosh; Physics Curriculum & Instruction (see above for address). Includes sections on standing waves in organ pipes. the Doppler effect, and sonic booms.
   c. *Physics Demonstrations*. See Chapter 2 SUGGESTIONS.
6. Computer Projects
   A spreadsheet or computer program can be used to add waves. Have students use it to investigate interference effects and beats.
7. Laboratory
   a. Meiners Experiment 12–2: *Velocity of Sound in Air* and Bernard Experiment 23: *Velocity of Sound in Air — Resonance-Tube Method.* Resonance of an air column is obtained by holding a tuning fork of known frequency at the open end of a tube with one closed end. The length of the column is changed by adjusting the amount of water in the tube. The wavelength and speed of sound are found.
   b. Meiners Experiment 12–3: *Velocity of Sound in Metals* and Bernard Experiment 24: *Velocity of Sound in a Metal — Kundt's-Tube Method.* A Kundt's tube is used to find the frequency of sound excited in a rod with its midpoint clamped and its ends free. The wavelength is known to be twice the rod length and λf is used to find the speed of sound. In another experiment, a transducer and oscilloscope are used to time a sound pulse as it travels the length of a rod and returns.
   c. Meiners Experiment 12–4: *Investigation of Longitudinal Waves.* The amplitude and phase of a sound wave are investigated as functions of distance from a speaker source. To do this, Lissajous figures are generated on an oscilloscope screen by the source signal and the signal picked up by a microphone. To eliminate noise, the speaker and microphone should be in a large sound-proof enclosure with absorbing walls. Use Meiners Experiment 10–10 to familiarize students with the oscilloscope and Lissajous figures.
   d. Probeware Activity 15: *Superposition.* Two sinusoidal sound waves with the same frequency are generated and the individual and superposed pressure waves are plotted on the computer monitor. The frequency and relative phase can be adjusted. Another part of the activity demonstrates the interference of four sound waves.
   e. Probeware Activity 16: *Interference — Beats.* Two sound waves with slightly different frequencies are generated. The individual and superposed pressure waves are shown on the computer monitor. The frequencies can be adjusted. The beat frequency is measured and the value compared with the theoretical value.

Chapter 18 TEMPERATURE, HEAT, AND THE FIRST LAW OF THERMODYNAMICS

BASIC TOPICS

I. The zeroth law of thermodynamics.
   A. Explain that if two bodies, not in thermal equilibrium, are allowed to exchange energy

then they will do so and one or more of their macroscopic properties will change. When no further changes take place, the bodies are in thermal equilibrium. Explain that two bodies in thermal equilibrium are said to have the same temperature.

B. For gases, the properties of interest include pressure, volume, internal energy, and the quantity of matter. Other properties may be included for other materials. The quantity of matter may be given as the number of particles or as the number of moles.

C. Explain what is meant by diathermal and adiabatic walls and remark that diathermal walls are used to obtain thermal contact without an exchange of particles. Adiabatic walls are used to thermally isolate a system.

D. State the zeroth law: if body A and body B are each in thermal equilibrium with body C, then A is in thermal equilibrium with B. Discuss the significance of the zeroth law. State that it is the basis for considering the temperature to be a property of an object. If it were not true, then, at best, an object might have a large number of temperatures, depending on what other objects were in thermal equilibrium with it.

E. Explain that the temperature of a body is measured by measuring some property of a thermometer in thermal equilibrium with it. Illustrate by reminding students that the length of the mercury column in an ordinary household thermometer is a measure of the temperature. Explain that the zeroth law guarantees that the same temperature, as measured by the same thermometer, will be obtained for two substances in thermal equilibrium with each other.

II. Temperature measurements.

A. Mention that the value of the temperature obtained depends on the substance used for the thermometer and on the property measured but that several techniques exist that allow us to define temperature independently of the thermometric substance and property.

B. Describe a constant-volume gas thermometer. If one is available, demonstrate its use. If not, show Fig. 18–5. The gas is placed in thermal contact with the substance whose temperature is to be measured and the pressure is adjusted so that the volume has some standard value (for that thermometer). After corrections are made, the temperature is taken to be proportional to the pressure: $T = ap$, where a is the constant of proportionality.

C. Describe the triple point of water and explain that water at the triple point is assigned the temperature $T = 273.16\,\text{K}$. Solve for a and show that $T = 273.16(p/p_3)$.

D. Point out that thermometers using different gases give different values for the temperature when used as described. Explain the limit used to obtain the Kelvin temperature. See Fig. 18–6.

E. Define the Celsius and Fahrenheit scales. Give the relationships between the degree sizes and the zero points. Give equations for conversion from one scale to another and give the temperature value for the ice and steam points in each system. Use Fig. 18–7 and Table 18–1.

F. Define the Kelvin scale and explain the kelvin as a unit of temperature. Give the relationship between the Celsius and Kelvin scales. Give the ice and steam points on the Kelvin scale.

III. Thermal expansion.

A. Describe linear expansion and define the coefficient of linear expansion: $\alpha = \Delta L/L\Delta T$. Point out Table 18–2. Obtain a bimetallic strip and use both a bunsen burner and liquid nitrogen (or dry ice) to show bending. After the students see the strip bend ask which of the metals has the greater coefficient of linear expansion. Explain that these devices are often used in thermostats.

B. Discuss area and volume expansion. Consider a plate and show that the coefficient of area expansion is 2α. Consider a rectangular solid and show that the coefficient of volume expansion is 3α. In each case apply the equation for linear expansion to each dimension of the object and find ΔA or ΔV to first order in ΔT.

C. Explain that the length of a scratch on the flat face of an object increases as the temperature increases. The area of a hole also increases. Carefully drill a 1/2 inch hole in a piece of aluminum, roughly $1\frac{1}{4}$ inch thick. Obtain a 13-mm diameter steel ball bearing and place it in the hole. It will not pass through. Heat the plate on a bunsen burner and the ball passes through easily.

D. Demonstrate volume expansion of a gas using a flat bottomed flask, a bulbed tube, a two hole stopper, and some colored water. Partially evacuate the bulb so the colored water stands in the tube somewhat above the stopper. Place your hand on the bulb to warm the air inside and the water in the tube drops in response.

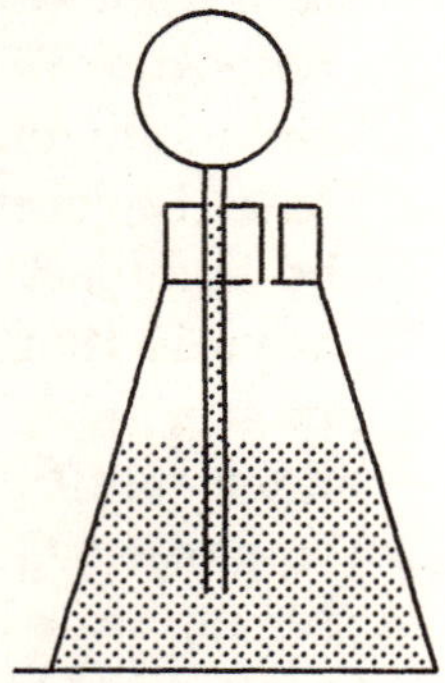

IV. Heat.

A. Explain that when thermal contact is made between two bodies at different temperatures, a net flow of energy takes place from the higher temperature body to the lower temperature body. The temperature of the hotter body decreases, the temperature of the cooler body increases, and the net flow continues until the temperatures are the same. Energy also flows from warmer to cooler regions of the same body. State that heat is energy that is transferred because of a temperature difference. Distinguish between heat and internal energy. Emphasize that the idea of a body having heat content is not meaningful. Also emphasize that heat is not a new form of energy. The energy transferred may be the kinetic energy of molecules or the energy in an electromagnetic wave. Examples: a bunsen burner flame, radiation across a vacuum. State that heat is usually measured in Joules but calories and British thermal units are also used. 1 kcal = 3.969 Btu = 4187 J. Remark that the unit used in nutrition, a Calorie (capitalized) is 1 kcal.

B. Remind students of the energy equation studied in Chapter 8. Tell them that for the systems considered here the center of mass remains at rest (or has a constant velocity) and changes in potential energy are ignored. Processes considered change only the internal energy. A new term, however, must be added since the environment can exchange energy as heat with the system, as well as do macroscopic work on the system. Write $\Delta E_{\text{int}} = Q - W$, where Q is the energy absorbed as heat *by* the system and W is the work done *by* the system.

C. Stress the sign convention for heat and work: Q is positive if the system takes in energy, W is positive if the system does positive work.

D. Stress that heat and work are alternate means of transferring energy and explain that, for example, temperature changes can be brought about by both heat and mechanical work. To demonstrate this, connect a brass tube, fitted with a rubber stopper, to a motor as shown. Make a wooden brake or clamp that fits tightly around the tube. Put a few drops of water into the tube, start the motor, and exert pressure on the tube with the clamp. Soon the stopper will fly off. Note that mechanical work was done and steam was produced.

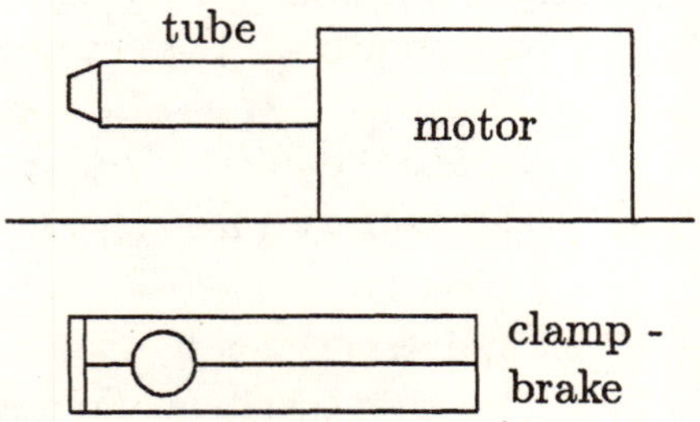

V. Heat capacity.

A. Define the heat capacity of a body as the amount of energy absorbed as heat per degree of temperature change: for a small temperature change $C = Q/\Delta T$. Point out that it depends on the temperature and on the constraints imposed during the transfer. The heat capacity at constant volume is different from the heat capacity at constant pressure because positive work is done by the system when the temperature is increased at constant pressure. More energy is therefore required as heat to obtain the same increase in internal energy and temperature.

B. Point out that the heat capacity depends on the amount of material. Define the specific heat c and the molar specific heat. Explain they are independent of the amount of material. Point out Table 18–3. You might use C' to denote a molar specific heat.

C. Do a simple calorimetric calculation (see Sample Problem 18–4). Stress the fundamental idea: the energy that leaves one body enters another, so the sum of the energies absorbed by all objects in a closed system vanishes.

D. Explain that energy must be transferred to or from a body when it changes phase (liquid to gas, etc.). The energy per unit mass is called the heat of transformation or latent heat. Point out Table 18–4. If time permits, work a calorimetric problem that involves a change in phase. Consider, for example, dropping ice into warm water and calculate the final temperature. Work a problem for which all the ice is melted and a problem for which only part of the ice is melted.

VI. Heat, work, and the first law of thermodynamics.

A. Describe a gas in a cylinder fitted with a piston. Remind students that as the piston moves the gas volume changes and the gas does work $W = \int p\,dV$ on the piston. Explain that the gas might exchange energy with its environment through both work and heat.

B. Draw a p-V diagram (such as Fig. 18–14) and mark initial and final states, with $V_f > V_i$. Explain that p and V are thermodynamic state variables and have definite, well defined values for a given thermodynamic state. They can be used to specify the state. Point out there are many paths from the initial to the final state. Define the term "quasi-static process" and explain that the various paths on the diagram represent quasi-static processes, for which the system is always infinitesimally close to equilibrium states. Point out that for different paths p is a different function of V and that the work is different for different paths. Also explain that the heat is different for different paths. Work and heat are not thermodynamic state variables.

C. Explain that $Q - W$ is independent of the process. Define the internal energy by $\Delta E_{\text{int}} = Q - W$ and point out that ΔE_{int} is the same for any two selected states regardless of the path used to get from one to the other. State that ΔE_{int} is the change in mechanical energy (kinetic and potential energy) of all the particles that make up the system. Stress that the first law $\Delta E_{\text{int}} = Q - W$ is an expression of the conservation of energy.

VII. Applications of the first law.

A. Adiabatic process. Explain that $Q = 0$ and $\Delta E_{\text{int}} = -W$. As an example, consider a gas in a thermally insulated cylinder and allow the volume to change by moving the piston. Explain that when the internal energy increases the temperature goes up for most materials. This can be achieved by compressing the gas. The opposite occurs when the piston is pulled out. Stress that no heat has been exchanged. Illustrate an adiabatic process on a p-V diagram.

B. Constant volume process. Explain that $W = 0$ and $\Delta E_{\text{int}} = Q$. Illustrate on a p-V diagram.

C. Isobaric process. Explain that $W = p(V_f - V_i)$ for a quasi-static isobaric process. For a change in phase, show that $\Delta E_{\text{int}} = mL - p\Delta V$. Illustrate on a p-V diagram.

D. Describe adiabatic free expansion and note that $\Delta E_{\text{int}} = 0$. Explain that this process is not quasi-static and cannot be shown on a p-V diagram. The end points, however, are well defined thermodynamic states and are points on a p-V diagram.

E. Cyclical process. Explain that all state variables return to their original values at the end of each cycle and, in particular, $\Delta E_{int} = 0$. Thus, $Q = W$. Illustrate on a p-V diagram. For later reference, stress that heat may be absorbed (or rejected) and work done during a cyclic process.

VIII. Transfer of heat.

A. Explain that steady state heat flow can be obtained if both ends of a slab are held at different temperatures. Define the thermal conductivity k of the material using $P_{cond} = -kA\,dT/dx$ for a slab of uniform cross section A. Here P_{cond} is the rate of heat flow. Emphasize that the negative sign appears because heat flows from hot to cold. Stress that P_{cond} and T are constant in time in the steady state. Explain that $P_{cond} = kA(T_H - T_C)/L$ for a uniform bar of length L, with the cold end held at temperature T_C and the hot end held at temperature T_H.

B. A demonstration that shows both thermal conductivity and heat capacity can be constructed from three rods of the same size, one made of aluminum, one made of iron, and one made of glass. Use red wax to attach small ball bearings at regular intervals along each rod. Clamp the rods so that each has one end just over a bunsen burner. The rate at which the wax melts and the ball bearings drop off is mostly dictated by the thermal conductivity of the rods, but it is influenced a bit by the specific heats.

C. For a practical discussion, introduce the idea of R value and discuss home insulation.

D. Qualitatively discuss radiation as a means of energy transfer. Place a heating element at the focal point of one spherical reflector and some matches, stuck in a cork, at the focal point of the another. Place the reflectors several meters apart and adjust the positions so that the heater is imaged at the matches. Use a 1 kW or so heater. The matches will ignite in about a minute.

E. Give $P_{rad} = \sigma\epsilon A T^4$ for the rate with which a surface with area A at Kelvin temperature T emits radiative energy. Here σ $(= 5.603 \times 10^{-8}\,\mathrm{W/m^2 \cdot K^4})$ is the Stefan-Boltzmann constant and ϵ is the emissivity of the surface. Explain that ϵ has a value between 0 and 1 and depends on the composition of the surface.

F. Qualitatively discuss convection as a means of heat transfer.

SUGGESTIONS

1. Assignments
   a. After discussing gas thermometers assign problems 1 and 3. Temperature scales are covered in problems 4 and 7.
   b. Use question 2 and one or two of problems 8, 12, and 15 to discuss thermal expansion. Use problem 17 in connection with the ball and hole demonstration.
   c. To test the fundamental concepts of heat capacity and heat of transformation assign question 3 and some of problems 22 through 41. Some of these problems involve phase changes. Include one or two of them.
   d. Problems 42 and 44 are good tests of understanding of the first law. Also assign questions 5 and 6 and problems such as 42, 43, 45, 48, and 49, which involve the interpretation of p-V diagrams. Tell students to pay attention to signs.
   e. Following the discussion of thermal conductivity, assign problems 51, 53, and 57 in connection with heat conduction.

2. Demonstrations
   a. Thermometers: Freier and Anderson Ha1 — 4.
   b. Thermal expansion: Freier and Anderson Ha5 — 12.
   c. Heat capacity and calorimetry: Freier and Anderson Hb1, 2.
   d. Work and heat: Freier and Anderson He1 — 6.

e. Heat transfer: Freier and Anderson Hc, Hd1 — 7, Hf.
f. p-V relations: Freier and Anderson Hg1 — 3.

3. Books and Monographs

   Resource Letters, Book Five; the American Association of Physics Teachers (AAPT, One Physics Ellipse, College Park MD 20740–3845; www.aapt.org). Contains an exhaustive list of journal articles on heat and thermodynamics.

   Resource Letters, Book Six; the AAPT (see above for address). Contains a list of journal articles on heat and thermodynamics.

4. Audio/Visual

   Heat and Temperature; Cinema Classics DVD 2: Mechanics (II) and Heat; available from Ztek Co. (PO Box 11768, Lexington, KY 40577–1768, www.ztek.com) and from the AAPT (see above for address).

   a. *Thermodynamics*; VHS video tape, DVD (part of a collection); Films for the Humanities and Sciences (PO Box 2053, Princeton, NJ 08543–2053; www.films.com).
   b. *Heat and Temperature*; VHS video tape; Films for the Humanities and Sciences (see above for address).
   c. *Heat*; VHS video tape, DVD; Films for the Humanities and Sciences (see above for address).
   d. *Heat*; VHS video tape, DVD (part of a collection); Films for the Humanities and Sciences (see above for address).
   e. *The Conduction of Heat*; VHS video tape; Films for the Humanities and Sciences (see above for address).
   f. *The Convection of Heat*; VHS video tape; Films for the Humanities and Sciences (see above for address).

5. Computer Software

   a. *Heat* from Exploration of Physics Volume I; Windows and Macintosh; Physics Curriculum & Instruction (22585 Woodhill Drive, Lakeville, MN 55044; www.PhysicsCurriculum.com). Simulated experiments with analysis. Includes sections on heat, heat conduction, calorimetry, and the first law of thermodynamics.
   b. *Thermodynamics Lecture Demonstrations*; Kurt Wick and Philip Johnson; Windows; available from Physics Academic Software (Centennial Campus, 940 Main Campus Drive, Suite 210, Raleigh, NC 27606–5212; www.aip.org/pas). Simulations of ten thermodynamic processes, including isochoric, adiabatic, and isothermal processes, the Carnot cycle, and the Otto and diesel engines. The processes are also diagramed on p-V and T-S diagrams.

6. Laboratory

   a. Meiners Experiment 9–3: *Linear Expansion* and Bernard Experiment 18: *Linear Coefficient of Expansion of Metals.* The length of a metal rod is measured at room temperature and at 100°C (in a steam jacket), then the data is used to compute the coefficient of thermal expansion. The experiment can be repeated for several different metals and the results compared.
   b. Probeware Activity 17: *Temperature and Heat.* Water is heated electrically and the temperature of the water is plotted as a function of time. The energy input for a given temperature change is computed and the values for two difference amounts of water are compared.
   c. Meiners Experiment 9–1: *Calorimetry — Specific Heat and Latent Heat of Fusion.* Students use a calorimeter to find the specific heat of water and a metal sample. They also measure the latent heat of fusion of ice. Since the specific heat of the stirring rod and the calorimeter must be taken into account, this is a good exercise in experimental design.
   d. Meiners Experiment 9–2: *Calorimetry — Mechanical Equivalent of Heat* and Bernard Experiment 30: *The Heating Effect of an Electric Current.* A calorimeter is used to find

the relationship between the energy dissipated by a resistive heating element and the temperature rise of the water in which it is immersed. Students must accept $P = i^2R$ for the power output of the heating element. With slight revision these experiments can also be used in conjunction with Chapter 27.

e. Bernard Experiment 19: *Specific Heat and Temperature of a Hot Body.* A calorimeter is used to obtain the specific heat of metal pellets. In a second part, a calorimeter and a metal sample with a known specific heat are used to find the temperature of a Bunsen burner flame.

f. Probeware Activity 18: *Specific Heat.* A small metal object of known mass, initially at room temperature, is placed in an ice bath of known mass and its temperature is monitored as it cools. The specific heat of the object, relative to that of water, is then computed.

g. Bernard Experiment 20: *Change of Phase — Heat of Fusion and Heat of Vaporization.* A calorimeter is used to measure the heat of fusion and heat of vaporization of water. If the lab period is long or writeups are done outside of lab, experiments 19 and 20 may be combined nicely.

h. Meiners Experiment 9–6: *Calorimetry Experiments* (with a microprocessor).

i. Meiners Experiment 9–4: *Thermal Conductivity.* The sample is sandwiched between a thermal reservoir and a copper block. The rate at which energy passes through the sample is found by measuring the rate at which the temperature of the copper increases. Temperature is monitored by means of a thermocouple.

j. Meiners Experiment 9–5: *Thermal Conductivity with Microprocessor.*

Chapter 19 THE KINETIC THEORY OF GASES

BASIC TOPICS

I. Macroscopic description of an ideal gas.

A. Explain that kinetic theory treats the same type problems as thermodynamics but from a microscopic viewpoint. It uses averages over the motions of individual particles to find macroscopic properties. Here it is used to clarify the microscopic basis of pressure and temperature.

B. Define the mole. Define Avogadro's number N_A and give its value, $6.02 \times 10^{23}\,\text{mol}^{-1}$. Explain the relationships between the mass of a molecule, the mass of the sample, the molar mass, the number of moles, the number of molecules, and Avogadro's number. These often confuse students.

C. Write down the ideal gas equation of state in the form $pV = nRT$ and in the form $pV = NkT$. Here N is the number of molecules and n is the number of moles. Give the values of R and k and state that $k = R/N_A$. Explain that for real gases at low density pV/T is nearly constant. Point out that the equation of state connects the thermodynamic variables n (or N), p, V, and T. Draw some ideal gas isotherms on a p-V diagram.

D. To show how the equation of state is used in thermodynamic calculations, go over Sample Problem 19–1. Also consider a problem in which the pressure and volume of an ideal gas are changed. Calculate the change in temperature.

E. Derive expressions for the work done by an ideal gas during an isothermal process and during an isobaric process.

II. Kinetic theory calculations of pressure and temperature.

A. Go over the assumptions of kinetic theory for an ideal gas. Consider a gas of molecules with only translational degrees of freedom. Assume the molecules are small and are free except for collisions of negligible duration. Also assume collisions with other molecules and with walls of the container are elastic. At the walls the molecules are specularly reflected.

B. Discuss a gas in a cubic container and explain that the pressure at the walls is due to the force of molecules as they bounce off. By considering the change in momentum at the wall per unit time, show that the pressure is given by $p = nMv_{\text{rms}}^2/3V$, where M is the molar mass. Define the rms speed. Use Table 20–1 to give some numerical examples of v_{rms}^2 and calculate the corresponding pressure. For many students, the rms value of a quantity needs clarification. Consider a system of five or so molecules and select numerical values for their speeds, then calculate v_{rms}^2 numerically.

C. Substitute $p = nMv_{\text{rms}}^2/3V$ into the ideal gas equation of state and show that $v_{\text{rms}} = \sqrt{3RT/M}$. Remark that this equation can be used to calculate the rms speed for a particular (ideal) gas at a given temperature.

D. Rearrange the equation for the rms speed to obtain $\frac{1}{2}Mv_{\text{rms}}^2 = \frac{3}{2}RT$ and use $M/m = N_A$ to show this can be written $\frac{1}{2}mv_{\text{rms}}^2 = \frac{3}{2}kT$, where m is the mass of a molecule. Remark that the left side is the mean kinetic energy of the molecules and point out that the temperature is proportional to the mean kinetic energy.

III. Internal energy and equipartition of energy.

A. Explain that the internal energy of a monatomic ideal gas is the sum of the kinetic energies of the molecules and write $E_{\text{int}} = \frac{1}{2}Nmv_{\text{rms}}^2 = \frac{3}{2}NkT = \frac{3}{2}nRT$, where N is the number of molecules and n is the number of moles. Stress that for an ideal gas the internal energy is a function of temperature alone, not the pressure and volume individually. This is an approximation for a real gas. Emphasize that the velocities used in computing the internal kinetic energy are measured relative to the center of mass and that the internal energy does not include the kinetic energy associated with motion of the system as a whole.

B. Point out that if adiabatic work W is done on the gas the internal energy increases by W and the temperature increases by $\Delta T = 2W/3nR$.

C. Point out that the expression obtained above for ΔT agrees closely with experimental values for monatomic gases but gives values that are too high for gases of diatomic and polyatomic molecules. Draw diagrams of these types of molecules and explain that they have two and three degrees of rotational freedom, respectively. Some of the energy goes into motions other than the translational motion of the molecules. Define the term degree of freedom and show how to count the number for monatomic, diatomic, and polyatomic molecules.

D. State the equipartition theorem: in thermal equilibrium the energy is distributed equally among all degrees of freedom, with each receiving $\frac{1}{2}kT$ for each molecule. Point out that this agrees with the previous result for monatomic gases: there are three degrees of freedom per molecule and an energy of $\frac{1}{2}kT$ is associated with each.

E. Discuss diatomic molecules and explain there are two new degrees of freedom, both rotational in nature. Show that $E_{\text{int}} = \frac{5}{2}nRT = \frac{5}{2}NkT$. Explain that $\frac{3}{2}nRT$ is in the form of translational kinetic energy and nRT is in the form of rotational kinetic energy.

F. Discuss polyatomic molecules. State that there are now three rotational degrees of freedom and show that $E_{\text{int}} = 3nRT = 3NkT$. Explain that $\frac{3}{2}nRT$ is in the form of translational kinetic energy and $\frac{3}{2}nRT$ is in the form of rotational kinetic energy.

G. Explain that vibrational motions may also contribute to the internal energy and that, since a vibration has both kinetic and potential energy, there are two degrees of freedom and energy kT associated with each vibrational mode. Explain that, in fact, for most materials vibrational modes generally do not contribute to the internal energy except at extremely high temperatures. Quantum mechanics is required to explain why vibrational modes are frozen out.

IV. Heat capacities of ideal gases.

A. Use equations previously derived for ΔE_{int} to obtain expressions for the molar specific heat at constant volume C_V. Point out the different results for monatomic, diatomic, and polyatomic molecules. Remark that C_V is used to denote *molar* specific heats in this and the next chapter. The symbol is not used for heat capacity as it was in the last chapter.

B. Show that the molar specific heat at constant pressure is related to the molar specific heat at constant volume by $C_p = C_v + R$ and derive the formulas for C_p for monatomic, diatomic, and polyatomic ideal gases.

C. For each type ideal gas, obtain the value for the ratio of molar specific heats $\gamma = C_p/C_v$. Point out these values are independent of T.

D. Derive $pV^\gamma =$ constant for an ideal gas undergoing an adiabatic quasi-static process. Also derive the expression $W = -(p_f V_f - p_i V_i)/(\gamma - 1)$ for the work done by the gas during an adiabatic change of state. Draw an ideal gas adiabat on a p-V diagram. Suppose the initial pressure and volume and the final volume are given. Show how to calculate the final pressure and temperature.

SUPPLEMENTARY TOPICS

1. Mean free path. This topic emphasizes the collisions of molecules and adds depth to the kinetic theory discussion but it is not crucial to subsequent chapters. Discuss as much as time allows.
2. Distribution of molecular speeds. This section deals with the Maxwell distribution and provides a deeper understanding of average speed and root-mean-square speed. Include it if you intend to cover thermonuclear fusion later in the course.

SUGGESTIONS

1. Assignments

   a. Assign a problem, such as 3, that is a straightforward application of the ideal gas law . Then, assign problems that show how the law is used to compute changes in various quantities when the gas changes state: 5 and 6, for example.

   b. Problem 13 provides an illustration of the work done by an ideal gas and problem 12 provides an example of heat exchange during a cycle. Also consider questions 1, 2, AND 3.

   c. Problems 14 and 15 deal with real-life applications. If possible, assign one or both. You may wish to discuss mixtures of gases and partial pressures; if so, consider problem 9.

   d. Use problem 21 in a discussion of the kinetic basis of pressure. Also assign problem 19.

   e. Assign problem 23 when you deal with the kinetic basis of temperature and the relationship between kinetic energy and temperature.

   f. After discussing the various specific heats, ask questions 5, 7, and 8 and assign problem 44. Assign problem 46 to emphasize the dependence of the heat capacity on the process.

   g. Consider using problem 54 to discuss adiabatic processes.

2. Demonstrations

   Kinetic theory models: Freier and Anderson Hh1, 2, 4, 5.

3. Audio/Visual

   a. *Boyle's Law*, *Equipartition of Energy*, *Maxwellian Speed Distribution*, *Random Walk and Brownian Motion*, *Diffusion*, *Gas Diffusion Rates*; from the AAPT collection 2 of single-concept films; DVD; available from Ztek Co. (PO Box 11768, Lexington, KY 40577–1768, www.ztek.com).

   b. *Gas Laws*; from Cinema Classics DVD 2: Mechanics (II) and Heat; available from Ztek Co. (see above for address) and from the American Association of Physics Teachers (AAPT, One Physics Ellipse, College Park MD 20740–3845; www.aapt.org).

4. Computer Software
   a. *Thermodynamics Lecture Demonstrations.* See Chapter 18 SUGGESTIONS.
   b. *Heat* from Exploration of Physics Volume I; Windows and Macintosh; Physics Curriculum & Instruction (22585 Woodhill Drive, Lakeville, MN 55044; www.PhysicsCurriculum.com). Simulated experiments with analysis. Includes a section on ideal gases.
   c. *Thermodynamics* from Exploration of Physics Volume II; Windows and Macintosh; Physics Curriculum & Instruction (see above for address). Includes sections on the ideal gas law, the molecular basis of internal energy, and the distribution of molecular speeds.
5. Laboratory
   a. Probeware Activity 19: *Ideal Gas Law.* Temperature and pressure sensors are used to verify the ideal gas law.
   b. Bernard Experiment 17: *Pressure and Volume Relations for a Gas.* The volume of gas in a tube is adjusted by changing the amount of mercury in the tube and a U-tube manometer is used to measure pressure. A logarithmic plot is used to determine the relationship between pressure and volume.
   c. Meiners Experiment 9–8: *Kinetic Theory Model.* The Fisher kinetic theory apparatus, consisting of a large piston-fitted tube of small plastic balls, is used to investigate relationships between pressure, temperature, and volume for a gas. A variable-speed impeller at the base allows changes in the average kinetic energy of the balls; the piston can be loaded to change the pressure. A variety of experiments can be performed.

Chapter 20 ENTROPY AND THE SECOND LAW OF THERMODYNAMICS

BASIC TOPICS

I. Entropy.
   A. Distinguish between reversible and irreversible processes. Remark that reversible processes are quasi-static but not all quasi-static processes are reversible (*i.e.* quasi-static processes involving friction). Also mention that for a gas the path of a reversible process can be plotted on a p-V diagram. As examples consider reversible and irreversible compressions of an ideal gas.
   B. Define the entropy difference between two infinitesimally close equilibrium states as $\mathrm{d}S = \mathrm{d}Q/T$ and between any two equilibrium states as $\Delta S = \int \mathrm{d}Q/T$. Explain that the integral is independent of path and that S is therefore a thermodynamic state function. Stress that a reversible path must be used to evaluate the integral but that entropy differences are defined regardless of whether the actual process is reversible or irreversible. The end points must be equilibrium states, however.
   C. Derive expressions for the change in entropy for an ideal gas undergoing processes at constant volume ($nC_V \ln(T_f/T_i)$), constant pressure ($nC_p \ln(T_f/T_i)$), and constant temperature ($nR \ln(V_f/V_i)$).
   D. Consider the adiabatic free expansion of an ideal gas. Point out that the process is irreversible, $Q = 0$, and $\Delta E_{\text{int}} = 0$. Since the gas is ideal, $T_f = T_i$. Find the change in entropy by evaluating $\int \mathrm{d}Q/T$ over a reversible isotherm through the initial and final states. Point out that the isothermal path does not represent the actual process. Show that $\Delta S = nR \ln(V_f/V_i)$ and state this is positive.
   E. Consider two identical rigid containers of ideal gas, at different temperatures, T_H and T_L. Place them in contact in an adiabatic enclosure. Show they reach equilibrium at temperature $T_m = (T_H + T_L)/2$. Then, consider a reversible, constant volume process that

connects the initial and final states and show that $\Delta S = C_V \ln(T_m^2/T_H T_L)$. Remark that this is positive.

II. The second law of thermodynamics.

A. State the second law: for processes that proceed from an initial equilibrium state to a final equilibrium state the total entropy of a closed system (or a system and its environment) does not decrease. State that if the process is reversible the total entropy does not change and if the process is irreversible it increases. Point out that the previous two examples are consistent with this statement.

B. Stress that the entropy change of the environment must be included. The entropy of a system can decrease but if it does the entropy of its environment increases by at least as much.

C. Remark that for reversible processes the total entropy of the system and its environment does not change because, for the combination of system and environment, the process is adiabatic and $dQ = 0$ for each segment of the reversible path. On the other hand, entropy increases for an adiabatic *irreversible* process.

III. Engines and refrigerators.

A. Discuss heat engines and refrigerators in general, from the point of view of the first law only. Explain that they run in cycles and that an engine absorbs energy as heat at a high temperature, rejects energy as heat at a low temperature, and does work. Describe a refrigerator in similar terms. Define the efficiency of an engine and the coefficient of performance of a refrigerator. Remark that heat engines and refrigerators may be reversible or irreversible.

B. Remind students that a cycle is a process for which the system starts and ends in the same equilibrium state and that $\Delta E_{\text{int}} = 0$, $\Delta p = 0$, $\Delta T = 0$, $\Delta V = 0$, and $\Delta S = 0$ for a cycle.

C. As an example, consider a gas undergoing a reversible cycle consisting of two isothermal processes at different temperatures and linked by two adiabatic processes (a Carnot cycle). Illustrate with a p-V diagram. Mention that, when run as a heat engine, energy enters the gas as heat during the isothermal expansion that energy leaves the gas as heat during the isothermal compression.

D. Over a cycle the change in the entropy of the working substance is zero, the change in the entropy of the high-temperature reservoir is $-Q_H/T_H$, and the change in entropy of the low temperature reservoir is $-Q_L/T_L$. Thus, the change in the total entropy of the system and its environment is $\Delta S = -(Q_L/T_L) - (Q_H/T_H)$. Since the process is reversible this must be zero. So $Q_L/Q_H = -T_L/T_H$ and the efficiency of the engine is $\varepsilon = W/Q_H = (Q_H + Q_L)/Q_H = 1 - (T_L/T_H)$. You may prefer to write these equations in terms of the absolute magnitudes of the quantities involved.

E. Remark that the efficiency is independent of the working substance. Also say that the second law of thermodynamics leads to the same expression for the efficiency of any reversible engine operating between those temperatures. The efficiencies of real engines, which are of necessity irreversible, are less.

F. Show that the second law forbids an engine with zero heat output. In particular, observe that the total entropy of the engine and reservoirs decreases if $|Q_L| < |Q_H|T_L/T_H$ and that this violates the second law.

G. Say that if the process is irreversible the total entropy must increase, so $|Q_L|/T_L > |Q_H|/T_H$ and the efficiency must be less than the ideal efficiency. Remark that no engine operating between two given temperatures can be more efficient than a reversible engine.

H. Carry out a similar analysis for an ideal refrigerator. Show that the coefficient of performance is given by $K = T_L/(T_H - T_L)$, independently of the working substance. State that all reversible refrigerators have the same coefficient of performance and that irreversible refrigerators have lower coefficients for the same reservoirs. Use an entropy argument to show that the second law forbids a refrigerator that operates with no work input.

IV. The statistical basis of entropy.

A. Explain that for any system composed of many molecules, there are many possible arrangements of the molecules. Illustrate by considering a small collection of molecules in a box. Say that each possible arrangement is called a microstate and that microstates can be grouped into configurations such that all the microstates in a given configuration are macroscopically equivalent. That is, the system has the same macroscopic properties. Use the molecules in the box to illustrate two equivalent and two non-equivalent microstates. State that the number of microstates associated with a configuration is called the multiplicity of the configuration.

B. Say that the fundamental assumption of statistical mechanics is that the system has the same probability of being in any microstate consistent with its macroscopic properties. Thus, the most likely configuration is the one with the largest multiplicity.

C. Show that if there are N molecules in the box, with n_R in the right half and n_L in the left half, the multiplicity is $W = N!/(n_R!)(n_L!)$.

D. Say that the entropy of a system when it has a given configuration is given by $S = k \ln W$, where W is the multiplicity of the configuration and k is the Boltzmann constant.

E. Use the statistical definition of entropy to show that the entropy changes by $\Delta S = nR \ln 2$ when the volume available to the molecules in the box is suddenly doubled. You will need to use Stirling's approximation ($\ln N! \approx N \ln N - N$).

SUGGESTIONS

1. Assignments
   a. To start students thinking about entropy changes as they occur in common processes, ask a few of the questions in the group 1 through 5. Assign problems 1, 4, 5, and 14. To include entropy changes in calorimetry experiments, ask problems 7, 9, and 16. Also consider problem 17.
   b. Use questions 8 and 9 to discuss real and ideal engines. Problems 27, 29, and 32 cover the fundamentals of cycles.
   c. Consider practical engines and their efficiencies by approximating their operation by reversible cycles. For a gasoline engine, $T_H \approx 1000°\text{F}$ and $T_L \approx 400°\text{F}$. Compare actual efficiencies with the ideal efficiency. Actual efficiencies can be obtained by considering the fuel energy available and the work actually obtained.
   d. Consider practical refrigerators. Look in a catalog for typical values of the coefficient of performance and compare with the ideal coefficient of performance. Also consider question 10 and assign some of problems 36, 38, 41, and 42.
   e. Ask question 11 in connection with the statistical interpretation of entropy.
2. Demonstrations

   Engines: Freier and Anderson Hm5, Hn.
3. Audio/Visual

   Entropy; from Physics Demonstrations in Heat, Part III; VHS video tape, DVD; ≈3 min; Physics Curriculum & Instruction (22585 Woodhill Drive, Lakeville, MN 55044; www.PhysicsCurriculum.com).
4. Computer Software
   a. *Thermodynamics* from Exploration of Physics Volume II; Windows and Macintosh; Physics Curriculum & Instruction Physics Curriculum & Instruction (see above for address). Includes sections on entropy and the second law of thermodynamics.

b. *Thermodynamics Lecture Demonstrations.* See Chapter 18 SUGGESTIONS.
c. *Physics Simulation Programs.* See Chapter 16 SUGGESTIONS.

Chapter 21 ELECTRIC CHARGE

BASIC TOPICS

I. Charge.
   A. Explain that there are two kinds of charge, called positive and negative, and that particles with like charges repel each other, particles with unlike charges attract each other. Give the SI unit (coulomb) and explain that it is defined in terms of current, to be discussed later. Optional: explain that current is the flow of charged particles and is measured in amperes. One coulomb of charge passes a cross section each second in a wire carrying a steady current of 1 A.
   B. Carry out the following sequence of demonstrations. They work best in dry weather.
      1. Suspend a pith ball by a string. Charge a rubber rod by rubbing it with fur, then hold the rod near the pith ball. The ball is attracted, touches the rod, then flies away after a short time. Use the rod to push the ball around without touching it. Explain that the rod and ball carry the same type charge. Hold the fur near the pith ball and explain that they are oppositely charged.
      2. Repeat using a second pith ball and a wooden rod charged by rubbing it on a plastic sheet (this replaces the traditional glass rod – silk combination and works much better). Place the two pith balls near each other and explain that they are oppositely charged.
      3. Suspend a charged rubber rod by a string. Use another charged rubber rod to push it around without touching it. Similarly, pull it with the charged wooden rod. Also show that only the rubbed end of the rubber rod is charged.

II. Conductors and insulators.
   A. Explain the difference between a conductor and an insulator as far as the conduction of charge is concerned. Explain that excess charge on a conductor is free to move and generally does so when influenced by the electric force of other charges. Excess charge on a conductor is distributed so the net force on any of it is zero. Any excess charge on an insulator does not move far from the place where it is deposited. Remind students of the demonstration that showed that only the rubbed end of the rubber rod remains charged. Metals are conductors. The rubber rod is an insulator. Mention semiconductors and superconductors.
   B. Use an electroscope to demonstrate the conducting properties of conductors. Charge the electroscope by contact with a charged rubber rod and explain why the leaves diverge. Discharge it by touching the top with your hand. Explain why the leaves converge. Recharge the electroscope with a charged wooden rod, then bring the charged rubber rod near the electroscope, but do not let it touch. Note the decrease in deflection and explain this by pointing out the attraction of the charged particles on the rod for the charged particles on the leaves. Throughout, emphasize the motion of the charged particles through the metal leaves and stem of the electroscope.
   C. Demonstrate charging by induction. Bring a charged rubber rod near to but not touching an uncharged electroscope. Touch your finger to the electroscope, then remove it. Remove the rubber rod and note the deflection of the leaves. Bring the rubber rod near again and note the decrease in deflection. Observe that the electroscope and rod are oppositely charged. Confirm this with the wooden rod. Explain the process.

III. Coulomb's law.

A. Assert that experimental evidence convinces us that there are only two kinds of charge and that the force between a pair of charged particles is along the line joining them, has magnitude proportional to the product of the magnitudes of the charges, and is inversely proportional to the square of the distance between them. Further, the force is attractive for particles with unlike charges and repulsive for particles with like charges.

B. Write down Coulomb's law for the magnitude of the electric force exerted by one point charged particle on another. Give the SI value for ϵ_0 and for $1/4\pi\epsilon_0$. Stress that the law holds for point charged particles. Note in detail that the mathematical form of the law contains all the qualitative features discussed previously in connection with gravitation. If Chapter 13 was covered, point out the similarity with Newton's law of gravity and mention that, unlike charge, there is no negative mass.

C. Explain that a superposition law holds for electric forces and illustrate by finding the resultant force on a charged particle due to two other charged particles. Use the analogy with Newton's law of gravity to show that the force of one spherical distribution of charge on another obeys the same law as two point charged particles and that the force on a charged particle inside a uniformly charged spherical shell is zero. If Chapter 13 was not covered, simply state the shell theorems.

IV. Quantization and conservation of charge.

A. State that all measured charge is an integer multiple of the charge on a proton: $q = ne$. Give the value of e: 1.60×10^{-19} C. State that the charge on the proton is $+e$, the charge on the electron is $-e$, and the neutron is neutral.

B. Remark that macroscopic objects are normally neutral; they have the same number of protons as electrons. Stress that the word "neutral" describes the algebraic sum of the charges and does not indicate the absence of charged particles. Remark that when an object is charged, the charge imbalance is usually slight but significant.

C. State that charge is conserved in the sense that for a closed system the sum of all charges before an event or process is the same as the sum after the event or process. Stress that the charges in the sum must have appropriate signs. Example: rubbing a rubber rod with fur. The rod and fur are oppositely charged afterwards and the magnitude of the charge is the same on both. Also discuss the conservation of charge in the annihilation and creation of fundamental particles and note that the identity of the particles may change in an event but charge is still conserved. Examples: beta decay, electron-positron annihilation.

SUGGESTIONS

1. Assignments
   a. Discuss question 10, perhaps in connection with demonstrations or lab experiments. Also see problems 4 and 9.
   b. Use questions 3 and 4 to test for understanding of the direction of an electrical force and the superposition of forces. Problems 8 and 13 deal with the addition of electric forces in one dimension and problems 6 and 15 deal with the addition of electric forces in two dimensions.
   c. Ask question 1 in connection with the shell theorems.
2. Demonstrations
   a. Charging, electroscopes: Freier and Anderson Ea1, 2, 11.
   b. Electric force: Freier and Anderson Ea5, 6, 8, 12, 15, 17, Eb3, 4, 9, 10, 12, Ec4 — 6.
   c. Induction: Freier and Anderson Ea12, 13, 14.
   d. Touch a grounded wire to several places within a small area of a wall. Rub a balloon with fur and place it in contact with that area. Ask students to explain why the balloon sticks.

3. Books and Monographs

   Teaching about Electrostatics; by Robert A. Morse; available from the American Association of Physics Teachers (AAPT, One Physics Ellipse, College Park MD 20740–3845; www.aapt.org). Describes reliable and inexpensive apparatus for demonstrations and student activities.

4. Audio/Visual

   a. *Electrostatics*; Cinema Classics DVD 4: Waves (II) & Electricity and Magnetism ; available Ztek Co. (PO Box 11768, Lexington, KY 40577–1768, www.ztek.com) and from the AAPT (see above for address).
   b. *Electricity*; VHS video tape; Films for the Humanities and Sciences (PO Box 2053, Princeton, NJ 08543–2053; www.films.com). Contains sections on conductors and insulators, charging and discharging, charging by induction, electrical currents, electric potential difference, and resistance.
   c. *Electricity and Magnetism*; interactive CD-ROM; Films for the Humanities and Sciences (PO Box 2053, Princeton, NJ 08543–2053; www.films.com).
   d. *Understanding Electricity*; VHS video tape, DVD; Films for the Humanities and Sciences (see above for address).
   e. *Electrostatics*; *Isolation of Charges*; from Physics Demonstrations in Electricity and Magnetism, Part I; VHS video tape, DVD; ≈3 min each; Physics Curriculum & Instruction (22585 Woodhill Drive, Lakeville, MN 55044; www.PhysicsCurriculum.com).

5. Computer Software

   a. *Electricity and Magnetism* from Exploration of Physics Volume I; Windows and Macintosh; Physics Curriculum & Instruction (see above for address). Simulated experiments with analysis. Includes a section on Coulomb's law.
   b. *Electricity and Magnetism* from Exploration of Physics Volume II; Windows and Macintosh; Physics Curriculum & Instruction (see above for address). Includes sections on pith ball experiments and electroscopes.
   c. *Electricity and Magnetism* from Exploration of Physics Volume II; Windows and Macintosh; Physics Curriculum & Instruction (see above for address). Includes sections on velocity and acceleration graphs and on free fall, with and without air resistance.

   Electric Field Hockey; Ruth W. Chabay; Windows, Macintosh; available from Physics Academic Software, North Carolina State University, PO Box 8202, Raleigh, NC 27690–0739. The user tries to score a goal by placing stationary charged particles so they guide a charged puck around obstacles and into the net. The force on the puck can be shown as the puck moves.

6. Laboratory

   Meiners Experiment 10–2: *The Electrostatic Balance.* A coulomb torsional balance is used to find the functional relationship between the electrostatic force of one small charged ball on another and the separation of balls. An electrostatic generator is used to charge the balls.

Chapter 22 ELECTRIC FIELDS

BASIC TOPICS

I. The electric field.

   A. Use a fluid to introduce the idea of a field. The temperature of the fluid $T(x, y, z, t)$ is an example of a scalar field and the velocity $\vec{v}(x, y, z, t)$ is an example of a vector field. Point out that these functions give the temperature and velocity at the place and time specified by the dependent variables.

B. Explain that charged particles may be thought to create an electric field at all points in space and that the field exerts a force on another charged particle, if present. The important questions to be answered are: Given the charge distribution, what is the field? Given the field, what is the force on a charged particle?

C. Consider two point charged particles and remark that each creates a field and that the field of either one exerts a force on the other. Explain that the two together produce a field that is the superposition of the individual fields and that this field exerts a force on a third charged particle, if present.

D. Define the field at any point as the force per unit charge on a positive test charge at the point, in the limit of a vanishingly small test charge. Mention that the limiting process eliminates the influence of the test charge on the charged particles creating the field. SI units: N/C.

E. Use Coulomb's law to obtain the expression for the field of a point charged particle. Explain that the field of a collection of charged particles is the vector sum of the individual fields.

II. Electric field lines.

A. Explain that field lines are useful for visualizing the field. Draw field lines for a point charged particle and explain that, in general, the field at any point is tangent to the line through that point and that the magnitude of the field is proportional to the number of lines per unit area that pass through a surface perpendicular to the lines.

B. By considering a sphere around a point charged particle and calculating the number of lines per unit area through the sphere, show that the $1/r^2$ law allows us to associate lines with a charged particle and to take the number of lines to be proportional to the charge. Explain that lines can be thought of as directed and that they originate at positively charged particles and terminate at negatively charged particles. Emphasize that they are not vectors.

C. Show Figs. 22-2, 22-3, 22-4, and 22-5 or similar diagrams that illustrate the field lines of some charge distributions.

D. Field lines can be illustrated by floating some long seeds in transformer oil in a shallow, flat-bottomed dish. Place two metal plates in the dish and connect them to an electrostatic generator. The seeds line up along the field lines. You can place the apparatus on an overhead projector and shadow project the seeds.

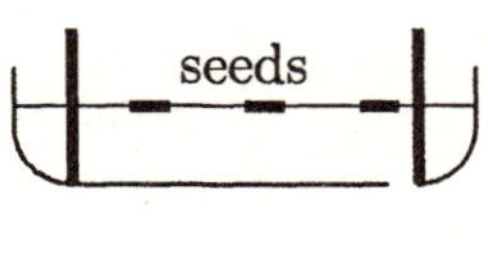

III. Calculation of the electric field.

A. Remind the students of the expression for the field of a point charged particle. State that the field is radially outward for a positively charged particle and radially inward for a negatively charged particle. Also remind them that the total field is the vector sum of the individual fields of the charged particles being considered.

B. Derive an expression for the field of an electric dipole by considering the field of two particles with charge of equal magnitudes and opposite signs. Consider a field point on a line perpendicular to the dipole moment, on a line along the dipole moment, or a general point. Evaluate the expression in the limit of vanishingly small separation and finite dipole moment. Define the dipole moment and stress that it points from the negative toward the positively charged particle. Point out that the field is proportional to $1/r^3$ for points far from the dipole.

C. Consider a small set of discrete charged particles and calculate the electric field by evaluating the vector sum of the individual fields. Example: the field at the center of a square with various charged particles on its corners.

D. As an introduction to the fields of continuous charge distributions, go over the ideas of linear and area charge densities. Graphically show how a line of charge is divided into infinitesimal segments and point out that a segment of length ds contains charge $dq = \lambda\, ds$. Explain that for purposes of calculating the field each segment can be treated as a point charged particle and that the fields of all segments are summed vectorially to find the total field.

E. Derive an expression for the field on the axis of a continuous ring of charge. Carefully explain how the integral is set up and how the vector nature of the field is taken into account by dealing with components. Explain in detail the symmetry argument used to show that the field is along the axis.

F. Extend the calculation to find an expression for the field on the axis of a charged disk and for an infinite sheet of charge. Remark that the field of a sheet is perpendicular to the sheet and is independent of distance from the sheet. This will be useful later when parallel plate capacitors are studied.

IV. Motion of a charged particle in an electric field.

A. Point out that the electric force on a charged particle is $q\vec{E}$ and explain that the electric field used is that due to all *other* charged particles (except q). Substitute the force into Newton's second law and remind the students that once the acceleration and initial conditions are known, kinematics can be used to find the subsequent motion of the charged particle.

B. Find the trajectory of a charged particle moving into a region of uniform field, perpendicular to its initial velocity. Compare to projectile motion problems studied in Chapter 4. See Sample Problem 22-5.

C. Show that the force on a dipole in a uniform field is zero and that the torque is $\vec{p} \times \vec{E}$. Also show that the potential energy of a dipole is $-\vec{p} \cdot \vec{E}$. Emphasize that the potential energy minimum occurs when the dipole moment is aligned with the field. To review oscillatory rotational motion calculate the angular frequency of small angle oscillations for a dipole with rotational inertia I in a uniform electric field. Assume no other forces act.

SUGGESTIONS

1. Assignments

   a. Center a qualitative discussion of electric field lines on question 1. Have students sketch field lines for various charge distributions. See problems 1, 3, and 8.

   b. Ask questions 2, 3, 4, and 5 and have students work some of problems 6, 8, 9, 11, and 12. These deal with the superposition of fields.

   c. Problems 18 and 19 are good tests of understanding of the derivation of the dipole field.

   d. Assign problems 24 and 25 in support of the calculation of the field of a ring of charge. Assign problems 27 and 29 to give students practice in deriving expressions for the field of a continuous charge distribution.

   e. Ask question 9 and assign problem 31 to support the discussion of the field of a uniformly charged disk.

   f. Assign problem 40 to help students with the motion of point charged particles in fields. Assign questions 10 and 11 and problems 50 (torque) and 53 (energy) in connection with the discussion of a dipole in a field.

   g. To include the Millikan oil drop experiment, assign problem 42.

2. Demonstrations

   Electric field lines: Freier and Anderson Eb1, Ec2 — 4.

3. Audio/Visual
   a. *Electrostatic Induction*; *The Van de Graaff Generator*; *Field as a Vector*; from the AAPT collection 2 of single-concept films; DVD; available from Ztek Co. (PO Box 11768, Lexington, KY 40577–1768, www.ztek.com).
   b. *Electric Fields*; Physics Demonstrations in Electricity and Magnetism, Part II; VHS video tape, DVD; ≈3 min; from Physics Curriculum & Instruction (22585 Woodhill Drive, Lakeville, MN 55044; www.PhysicsCurriculum.com).
4. Computer Software
   a. *Electricity and Magnetism* from Exploration of Physics Volume I; Windows and Macintosh; Physics Curriculum & Instruction (22585 Woodhill Drive, Lakeville, MN 55044; www.PhysicsCurriculum.com). Simulated experiments with analysis. Includes a section on electric fields.
   b. *Electricity and Magnetism* from Exploration of Physics Volume II; Windows and Macintosh; Physics Curriculum & Instruction (see above for address). Includes sections on electric field lines and the trajectory of a charged particle in an electric field.
   c. *Electric Field Plotter*; Windows; Bob Nelson; available from Physics Academic Software (Centennial Campus, 940 Main Campus Drive, Suite 210, Raleigh, NC 27606–5212; www.aip.org/pas). Draws electric field lines and equipotential lines. Students can place up to nine charged particles anywhere on the screen. The program also searches for points where the electric field vanishes.
   d. *EM Field*; David Trowbridge; Windows, Macintosh; available from Physics Academic Software (see above for address) and the American Association of Physics Teachers (AAPT, One Physics Ellipse, College Park MD 20740–3845; www.aapt.org). Plots electric field lines and equipotential surfaces of point and line charges, set up by the user. The electric field vectors can be shown. Plots magnetic field lines of user-selected current distributions. Users can draw Gaussian surfaces or Amperian paths. The program gives the flux through the surface or the contribution to $\int \vec{B} \cdot d\vec{s}$. When the surface or line is closed the program gives the charge or current enclosed. A game asks the user to find hidden charge or current.
   e. *Virtual E-Field Lab*; Gregory Marlow; Windows; available from Physics Academic Software (see above for address). Shows field and equipotential lines for continuous charge distributions as well as point charged particles. Gives numerical values of the electric potential and field at the position of the curser.
   f. *Motion in Electromagnetic Fields*; WhistleSoft, Inc.; Windows; available from Physics Academic Software (see above for address). Interactive tutorial with real-life examples, chiefly from scientific research.
   g. *Electric Field Hockey*; Windows; available from Physics Academic Software and from the AAPT (see above for addresses).
   h. *Dynamic Analyzer*. See Chapter 2 SUGGESTIONS.
   i. *Forces*. See Chapter 5 SUGGESTIONS.
5. Computer Projects
   a. Have students use a commercial math program or write their own programs to calculate the electric fields of discrete charge distributions. Have them use the programs to plot the magnitude of the field at various distances from a dipole, along lines that are perpendicular and parallel to the dipole moment.
   b. Have students write programs to trace field lines for discrete charge distributions.

Chapter 23 GAUSS' LAW

BASIC TOPICS

I. Electric flux.

A. Start by discussing some of the important concepts in a general way. Define a vector surface element. Define the flux of a vector field through a surface. Distinguish between open and closed surfaces and explain that for the latter the surface normal is taken to be *outward.* Interpret the surface integral for the flux as a sum over surface elements. If you covered Chapter 14, use the velocity field of a fluid as an example.

B. Define electric flux. Point out that it is the normal component of the field that enters. Also point out that the sign of the contribution of any surface element depends on the choice for the direction of $d\vec{A}$.

C. Interpret electric flux as a quantity that is proportional to the net number of field lines penetrating the surface. Remind students that the number of lines through a small area perpendicular to the field is taken to be proportional to the magnitude of the field. By considering surfaces with the same area but different orientations, show that the net number of penetrating lines is proportional to the cosine of the angle between the field and the normal to the surface. Conclude that $\vec{E}\cdot d\vec{A}$ is proportional to the number of lines through $d\vec{A}$.

D. Stress that lines roughly in the same direction as the normal contribute positively to the flux, lines roughly in the opposite direction contribute negatively, and lines that pass completely through a volume do not contribute to the flux through its boundary. Point out that zero flux through a surface does not imply zero field at points on the surface.

E. As an example, calculate the flux through each side of a cube in a uniform electric field. Also consider Sample Problem 23–2, which deals with a nonuniform field.

II. Gauss' law.

A. Write down the law. Stress that the surface is closed and that the charge appearing in the law is the net charge enclosed. Interpret the law as a statement that the number of (signed) lines crossing the surface is proportional to the net charge inside, and make the statement plausible by reminding students that the field of each charge is proportional to the charge and its direction depends on the sign of the charge.

B. Illustrate by considering the surface of a sphere with positively charged particles inside, with negatively charged particles inside, with both positively and negatively charged particles inside, and with charged particles outside. In each case draw representative field lines with the number of lines proportional to the net charge. Stress that the position of the charged particles inside is irrelevant for the flux through the surface. Also use Gauss' law to calculate the flux.

C. Use Gauss' law and symmetry arguments to obtain an expression for the electric field of a point charged particle.

III. Gauss' law and conductors.

A. Argue that the electrostatic field vanishes inside a conductor and use Gauss' law to show that there can be no net charge at interior points under static conditions. Point out that exterior charged particles and charged particles on the surface separately produce fields in the interior but that the resultant field vanishes. For contrast, point out that an insulator may have charge distributed throughout.

B. Demonstrate that any excess charge on a conductor resides on the exterior surface. Use a hollow metal sphere with a small hole cut in it. As an alternative, solder shut the top of an empty metal can and drill a small hole in it. This will not work as well because of the sharp edges. Charge a rubber rod by rubbing it with fur and touch it to the inside of the sphere, being careful not to touch the edge of the hole. Repeat several times to build up charge. Now scrape at the interior with a metal transfer rod, again being careful not to touch the edge of the hole. Touch the transfer rod to an uncharged electroscope and note the lack of deflection. Scrape the exterior of the sphere with the transfer rod and touch the electroscope. Note the deflection.

C. Show how to calculate the charge on the inner and outer surfaces of neutral and charged conducting spherical shells when charge is placed in the cavities. See Sample Problem 23–4. Also see Checkpoint 4.

D. Use Gauss' law to show that the magnitude of the field just outside a charged conductor is given by $E = \sigma/\epsilon_0$, where σ is the surface charge density.

IV. Applications of Gauss' law.

A. Derive expressions for the electric field at various points for a uniformly charged sphere and for a uniformly charged thick spherical shell. Remark that such distributions are possible if the sphere or shell is not conducting. Carefully give the symmetry argument to show the field is radial and has the same magnitude at all points on a concentric sphere.

B. Derive an expression for the electric field at a point outside an infinite sheet with a uniform charge distribution. Contrast with the field outside an infinite conducting sheet with the same area charge density on one surface. Point out that for the conductor the field is not due only to the charge on the surface being considered. Another field must be present to produce a net field of zero in the interior and this doubles the field in the exterior.

C. Consider a point charged particle at the center of a neutral spherical conducting shell and derive expressions for the electric field in the various regions. Repeat for a charged shell.

D. Work one problem with cylindrical symmetry. For example, consider charge distributed uniformly throughout a cylinder and find the field in all regions.

E. Note that Gauss' law can be used to find $\vec{E}$ only if there is adequate symmetry.

SUGGESTIONS

1. Assignments
   a. Use questions 1, 4, 6, and 7 to help students understand the flux integral and charge that appear in Gauss' law. Use problems 1 and 2 to introduce electric flux. The latter problem also demonstrates the vanishing of the total flux for a closed surface in a uniform field.
   b. Problem 5 illustrates the fundamental idea of Gauss' law. Problems 4 and 13 are also instructive.
   c. Use questions 8 and 9 and problem 19 to discuss the electrostatic properties of conductors.
   d. Assign a variety of problems dealing with applications: 29 (cylinder of charge); 32, 38, and 41 (plane of charge); 45 and 49 (sphere of charge). Assign problem 47 or 51 to challenge good students.
2. Demonstrations

   Charges on conductors: Freier and Anderson Ea7, 18, 23, Eb7.
3. Audio/Visual

   Charge Distribution — Faraday Ice Pail Experiment; from the AAPT collection 2 of single-concept films; video tape; available from Ztek Co. (PO Box 11768, Lexington, KY 40577–1768, www.ztek.com).
4. Computer Software

   EM Field. See Chapter 22 SUGGESTIONS.
5. Computer Project

   Have students use a commercial math program or their own programs to evaluate the flux integral in Gauss' law. Have them separately calculate the flux through each face of a cube containing a point charged particle. Consider various positions of the particle within the cube to show that the flux through individual faces may change as the particle changes position but the total flux remains the same and obeys Gauss' law. Repeat for a point charged particle outside the cube.

Chapter 24 ELECTRIC POTENTIAL

BASIC TOPICS

I. Electric potential.

A. Define the potential difference of two points as the negative of the work per unit charge done by the electric field when a positive test charge moves from one point to the other. Stress the sign of the potential: the potential of the end point is higher than that of the initial point if the work is negative. The electric field points from regions of high potential toward regions of low potential and positively charged particles tend to be repelled from regions of high potential. The region near an isolated positively charged particle has a higher potential than regions far away. The opposite is true for a negatively charged particle. Unit: volt. Define electron volt as a unit of energy.

B. If you covered Chapter 14, use the similarity of Coulomb's law and Newton's law of gravity to argue that the electrostatic force is conservative and that the work is independent of path. If you did not cover Chapter 14, either derive or state these results.

C. Show that the definition is equivalent to $V_b - V_a = -\int \vec{E} \cdot d\vec{s}$, where the integral is along a path from a to b. Point out that the potential is constant in regions of zero field. Note that the unit N/C is the same as V/m and the latter is a more common unit for $\vec{E}$.

D. Point out that the potential is a scalar and that only potential differences are physically meaningful. One point can be chosen arbitrarily to have zero potential and the potential at other points is measured relative to the potential there. Often the potential is chosen to be zero where the field (or force) is zero. For a finite distribution of charge, the potential is usually chosen to be zero at a point far away (infinity). Show a voltmeter and remark that the meter reads the potential difference between the leads.

E. Show that the potential a distance r from an isolated point charged particle is given by $V = q/4\pi\epsilon_0 r$. Remark that this is the potential energy per unit test charge of a system consisting of the point particle with charge q and the test charge. Explain that the equation is valid for both positively and negatively charged particles. Show how to calculate the potential due to a collection of point charged particles. Derive the expression for the potential of an electric dipole.

F. Give some examples of calculations of the potential from the electric field. Start with a uniform electric field, like that outside a uniform plane distribution of charge, and show that potential is given by $-Ex+C$, where C is a constant. Since the distribution is infinite the point at infinity cannot be picked as the zero of potential.

F. As a more complicated example, consider one of the configurations discussed in the last chapter, a point charged particle at the center of a spherical conducting shell, say. Take the potential to be zero at infinity and compute its value at points outside the outer surface, within the shell, and inside the inner surface. As an alternative you might find expressions for the potential in various regions around and inside a nonconducting sphere with a uniform charge distribution.

G. Write down the integral expressions for the potential due to a line of charge and for a surface of charge, in terms of the linear and area charge densities. Work an example, such as the potential of a uniform finite line of charge or a uniform disk of charge.

II. Equipotential surfaces.

A. Define the term equipotential surface. Show diagrams of equipotential surfaces for an isolated point charged particle and for the region between two uniformly charged plates. Equipotential surfaces of a dipole are shown in Fig. 24–3(c).

B. Point out that the field does zero work if a test charge is carried between two points on the same equipotential surface and note that this means that the force, and hence $\vec{E}$, is perpendicular to the equipotential surfaces. Note further that the work done by the field when a charged particle is carried from any point on one surface to any point on another is the product of the charge and the negative of the potential difference.

III. Calculation of $\vec{E}$ from V.

A. Remind students that $\Delta V = -E\Delta x$ for a uniform field in the positive x direction. Note that E has the form $-\Delta V/\Delta x$ and $\vec{E}$ is directed from high to low potential. Use this result to reenforce the idea of an equipotential surface and the fact that $\vec{E}$ is perpendicular to equipotential surfaces.

B. Generalize the result to $E = -\mathrm{d}V/\mathrm{d}s$, where s is the distance along a normal to an equipotential surface. Then specialize this to $E_x = -\partial V/\partial x$, $E_y = -\partial V/\partial y$, and $E_z = -\partial V/\partial z$. Verify that the prescription works for a point charged particle and for a dipole.

IV. Electrostatic potential energy.

A. Remark that when a particle with charge Q moves from point a to point b the potential energy of the system changes by $Q(V_b - V_a)$, where V is the potential due to the other charged particles. When a particle with charge Q is brought into position from infinity (where the potential is zero), the potential energy changes by QV, where V is the potential at the final position of Q due to charged particles already in place.

B. Show that the potential energy of two point charged particles is given by $q_1q_2/4\pi\epsilon_0 r$, where r is their separation and the zero of potential energy is taken to be infinite separation. Point out that the potential energy is positive if the charges on the particles have like signs and negative if they have opposite signs. Explain that the potential energy decreases if particles with charge of the same sign move apart or if two particles with charges of opposite sign move closer together.

C. Remind students that potential energy can be converted to kinetic energy. Explain what happens if the particles used in the last example are released from their positions. Consider a proton fired directly at a heavy nucleus with charge Ze and find the distance of closest approach in terms of the initial speed.

D. Calculate the potential energy of a simple system: charged particles at the corners of a triangle or square, for example. Assume the particles are brought in from infinity one at a time and sum the potential energies. Explain that the total is the sum over particle pairs. Show how to calculate the potential energy of any collection of point charged particles.

E. Explain that the potential energy of a system of charged particles is the work an agent must do to assemble the system from rest at infinite separation. This is the negative of the work done by the field.

V. An isolated conductor.

A. Recall that the electric field vanishes at points in the interior of a conductor. Argue that the surface must be an equipotential surface and that V at all points inside must have the same value as on the surface. State that this is true if the conductor is charged or not and if an external field exists or not.

B. Consider two spherical shells of different radii, far apart and connected by a very fine wire. Explain that $V_1 = V_2$ and show that $q_1/R_1 = q_2/R_2$. Then show that the surface charge density varies inversely with the radius: $\sigma_1/\sigma_2 = R_2/R_1$. Recall that E is proportional to σ just outside a conductor and argue that σ and E are large near places of small radius of curvature and small near places of large radius of curvature. Use an electrostatic generator to show discharge from a sharp point and from a rounded (larger radius) ball. Discuss the function of lightning rods and explain their shape.

SUGGESTIONS

1. Assignments
   a. Questions 1, 2, 3, 4, and 6 can be used to help students think about some qualitative aspects of electric potential.
   b. Use question 5 and problem 3 to test for understanding of equipotential surfaces.
   c. Ask students to calculate potential differences for various situations: see problems 5, 7, 15, 19, and 28.
   d. Use questions 9 and 10 and problems 37 and 41 in connection with the discussion of electrostatic potential energy and the work done by an electric field or an external agent. Also assign some conservation of energy problems, such as 43 and 45.
   e. Assign one or two of problems 53, 56, and 58 to aid in a discussion of the field and potential of a conductor.
2. Demonstrations
   Electrostatic generators: Freier and Anderson Ea22, Ec1.
3. Audio/Visual
   Electrical Energy; VHS video tape, DVD; Films for the Humanities and Sciences (PO Box 2053, Princeton, NJ 08543–2053; www.films.com).
4. Computer Software
   a. *Electric Field Plotter* See Chapter 22 SUGGESTIONS.
   b. *EM Field*. See Chapter 22 SUGGESTIONS.
5. Computer Project
   Have students use a commercial math program or their own root finding programs to plot equipotential surfaces for a discrete charge distribution. It is instructive to consider two particles with unequal charges (any combination of signs).
6. Laboratory
   Meiners Experiment 10–1: *Electric Fields* and Bernard Experiment 25: *Mapping of Electric Fields.* Students map equipotential lines on sheets of high resistance paper with metallic electrodes at two sides. In the Meiners experiment an audio oscillator generates the field and an oscilloscope or null detecting probe is used to find points of equal potential. If students are not familiar with oscilloscopes, you might want to preface this experiment with Part A of Meiners Experiment 10–10. In the Bernard experiment the field is generated by a battery and a galvanometer is used as a probe.

Chapter 25 CAPACITANCE

BASIC TOPICS

I. Capacitance.
   A. Describe a generalized capacitor. Draw a diagram showing two separated, isolated conductors. Assume they carry charge q and $-q$, respectively, draw representative field lines, and point out that all field lines start on one conductor and terminate on the other. Explain that there is a potential difference V between the conductors and that the positively charged conductor is at the higher potential. Define capacitance as $C = q/V$. Explain that V is proportional to q and that C is independent of q and V. C does depend on the

shapes, relative positions, and orientations of the conductors and on the medium between them. Unit: 1 farad = 1 C/V.

B. Show a radio tuning capacitor and some commercial fixed capacitors. Mention that one usually encounters μF and pF capacitors. Capacitors on the order of 1 F have been developed for the electronics industry.

C. Remark that in circuit drawings a capacitor is denoted by —| |—.

D. State that a battery can be used to charge a capacitor. The battery transfers charged particles from one plate to the other until the potential difference of the plates is the same as the terminal potential difference of the battery. Calculate the charge, given the battery potential difference and the capacitance.

E. In a general way, give the steps required to calculate capacitance: put charge q on one conductor, $-q$ on the other, and calculate the electric field due to the charge, then calculate the potential difference V between the conductors, and finally use $q = CV$ to find the capacitance. Except for highly symmetric situations, the charge is not uniformly distributed over the surfaces of the conductors and fairly sophisticated means must be used to calculate V. The text deals with symmetric situations for which Gauss' law can be used to calculate the electric field.

F. Examples: derive expressions for the capacitance of two parallel plates (neglect fringing) and two coaxial cylinders or two concentric spherical shells. Use Gauss' law to find the electric field, then evaluate the integral for the potential difference. Emphasize that the field is due to the charged particles on the plates.

G. Large demonstration parallel plate capacitors with variable plate separations are available commercially. You can also make one using two $\approx$ 1-ft diameter circular plates of 1/8 inch aluminum sheeting. Attach an aluminum disk to the center of each with a hole drilled for a support rod. Use an insulating rod on one and a metal rod on the other. By sliding the two conductors closer together, you can show the effect of changing d while holding q constant. An electroscope serves as a voltmeter.

H. Explain how the equivalent capacitance of a device can be measured. Consider a black box with two terminals. State that a potential difference V is applied and the total charge q deposited is measured. The capacitance is q/V.

I. Derive $1/C_{eq} = 1/C_1 + 1/C_2$ for the equivalent capacitance of two capacitors in series and $C_{eq} = C_1 + C_2$ for the equivalent capacitance of two capacitors in parallel. Emphasize that two capacitors in parallel have the same potential difference and that two in series have the same charge. Explain the usefulness of these equations for circuit analysis.

II. Energy storage.

A. Derive the expression $W = \frac{1}{2}q^2/C$ for the work required to charge a capacitor. Explain that, as an increment of charge is transferred, work is done by an external agent (a battery, for example) against the electric field of the charged particles already on the plates. Show that this expression is equivalent to $W = \frac{1}{2}CV^2$. Interpret the result as the potential energy stored in the charge system and explain that it can be recovered when the capacitor is discharged.

B. Remark that if two capacitors are in parallel the larger stores the greater energy. If two capacitors are in series, the smaller stores the greater energy.

C. Show that the energy density in a parallel plate capacitor is $\frac{1}{2}\epsilon_0 E^2$. State that this result is quite general and that its volume integral gives the work required to assemble charged particles to create the electric field E. Explain that the energy may be thought to reside in the field or it may be considered to be the potential energy of the charged particles.

D. Integrate the energy density to find an expression for the energy stored in the electric field of a charged spherical capacitor or a charged cylindrical capacitor. Compare the result with $\frac{1}{2}q^2/C$.

III. Dielectrics.

A. Explain that when the region between the conductors of a capacitor is occupied by insulating material the capacitance is multiplied by a factor $\kappa > 1$, called the dielectric constant of the material. Remark that $\kappa = 1$ for a vacuum.

B. Use a large commercial or homemade capacitor to show the effect of a dielectric. Charge the capacitor, then isolate it and insert a glass plate between the plates. The electroscope shows that V decreases and, since q is fixed, the capacitance increases.

C. Calculate the change in stored energy that occurs when a dielectric slab is inserted between the plates of an isolated parallel plate capacitor (see Sample Problem 25–6). Also calculate the change in stored energy when the slab is inserted while the potential difference is maintained by a battery. Explain that the battery now does work in moving charged particles from one plate to the other.

D. Explain that dielectric material between the plates becomes polarized, with the positively charged ends of the dipoles attracted toward the negative conductor. The field of the dipoles opposes the external field, so the electric field is weaker between the plates than it would be if the material were not there. This reduces the potential difference between the conductors for a given charge on them. Since the potential difference is less for the same charge on the plates, the capacitance is greater.

E. Explain that if the polarization is uniform, the material behaves like neutral material with charge on its surfaces.

F. Optional: Show how Gauss' law can be written in terms of $\kappa\vec{E}$ and the free charge. Show how to compute the polarization charge for a parallel plate capacitor with dielectric material between its plates.

SUGGESTIONS

1. Assignments
   a. Use question 1 to emphasize the dependence of capacitance on geometry.
   b. The fundamental idea of capacitance is illustrated by problem 2. Assign problem 6 to have students compare spherical and plane capacitors. Problem 4 covers the dependence of the capacitance of a parallel plane capacitor on area and separation.
   c. Include some of questions 5 through 9 in the discussion of series and parallel connections of capacitors. Problems 8 and 10 cover equivalent capacitance, charge, and potential difference for series and parallel combinations. Also consider assigning some problems in which students must find the equivalent capacitance of more complicated combinations. See problems 7 and 9, for example. Problem 23 is more challenging.
   d. Problem 26 covers most of the important points discussed in connection with energy storage. Also assign problem 31, which deals with the energy needed to separate the plates of a parallel plate capacitor, and problem 32, which deals with the energy density around a charged metal sphere.
   e. Include question 11 in the discussion of the influence of a dielectric on capacitance. Assign problems 36 and 40.
   f. To test understanding of induced polarization charge, assign problem 45 or 47.

2. Demonstrations
   a. Charge storage: Freier and Anderson Eb8, Ed3, 7.
   b. Capacitance and voltage: Freier and Anderson Ed1.
   c. Energy storage: Freier and Anderson Ed8
   d. Dielectrics: Freier and Anderson Ed2, 4.

3. Audio/Video
   a. *Magnetic Fields*; VHS video tape; Films for the Humanities and Sciences (PO Box 2053, Princeton, NJ 08543–2053; www.films.com).
4. Computer Software
   a. *Electricity and Magnetism* from Exploration of Physics Volume I; Windows and Macintosh; Physics Curriculum & Instruction (22585 Woodhill Drive, Lakeville, MN 55044; www.PhysicsCurriculum.com). Simulated experiments with analysis. Includes a section on capacitors.
5. Laboratory
   a. Meiners Experiment 10–7 (Part B): *Measuring Capacitance with a Ballistic Galvanometer.* A ballistic galvanometer is used to measure the capacitance of individual capacitors and capacitors in series and parallel. Students must temporarily accept on faith that the deflection of the galvanometer is proportional to the total charge that passes through it.
   b. Meiners Experiment 11–2 (Part C): *Coulomb Balance Attachment (to the current balance).* Students use gravitational force to balance the force of one capacitor plate on the other. The voltage and plate separation are used to find the charge on the plates, then ϵ_0 is calculated.

Chapter 26 CURRENT AND RESISTANCE

BASIC TOPICS

I. Current and current density.
   A. Explain that an electric current is moving charged particles. Draw a diagram of a long straight wire with positively charged particles moving in it. Consider a cross section and state that the current is dq/dt if charge dq passes the cross section in time dt. Give the sign convention: both positively charged particles moving to the right and negatively charged particles moving to the left constitute currents to the right. Early on, use the words "conventional current" quite often. Later "conventional" can be dropped. Some high school courses now take the current to be in the direction of electron flow and it is worthwhile making the effort to reduce confusion in students' minds. Unit: $1\,\text{ampere} = 1\,\text{C/s}$.
   B. Explain that under steady state conditions, in which no charge is building up or being depleted anywhere in the wire, the current is the same for every cross section. Remark that current is a scalar, but arrows are used to show the direction of positive charge flow.
   C. Explain that current is produced when charged particles are free to move in an electric field. For most materials, it is the negative electrons that move and their motion is opposite to the direction of the electric field. Current is taken to be in the direction opposite to that of electron drift, in the direction of the field.
   D. Distinguish between the drift velocity and the velocities of individual charged particles. Note that the drift velocity of electrons in an ordinary wire is zero unless an electric field is turned on. Also note that the drift speed is many orders of magnitude smaller than the average electron speed.
   E. Explain that current density is a microscopic quantity used to describe current flow at a point. Use the same diagram but now consider a small part of the cross section and state that $J = i/A$ in the limit as the area diminishes to a point. State that current density is a vector in the direction of the drift velocity for positively charged particles and opposite the drift velocity for negatively charged particles. Explain that $i = \int \vec{J} \cdot d\vec{A}$ is the current

through a finite surface, where $d\vec{A}$ is normal to the surface. This reduces to $J = i/A$ for uniform current density and an area that is perpendicular to the current. Unit: $\mathrm{A/m^2}$.

F. Derive $\vec{J} = en\vec{v}_d$ and show how to calculate the drift speed from the free-electron concentration and current in the wire, assuming uniform current density. You may want to go over the calculation of the free-electron concentration from the mass density of the sample and the molar masses of its constituents.

II. Resistance and resistivity.

A. Define resistance by $R = V/i$ and point out that R may depend on V. Unit: $1\,\text{ohm} = 1\,\text{V/A}$; abbreviation: Ω. Also define resistivity ρ and conductivity σ. Point out Table 26–1. Explain that the latter quantities are characteristic of the material while resistance also depends on the sample shape and the positions of the current leads.

B. Make a sketch similar to the one shown here. Indicate that $V_a - V_b = iR$ is algebraically correct, even if i is negative, and effectively defines the resistance of the sample with the leads connected at a and b. Emphasize that the point at which the current enters is iR higher in potential than the point at which it leaves.

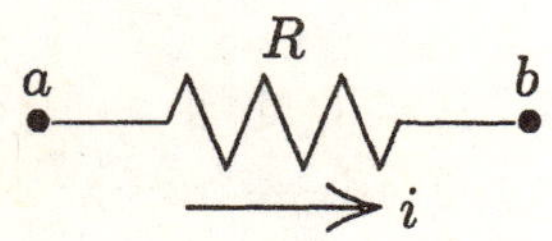

C. Show that $R = \rho L/A$ for a conductor with uniform cross section A and length L, carrying a current that is uniformly distributed over the cross section.

D. Point out that for many samples the current is proportional to the potential difference and the resistance is independent of the voltage applied. These materials are said to obey Ohm's law. Also point out that many important materials do not obey Ohm's law. Show Fig. 26–12.

E. Use a variable-voltage power supply and connect, in turn, samples of ohmic (carbon resistor) and non-ohmic (solid state diode) material across the terminals. Use analog meters to display the current and potential difference and vary the supply smoothly and fairly rapidly. For the ohmic material, it will be apparent that i is proportional to V, while for the non-ohmic material, it will be apparent that i is not proportional to V.

F. Give a qualitative description of the mechanism that leads to Ohm's law behavior. Explain that collisions with atoms cause the drift velocity to be proportional to the applied field. Assume the electrons have zero velocity after each collision and that they accelerate for a time τ between collisions. Show that an electron goes the same distance on the average during the first five collisions as it does during the second five so the drift velocity is proportional to the field even though the electron accelerates between collisions. Now consider the quantitative aspects: derive the expression for the drift velocity in terms of $\vec{E}$ and the mean free time τ, then derive $\rho = m/ne^2\tau$. Emphasize that the mean free time is determined by the electron speed and, since drift is an extremely small part of the speed, τ is essentially independent of the electric field. Point out that a long mean free time means a small resistivity because the electrons accelerate for a longer time between collisions and thus have a higher drift speed.

G. Remark that the resistivity of a sample depends on the temperature. Define the temperature coefficient of resistivity and point out the values given in Table 26–1.

III. Energy considerations.

A. Point out that when current flows from the high to the low potential side of any device, energy is transferred from the current to the device at the rate $P = iV$. Reproduce Fig. 26–14 and note that $P = i(V_a - V_b)$ is algebraically correct if P is the power supplied to the device. Note that if P is negative the device is supplying energy at the rate $-P$.

B. Give examples: Energy may be converted to mechanical energy (a motor), to chemical energy (a charging battery), or to internal energy (a resistor). Also note the converse:

mechanical energy (a generator), chemical energy (a discharging battery), and internal energy (a thermocouple) may be converted to electrical energy.

C. Explain that in a resistor the electrical potential energy of the free electrons is converted to kinetic energy as the electric field does work on them and that the kinetic energy is lost to atoms in collisions. This increases the thermal motion of the atoms. Show that the rate of energy loss in a resistor is given by $P = i^2R = V^2/R$.

SUPPLEMENTARY TOPICS

1. Semiconductors
2. Superconductors

Both topics are important for modern physics and technology. Say a few words about them if you have time or encourage students to read about them on their own.

SUGGESTIONS

1. Assignments
   a. Discuss question 2 to emphasize the sign convention for current.
   b. Use questions 5 and 6 in a discussion of current density, resistivity, and drift velocity. Definitions are covered in problems 1 (current), 5 (current density), and 6 (drift speed).
   c. Use questions 3 and 5 when you discuss the calculation of resistance. Assign problems 18, 21, and 22. For a greater challenge assign problem 33.
   d. As part of the coverage of energy dissipation by a resistor, assign problems 37 and 41.
2. Demonstrations
   a. Model of resistance: Freier and Anderson Eg1.
   b. Thermal dissipation by resistors: Freier and Anderson Eh3.
   c. Fuses: Freier and Anderson Eh5.
   d. Ohm's law: Freier and Anderson Eg2, Eo1.
   e. Measurement of resistance, values of resistance: Freier and Anderson Eg3, 6.
   f. Temperature dependence of resistance: Freier and Anderson Eg4, 5.
3. Audio/Visual
   a. *Electric Currents*; from Cinema Classics DVD 4: Waves (II) & Electricity and Magnetism; available from Ztek Co. (PO Box 11768, Lexington, KY 40577–1768, www.ztek.com) and from the American Association of Physics Teachers (AAPT, One Physics Ellipse, College Park MD 20740–3845; www.aapt.org).
   b. *Electric Current*; VHS video tape, DVD; Films for the Humanities and Sciences (PO Box 2053, Princeton, NJ 08543–2053; www.films.com).
   c. *Temperature and Resistance*; from Physics Demonstrations in Electricity and Magnetism, Part II; VHS video tape, DVD; ≈3 min; Physics Curriculum & Instruction (22585 Woodhill Drive, Lakeville, MN 55044; www.PhysicsCurriculum.com).
4. Laboratory
   a. Probeware Activity 20: *Ohm's Law.* Voltage and current probes, connected to a computer, are used to plot the current versus the potential difference for several resistors and a diode.
   b. Meiners Experiment 10–3: *Electrical Resistance.* An ammeter and voltmeter are used to find the resistance of a light bulb and wires of various dimensions, made of various materials. The dependence of resistance on length and cross section is investigated. Resistivities of the substances are calculated and compared.
   c. Bernard Experiment 29: *A Study of the Factors Affecting Resistance.* A Wheatstone bridge and a collection of wire resistors are used to investigate the dependence of resistance on length, cross section, temperature, and resistivity.

d. Meiners Experiment 10–8: *Temperature Coefficient of Resistors and Thermistors.* A Wheatstone bridge is used to measure the resistances of a resistor and thermistor in a water-filled thermal reservoir. The temperature is changed by an immersion heater. Students see two different behaviors. A voltmeter-ammeter technique can replace the bridge if desired.

e. Also see Meiners Experiment 9–2 and Bernard Experiment 30, described in the Chapter 20 notes. These experiments can be revised to emphasize the power dissipated by a resistor. In several runs the students measure the power dissipated for different applied voltages.

Chapter 27 CIRCUITS

BASIC TOPICS

I. Emf devices.

A. Explain that an emf device moves positive charged particles inside from its negative to its positive terminal or negatively charged particles in the opposite direction and maintains the potential difference between its terminals. Emf devices are used to drive currents in circuits. Example: a battery is an emf device with an internal resistance. Note the symbol used in circuit diagrams to represent an ideal emf device (no internal resistance).

B. Explain that a direction is associated with an emf and that it is from the negative to the positive terminal, inside the device. This is the direction current would flow if the device acted alone in a completed circuit. Point out that when current flows in this direction the device does positive work on the charged particles and define the emf of an ideal device as the work per unit positive charge: $\mathcal{E} = dW/dq$. Also point out that the positive terminal of an ideal device is $\mathcal{E}$ higher in potential than the negative terminal, regardless of the direction of the current. Unit: volt.

C. Point out that the rate at which energy is supplied by an ideal device is $i\mathcal{E}$. State that for a battery the energy comes from a store of chemical energy. Mention that a battery is charging if the current and emf are in opposite directions.

II. Single loop circuits.

A. Consider a circuit containing a single ideal emf and a single resistor. Use energy considerations to derive the steady state loop equation (Kirchhoff's loop rule): equate the power supplied by the emf to the power loss in the resistor.

B. Derive the loop equation by picking a point on the circuit, selecting the potential to be zero there, then traversing the circuit and writing down expressions for the potential at points between the elements until the zero potential point is reached again. Tell the students that if the current is not known a direction must be chosen for it and used to determine the sign of the potential difference across the resistor. When the circuit equation is solved for i, a negative result will be obtained if the current is actually opposite in direction to the arrow. As you carry out the derivation remind students that current enters a resistor at the high potential end and that the positive terminal of an emf is at a higher potential than the negative terminal.

C. Consider slightly more complicated single loop circuits. Include the internal resistance of the battery and solve for the current. Place two batteries in the circuit, one charging and the other discharging. Once the current is found, calculate the power gained or lost in each element.

D. For the circuits considered, show how to calculate the potential difference between two points on the circuit and point out that the answer is independent of the path used for the calculation. Explain the difference between the closed and open circuit potential difference across a battery.

III. Multiloop circuits.

A. Explain Kirchhoff's junction rule for steady state current flow. State that it follows from the conservation of charge and the fact that charge does not build up anywhere when the steady state is reached.

B. Using an example of a two-loop circuit, go over the steps used to write down the loop and junction equations and to solve for the currents. Explain that if the current directions are unknown an arbitrary choice must be made in order to write the equations and that if the wrong choice is made, the values obtained for the current will be negative.

C. Warn students not to write duplicate junction equations. Define a branch and state that different symbols must be used for currents in different branches. State that the total number of equations will be the same as the number of branches, that the number of independent junction equations equals one less than the number of junctions, and that the remaining equations are loop equations. Also state that each current must appear in at least one loop equation.

D. Derive expressions for the equivalent resistance of two resistors in series and in parallel. Contrast with the expressions for the equivalent capacitance of two capacitors in series and in parallel. Show how to calculate potential differences across resistors in series and currents in resistors in parallel. Show how series and parallel combinations can sometimes be used to solve complicated circuits. Mention that not all circuits can be considered combinations of series and parallel connections.

E. State that the current is the same in all resistors of a series combination and the potential difference is the same for all resistors of a parallel combination.

IV. *RC* circuits.

A. Consider a series circuit consisting of an emf device, a resistor, a capacitor, and a switch. Suppose the switch is closed at time $t = 0$ with the capacitor uncharged. Use the loop rule and $i = dq/dt$ to show that $R(dq/dt) + (q/C) = \mathcal{E}$. By direct substitution, show that $q(t) = C\mathcal{E}[1 - e^{-t/RC}]$ satisfies this equation and yields $q = 0$ for $t = 0$. Also find expressions for the potential differences across the capacitor and across the resistor. Plot the expressions for q and i. Show that $q = C\mathcal{E}$ for times long compared to RC. State that $i = dq/dt$ only if the current arrow is into the positive plate of the capacitor. If it is into the negative plate, then $i = -dq/dt$.

B. Explain that $\tau = RC$ is called the time constant for the circuit and that it is indicative of the time required to charge the capacitor. If RC is large, the capacitor takes a long time to charge. Show that $q/C\mathcal{E} \approx 0.63$ when $t = \tau$ for a charging capacitor.

C. Show that the current is given by $i(t) = (\mathcal{E}/R)e^{-t/RC}$. Point out that $i = \mathcal{E}/R$ for $t = 0$ and that the potential difference across the capacitor is zero at that time because the capacitor is uncharged. Thus the potential difference across the resistor is $\mathcal{E}$. Also point out that the current tends toward zero for times that are long compared to τ. Then, the potential difference across the resistor is zero and the potential difference across the capacitor is $\mathcal{E}$.

D. Derive the loop equation for a series circuit consisting of a capacitor and resistor. Suppose the capacitor has charge q_0 at time $t = 0$ and show that $q = q_0e^{-t/RC}$. Again find expressions for the potential differences across the capacitor and resistor. Plot q and i. Point out that RC is indicative of the time for discharge. Show that $q/q_0 \approx 0.37$ when $t - \tau$ for a discharging capacitor.

E. Write the expression for the energy initially stored in the capacitor: $U = \frac{1}{2}q_0^2/C$. Evaluate $\int_0^\infty i^2R\,dt$ to find the energy dissipated in the resistor as the capacitor discharges. Show that these energies are the same.

SUPPLEMENTARY TOPIC

Electrical measuring instruments (voltmeters and ammeters). This material can be covered as needed in conjunction with the laboratory.

SUGGESTIONS

1. Assignments
   a. Problem 2 covers the fundamental idea of emf. Use problems 7 and 14 to discuss the distinction between the emf and terminal potential difference of a battery.
   b. Assign some single-loop problems, such as 5.
   c. Discuss some of questions 2, 3, 4, 5, 7, and 9 in connection with parallel and series combinations of resistors. Assign problems 11, 15, and 29.
   d. Assign some problems dealing with multiloop circuits. Consider problems 19, 20, 32, and 33.
   e. Assign problem 39 if voltmeters and ammeters are discussed in lecture or lab. Also consider problems 41 and 42.
2. Demonstrations
   a. Seats of emf: Freier and Anderson Ee2, 3, 4.
   b. Measurement of emf: Freier and Anderson Eg7.
   c. Resistive circuits: Freier and Anderson Eh1, 2, 4, Eo2 — 8.
3. Computer Software
   a. *DC Circuits*; Windows; Miky Ronen, Matzi Eliahu, and Igal Yastrubinezky; available from Physics Academic Software (Centennial Campus, 940 Main Campus Drive, Suite 210, Raleigh, NC 27606–5212; www.aip.org/pas). Circuit elements can be put together to form circuits, values of the parameters can be selected, and the circuits can then be analyzed.
   b. *Electricity and Magnetism* from Exploration of Physics Volume I; Windows and Macintosh; Physics Curriculum & Instruction (22585 Woodhill Drive, Lakeville, MN 55044; www.PhysicsCurriculum.com). Simulated experiments with analysis. Includes sections on resistive circuits and *RC* circuits.
   c. *Electricity and Magnetism* from Exploration of Physics Volume II; Windows and Macintosh; Physics Curriculum & Instruction (see above for address). Includes sections on electric circuits, Ohm's law, parallel and series connects of resistors.
4. Computer Projects
   A computer can easily be programmed to solve simultaneous linear equations. Have students use such a program to solve multiloop circuit problems.
5. Laboratory
   a. Probeware Activity 21: *Resistance in Series* and Probeware Activity 22: *Resistance in Parallel.* Voltage and current probes, connected to a computer, are used to find the equivalent resistance of several resistors in series and in parallel.
   b. Meiners Experiment 10–7 (Part A): *Measuring Current with a d'Arsonval Galvanometer.* Students determine the characteristics and sensitivity of a galvanometer. To expand this lab, ask the students to design an ammeter and a voltmeter with full scale deflections prescribed by you. Students practice circuit analysis while trying to understand design considerations.
   c. Meiners Experiment 10–9: *The EMF of a Solar Cell.* Students study a slide wire potentiometer and use it to measure the emf of a solar cell. This is another experiment that gives them practice in circuit analysis.
   d. Bernard Experiment 28: *Measurements of Potential Difference with a Potentiometer.* Students study a slide wire potentiometer and use it to investigate the emf and terminal voltage of a battery and the workings of a voltage divider.

e. Bernard Experiment 26: *A Study of Series and Parallel Electric Circuits.* Students use ammeters and voltmeters to verify Kirchhoff's laws and investigate energy balance for various circuits. They also experimentally determine equivalent resistances of resistors in series and parallel. This experiment can be extended somewhat by having them consider a network of resistors that cannot be reduced by applying the rules for series and parallel resistors. Also see Bernard Experiment 27: *Methods of Measuring Resistance.* Two voltmeter-ammeter methods and a Wheatstone bridge method are used to measure resistance and to check the equivalent resistance of series and parallel connections.

f. Bernard Experiment 31: *Circuits Containing More Than One Potential Source.* Similar to Bernard Experiment 26 described above except circuits with more than one battery are considered. The two experiments can be done together, if desired.

g. Probeware Activity 23: *RC Circuit.* Voltage and current sensors, connected to a computer, are used to plot the potential difference across a charging and then discharging capacitor as a function of time. The data is used to compute the capacitance.

h. Meiners Experiment 10–4: *The R-C Circuit.* Students connect an unknown resistor to a known capacitor, charged by a battery. The battery is disconnected and a voltmeter and timer are used to measure the time constant. The value of the resistance is calculated. In a second part an unknown capacitor is charged by means of a square wave generator and the decay is monitored on an oscilloscope. Again the time constant is measured, then it is used to calculate the capacitance. A third part explains how to use a microprocessor to collect data. Also see Bernard Experiment 32: *A Study of Capacitance and Capacitor Transients.*

Chapter 28 MAGNETIC FIELDS

BASIC TOPICS

I. Definition of the field and the magnetic force on a moving charged particle.

A. Explain that moving charged particles create magnetic fields and that a magnetic field exerts a force on a moving charged particle. Both the field of a moving charged particle and the force exerted by a field depend on the velocity of the particle involved. The latter property distinguishes it from an electric field. Also say that many particles, among them the electron, proton, and neutron, have intrinsic magnetic fields associated with them, even when they are not moving.

B. Define the magnetic field: the force on a moving test charge is $q_0\vec{v} \times \vec{B}$ after the electric force is taken into account. Review the rules for finding the magnitude and direction of a vector product. Point out that the force must be measured for at least two directions of $\vec{v}$ since the component of $\vec{B}$ along $\vec{v}$ cannot be found from the force. The direction of $\vec{B}$ can be found by trying various directions for $\vec{v}$ until one is found for which the force vanishes. The magnitude of $\vec{B}$ can be found by orienting $\vec{v}$ perpendicular to $\vec{B}$. Units: 1 tesla = 1 N/A·m, 1 gauss = 10^{-4} T. Point out the magnitudes of the fields given in Table 28–1.

C. Explain that the magnetic force on any moving charged particle is given by $\vec{F}_B = q\vec{v} \times \vec{B}$. Point out that the force is perpendicular to both $\vec{v}$ and $\vec{B}$ and is zero for $\vec{v}$ parallel or antiparallel to $\vec{B}$. Also point out that the direction of the force depends on the sign of q. Remark that the field cannot do work on the charged particle and so cannot change its speed or kinetic energy. A magnetic field can change the direction of travel of a moving charged particle. It can, for example, be used to produce a centripetal force and can cause a charged particle to move in a circular orbit.

D. To show a magnetic force qualitatively, slightly defocus an oscilloscope so the central spot is reasonably large. Move a bar magnet at an angle to the face of the scope and note the movement of the beam.

E. Point out that the total force on a charged particle is $q(\vec{E} + \vec{v} \times \vec{B})$ when both an electric and a magnetic field are present.

II. Magnetic field lines.

A. Explain that field lines can be associated with a magnetic field. At any point the field is tangent to the line through that point and the number of lines per unit area that pierce a plane perpendicular to the field is proportional to the magnitude of the field.

B. To show field lines project Figs. 28–4 and 28–5 or place a sheet of clear plastic over a bar magnet and place iron filings on the sheet. Place the arrangement on an overhead projector. Explain that the filings line up along field lines.

C. Point out that magnetic field lines form closed loops; they continue into the interior of the magnet, for example. Contrast with electric field lines and remark that no magnetic charge has yet been found. Mention that magnetic field lines would start and stop at magnetic monopoles, if they exist. Remark that lines enter at the south pole of a magnet and exit at the north pole.

III. Motions of charged particles in magnetic fields.

A. Derive $v = E/B$ for the speed of a charged particle passing through a velocity selector.

B. Outline the Thompson experiment and derive Eq. 28–8 for the mass-to-charge ratio.

C. Show how the Hall effect can be used to determine the sign and concentration of charge carriers in a conductor. Mention that these measurements are important for the semiconductor industry. Also mention that the Hall effect is used to measure magnetic fields. Show a Hall effect teslameter.

D. Consider a charged particle with velocity perpendicular to a constant magnetic field. Show that the orbit radius is given by $r = mv/qB$ and the period of the motion is given by $T = 2\pi m/qB$ (independently of v) for non-relativistic speeds. If you are covering modern topics, state that $r = p/qB$ is relativistically correct but $p = mv/\sqrt{1 - v^2/c^2}$, where c is the speed of light, must be used for the momentum. Remark that the orbit is a helix if the velocity of the charged particle has a component along the field. Show how to calculate the pitch of the helix, given the velocity components parallel and perpendicular to the field. Mention that cyclotron motion is used in cyclotrons and synchrotrons. If you have time, explain how a cyclotron works.

IV. Force on a current-carrying wire.

A. Run a flexible non-magnetic wire near a strong permanent magnet. Observe that the wire does not move. Turn on a power supply so about 1 A flows in the wire and watch the wire move. Remark that magnetic fields exert forces on currents. A car battery and jumper cables can be used. To avoid an explosion, place a heavy-duty switch in the circuit, far from the battery.

B. Consider a thin wire carrying current, with all charge carriers moving with the drift velocity. Start with the force on a single charged particle and derive $d\vec{F}_B = i\,d\vec{L} \times \vec{B}$ for an infinitesimal segment and $\vec{F}_B = i\vec{L} \times \vec{B}$ for a finite straight segment in a uniform field. Stress that $d\vec{L}$ and $\vec{L}$ are in the direction of the current.

C. Consider an arbitrarily shaped segment of wire in a uniform field. Show that the force on the segment between a and b is $\vec{F}_B = i\vec{L} \times \vec{B}$, where $\vec{L}$ is the vector joining the ends of the segment. This expression is valid only if the field is uniform.

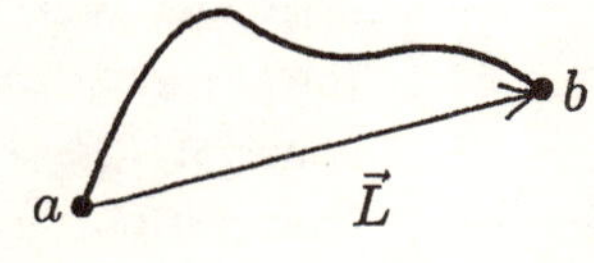

D. Point out that the force on a closed loop in a uniform field is zero since $\vec{L} = 0$.

E. Calculate the force of a uniform field on a semicircular loop of wire, in the plane perpendicular to $\vec{B}$. Do this by evaluating the integral $i\int d\vec{L} \times \vec{B}$ along the wire, then repeat using the result given in C above.

V. Torque on a current loop.
   A. Calculate the torque exerted by a uniform field on a rectangular loop of wire arbitrarily oriented with two opposite sides perpendicular to $\vec{B}$. See Fig. 28–21.
   B. Define the magnetic dipole moment of a current loop ($\mu = NiA$) and give the right hand rule for determining its direction. For a rectangular loop in a uniform field, show that $\vec{\tau} = \vec{\mu} \times \vec{B}$. State that the result is generally valid for any loop in a uniform field. Mention that other sources of magnetic fields, such as bar magnets and Earth, have dipole moments. Mention that many fundamental particles have intrinsic dipole moments and that these are the sources of their magnetic fields. See Table 29–2.
   C. Note that this is a restoring torque and that if the dipole is free to rotate it will oscillate about the direction of the field. If damping is present, it will line up along the field direction. Remark that this is the basis of magnetic compasses.
   D. Explain how analog ammeters and voltmeters work. To demonstrate the torque on a current-carrying coil, remove the case from a galvanometer and wire it to a battery and resistor so that it fully deflects.
   E. Remark that a potential energy cannot be associated with a moving charged particle in a magnetic field but can be associated with a magnetic dipole in a magnetic field. Show that $U = -\vec{\mu} \cdot \vec{B}$. Find the work required to turn a dipole through 90° and 180°, starting with it aligned along the field. Point out that U is a minimum when $\vec{\mu}$ and $\vec{B}$ are parallel and is a maximum when they are antiparallel.

SUGGESTIONS

1. Assignments
   a. Use question 1 to help in understanding the magnetic force. The dependence of magnetic force on velocity and charge is emphasized in problem 2.
   b. Use questions 5 through 8 to test for understanding of the motion of charge particles in magnetic fields. Problems 17 and 20 deal with the circular orbit of a charged particle in a uniform magnetic field. Crossed electric and magnetic fields, used as a velocity filter, are explored in questions 3 and 4 and in problems 7 and 9. Problem 25 deals with a mass spectrometer. Problem 29 deals with cyclotrons. Use some of these problems to include practical applications.
   c. Use problem 12 to help students study the Hall effect.
   d. Use problems 32 and 36 to stress the importance of the angle between the magnetic field and the current carrying wire on which it exerts a force. Use problem 40 to emphasize that the force of a uniform magnetic field on a closed loop is zero. Problem 45 asks students about the dynamics of current-carrying wires in magnetic fields. Assign problems 55 in support of the discussion of magnetic torques on current-carrying loops.
   e. Magnetic dipoles and the torques exerted on them by magnetic fields are explored in problems 47 and 48. Question 10 and problem 53 deal with the energy of a dipole in a field. Also consider problem 52.
2. Demonstrations
   a. Force on an electron beam: Freier and Anderson Ei18, Ep8, 11.
   b. Forces and torques on wires: Freier and Anderson Ei7, 12, 13 — 15, 19, 20.
   c. Meters: Freier and Anderson Ej1, 2.
   d. Hall effect: Freier and Anderson Ei16.
3. Books and Monographs

   Teaching about Magnetism; by Robert J. Reiland; available from the American Association of Physics Teachers (AAPT, One Physics Ellipse, College Park MD 20740–3845;

www.aapt.org). A PTRA workshop manual containing a collection of demonstrations and student activities.

4. Computer Software

   Electricity and Magnetism from Exploration of Physics Volume II; Windows and Macintosh; Physics Curriculum & Instruction Physics Curriculum & Instruction (22585 Woodhill Drive, Lakeville, MN 55044; www.PhysicsCurriculum.com). Includes sections on the motion of a charged particle in a magnetic field and the magnetic force on a current-carrying wire.

5. Audio/Visual
   a. *The Force on a Current*; from the AAPT collection 2 of single-concept films; DVD; available from Ztek Co. (PO Box 11768, Lexington, KY 40577–1768, www.ztek.com).
   b. *Magnetism and Magnetic Fields*; from Cinema Classics DVD 4: Waves (II) & Electricity and Magnetism; available from Ztek Co. and from the AAPT (see above for addresses).
   c. *Understanding Magnetism*; VHS video tape, DVD; Films for the Humanities and Sciences (PO Box 2053, Princeton, NJ 08543–2053; www.films.com).
   d. *Magnets*; VHS video tape, DVD; Films for the Humanities and Sciences (see above for address).
   e. *Magnetism and Static electricity*; VHS video tape, DVD; Films for the Humanities and Sciences (see above for address).
   f. *Magnetic Fields*; Physics Demonstrations in Electricity and Magnetism, Part III; VHS video tape, DVD; ≈3 min; Physics Curriculum & Instruction (see above for address).

6. Computer Software
   a. *Electricity and Magnetism* from Exploration of Physics Volume I; Windows and Macintosh; Physics Curriculum & Instruction (see above for address). Simulated experiments with analysis. Includes a section on magnetic fields.
   b. *Dynamic Analyzer*. See Chapter 2 SUGGESTIONS.

7. Computer Project

   Have students use numerical integration of Newton's second law to investigate the orbits of charged particles in magnetic and electric fields.

8. Laboratory
   a. Bernard Experiment 33: *A Study of Magnetic Fields.* A small magnetic compass is used to map field lines of various permanent magnets, a long straight current-carrying wire, a single loop of current-carrying wire, a solenoid, and Earth. Parts of this experiment might be performed profitably in connection with Chapter 31.
   b. Meiners Experiment 11–3: *Determination of* e/m. Students use the accelerating potential and the radius of the orbit in a magnetic field to calculate the charge-to-mass ratio for the electron.
   c. Meiners Experiment 11–5: *The Hall Effect.* Students measure the Hall voltage and use it to calculate the drift speed and carrier concentration for a bismuth sample. The influence of the magnetic field on the Hall voltage is also investigated. Values of the magnetic field are given to them by the instructor.

Chapter 29 MAGNETIC FIELDS DUE TO CURRENTS

BASIC TOPICS

I. Magnetic field of a current.
   A. Place a magnetic compass near a wire carrying a dc current of several amperes, if possible. Turn the current on and off and reverse the current. Note the deflection of the compass

needle and remark that the current produces a magnetic field and that the field reverses when the current reverses.

B. Write the Biot-Savart law for the field produced by an infinitesimal segment of a current-carrying wire. Give the value for μ_0. Draw a diagram to show the direction of the current, the displacement vector from the segment to the field point, and the direction of the field. Explain that $d\vec{B}$ is in the direction of $i\,d\vec{s}\times\vec{r}$. Point out the angle between $\vec{r}$ and $d\vec{s}$. Mention that the integral for the field of a finite segment must be evaluated one component at a time. Point out that the angle between $d\vec{B}$ and a coordinate axis must be used to find the component of $d\vec{B}$.

C. Example: Show how to calculate the magnetic field of a straight finite wire segment. See the text, but use finite limits of integration. State that magnetic fields obey a superposition principle and point out that the result of the calculation can be used to find the field of a circuit composed of straight segments. Specialize the result to an infinite straight wire. Demonstrate the right-hand rule for finding the direction of $\vec{B}$ due to a long straight wire.

D. Explain that the field lines around a straight wire are circles in planes perpendicular to the wire and are centered on the wire. Draw a diagram to illustrate. Use symmetry to argue that the magnitude of the field is uniform on a field line. Point out that for other current configurations B is not necessarily uniform on a field line.

E. Show how to find the force per unit length of one long straight wire on another. Treat currents in the same and opposite directions. Lay two long automobile starter cables on the table. Connect them in parallel to an auto battery, with a $0.5\,\Omega$, 500 W resistor and an "anti-theft" switch or starter relay in each circuit. Close one switch and note that the wires do not move. Close the other switch and note the motion. Show parallel and antiparallel situations. It is better to reconnect the wires or rearrange them rather than to use a reversing switch.

F. Give the definition of the ampere and remind students of the definition of the coulomb.

G. Consider a circular arc of radius R, subtending an angle ϕ, and carrying current i. Use the Biot-Savart law to show that the magnetic field at the center is given by $B = \mu_0 i\phi/4\pi R$. Note that ϕ must be in radians. Specialize to the cases of a semicircle and a full circle.

II. Ampere's law.

A. Write the law in integral form. Explain that the integral is a line integral around a closed contour and interpret it as a sum over segments. Point out that it is the tangential component of $\vec{B}$ that enters. Explain that the current that enters is the net current through the contour. Two currents in opposite directions tend to cancel, for example. Illustrate by considering a contour that encircles five or six wires, with some currents in each direction. Also consider a wire passing through the plane of the contour but outside the contour. Mention that this current produces a magnetic field at all points on the contour but the integral of its tangential component is zero.

B. Explain the right-hand rule that relates the direction of integration around the contour and the direction of positive current through the contour.

C. Pick a functional form for the magnetic field ($B_x = 2axy$, $B_y = -ay^2$, $B_z = 0$, for example). Be sure the divergence is zero and the curl is not. Now consider a simple contour, such as a square in the xy plane. Integrate the tangential component of the field around the contour and calculate the net current through it.

D. Use Ampere's law to calculate the magnetic field *outside* a long straight wire. Either use without proof the circular nature of the field lines or give a symmetry argument to show that $\vec{B}$ at any point is tangent to a circle through the point and has constant magnitude around the circle. Point out that the integration contour is taken tangent to $\vec{B}$ in order to evaluate the integral in terms of the unknown magnitude of $\vec{B}$.

E. Use Ampere's law to calculate the field *inside* a long straight wire with a uniform current distribution. Note that the use of Ampere's law to find B has the same limitations as Gauss' law when used to find E: there must be sufficient symmetry.

F. Use Ampere's law to calculate the field inside a solenoid. First argue that, for a long tightly wound solenoid, the field at interior points is along the axis and nearly uniform while the field at exterior points is nearly zero.

G. Similarly, use Ampere's law to calculate the field inside a toroid.

III. Magnetic dipole field.

A. Use the Biot-Savart law to derive an expression for the field of a circular current loop at a point on its axis. Stress the resolution of $d\vec{B}$ into components.

B. Take the limit as the radius becomes much smaller than the distance to the field point and write the result in terms of the dipole moment. Explain that the result is generally true for loops of any shape as long as the field point is far from the loop. Remind students that the dipole moment of a loop is determined by its area and the current it carries.

SUGGESTIONS

1. Assignments
   a. Use questions 1 and 2 and problem 1 as part of the discussion of the magnetic field due to a long straight wire. Problems 17 and 19 deal with the field of a finite straight wire. Assign them, then problem 25, which asks students to superpose the fields of finite wires.
   b. Question 3 deals with the field of a circular arc. Problems 4, 5, 6, and 13 deal with circuits consisting of straight line and circular segments. Assign one or two of them.
   c. Ask questions 5 and 6 in association with the magnetic forces exerted by wires on each other. Assign problem 29.
   d. Use questions 7 through 9 in your discussion of Ampere's law. After discussing line integrals around closed loops, assign problems 35 and 36 to test the fundamentals; problem 37 gives an application. Assign problem 39 if you want to include the field of a wire with nonuniform current density.
   e. Problems 40 and 43 can be assigned to support the discussion of solenoids and toroids.
   f. Problems 48 and 53 deal with the magnetic fields of coils and dipole loops. Assign problem 50 if you cover Helmholtz coils or use them in lab.

2. Demonstrations
   a. Magnetic fields of wires: Freier and Anderson Ei8 — 11.
   b. Magnetic forces between wires: Freier and Anderson Ei1 — 6.
   c. *Magnetic Fields*; VHS video tape; Films for the Humanities and Sciences (PO Box 2053, Princeton, NJ 08543–2053; www.films.com).

3. Computer Software
   a. *Electricity and Magnetism* from Exploration of Physics Volume II; Windows and Macintosh; Physics Curriculum & Instruction (22585 Woodhill Drive, Lakeville, MN 55044; www.PhysicsCurriculum.com). Includes sections on the magnetic field of a current-carrying wire, the magnetic field of a solenoid, and the magnetic field of a bar magnet.
   b. *EM Field*; David Trowbridge. See Chapter 22 SUGGESTIONS.

4. Books and Monographs

   The Solenoid; by Carl R. Stannard, Arnold A. Strassenberg, and Gabriel Kousourou; available from the American Association of Physics Teachers (AAPT, One Physics Ellipse, College Park MD 20740–3845; www.aapt.org). Covers the magnetic field of a solenoid and practical applications as a mechanical switch.

5. Computer Projects
   a. Have students use the Biot-Savart law and numerical integration to calculate the magnetic field due to a circular current loop at off-axis points. They can use a commercial math program or their own programs.
   b. Use numerical integration to verify Ampere's law for several long straight wires passing through a square contour. Have them show the result of the integration is independent of the positions of the wires, as long as they are inside the square. Also have them consider a wire outside the square.
6. Laboratory
   a. Probeware Activity 25: *Magnetic Field of a Solenoid.* A magnetic field probe, connected to a computer, is used to plot the magnitude of the magnetic field of a solenoid as a function of radial position, both outside and inside the solenoid.
   b. Meiners Experiment 11–1: *The Earth's Magnetic Field.* A tangent galvanometer is used to measure Earth's magnetic field. The dip angle is calculated.
   c. Meiners Experiment 11–2: *The Current Balance.* The gravitational force on a current-carrying wire is used to balance the magnetic force due to current in a second wire. The data can be used to find the value of μ_0 or to find the current in the wires. The second version essentially defines the ampere. Part B describes how a microprocessor can be used to collect and analyze the data.
   d. Bernard Experiment 34: *Measurement of the Earth's Magnetic Field.* The oscillation period of a small permanent magnet suspended inside a solenoid is measured with the solenoid and Earth's field aligned. The reciprocal of the period squared is plotted as a function of the current in the solenoid, and the slope, along with calculated values of the solenoid's field, is used to find Earth's field.
   e. Meiners Experiment 11–3: *Determination of e/m.* Students find the speed and orbit radius of an electron in the magnetic field of a pair of Helmholtz coils and use the data to calculate e/m. Information from this chapter is used to compute the field, given the coil radius and current. If you are willing to postulate the field for the students, this experiment can be performed in connection with Chapter 30.

Chapter 30 INDUCTION AND INDUCTANCE

BASIC TOPICS

I. The law of induction.
   A. Connect a coil (50 to 100 turns) to a sensitive galvanometer and move a bar magnet in and out of the coil. Note that a current is induced only when the magnet is moving. Show all possibilities: the north pole entering and exiting the coil and the south pole entering and exiting the coil. In each case point out the direction of the induced current. With a little practice you might also demonstrate effectively that the deflection of the galvanometer depends on the speed of the magnet.
   B. To show the current produced by changing the orientation of a loop, align the coil axis with Earth's magnetic field and rapidly rotate the coil once through 180°. Note the deflection of a galvanometer in series with the coil. Explain that this forms the basis of electric generators.
   C. Connect a coil to a switchable dc power supply. Connect a voltmeter (digital, if possible) to the supply to show when it is on. Place a second coil, connected to a sensitive galvanometer, near the first. Show that when the switch is opened or closed, current is induced in the second coil, but that none is induced when the current in the first coil is steady.

D. Define the magnetic flux through a surface. Unit: 1 weber = 1 T·m^2. Point out that Φ_B measures the number of magnetic field lines that penetrate the surface. Remark that $\Phi_B = BA\cos\theta$ when $\vec{B}$ is uniform over the surface and makes the angle θ with its normal.

E. Give a qualitative statement of the law: an emf is generated around a closed contour when the magnetic flux through the contour changes. Stress that the law involves the flux through the surface bounded by the contour. Point out the surface and contour for each of the demonstrations done, then remark that the contour may be a conducting wire, the physical boundary of some material, or a purely geometric construction. Remark that if the contour is conducting, then charge flows.

F. Give the equations for Faraday's law: $\mathcal{E} = -\mathrm{d}\Phi_B/\mathrm{d}t$ for a single loop and $\mathcal{E} = -N\,\mathrm{d}\Phi_B/\mathrm{d}t$ for N tightly packed loops. Note that the emf's add.

II. Lenz's law.

A. Explain Lenz's law in terms of the magnetic field produced by the current induced if the contour is a conducting wire. Stress that the induced field must re-enforce the external field in the interior of the loop if the flux is decreasing and must tend to cancel it if the flux is increasing. This gives the direction of the induced current, which is the same as the direction of the emf. Review the right-hand rule for finding the direction of the field produced by a loop of current-carrying wire. State that Lenz's law can be used even if the contour is not conducting. The current must then be imagined.

B. Optional: Give the right-hand rule for finding the direction of positive emf. When the thumb points in the direction of $\mathrm{d}\vec{A}$, then the fingers curl in the direction of positive emf. If Faraday's law gives a negative emf, then it is directed opposite to the fingers. Stress that the negative sign in the law is important if the equation, with the right-hand rule, is to describe nature.

C. Consider a rectangular loop of wire placed perpendicular to a magnetic field. Assume a function $B(t)$ and calculate the emf and current. Show how the directions of the emf and current are found. Point out that an *area* integral is evaluated to find Φ_B and a *time* derivative is evaluated to find the emf. Some students confuse the variables and integrate with respect to time.

III. Motional emf.

A. Consider a rectangular loop being pulled with constant velocity past the boundary of a uniform magnetic field. Calculate the emf and current.

B. Consider a rod moving with a constant velocity that is perpendicular to a uniform magnetic field. Show how to complete the loop and calculate the emf. Mention that the emf exists only in the moving rod, regardless of whether the rest of the contour is conducting.

C. Consider a rectangular loop of wire rotating with constant angular velocity about an axis that is in the plane of the loop and through its center. Take the magnetic field to be uniform and point out that now the flux is changing because the angle between the field and the normal to the loop is changing. Derive the expression for the emf and point out it is time dependent.

IV. Energy considerations.

A. Point out that an emf does work at the rate $\mathcal{E}i$, where i is the current. Explain that for a current induced by motion, the energy comes from the work done by an external agent or from the kinetic energy of the moving portion of the loop.

B. Consider four conducting rails that form a rectangle, three fixed and the fourth riding on two of them. Take the magnetic field to be uniform and normal to the loop. Assume that essentially all of the electrical resistance of the loop is associated with the moving rail. First, suppose the moving rail has constant velocity and derive expressions for the emf, current, and magnetic force on the rail. Next, derive expressions for the rate at which an external agent must do work to keep the velocity constant and for the rate at which energy is dissipated by the resistance of the loop. Point out that all the energy supplied by the agent is dissipated.

C. Now suppose the rail is given an initial velocity and, thereafter, it is acted on by the magnetic field alone. Use Newton's second law to derive an expression for the velocity as a function of time. Compare the rate at which the kinetic energy is decreasing with the rate of energy dissipation in the resistance. Remark that this phenomenon finds practical application in magnetic braking.

D. Mention that energy is also dissipated when a current is induced by a changing magnetic field and it comes from the agent that is changing the field.

V. Induced electric fields.

A. Explain that a changing magnetic field produces an electric field, which is responsible for the emf. The emf and electric field are related by $\mathcal{E} = \oint \vec{E} \cdot d\vec{s}$, where the integral is around the contour. Remind students that this integral is the work per unit charge done by the field as a charge goes around the contour. Write Faraday's law as $\oint \vec{E} \cdot d\vec{s} = -\frac{d}{dt} \int \vec{B} \cdot d\vec{A}$. Note that $d\vec{s}$ and $d\vec{A}$ are related by a right-hand rule: fingers along $d\vec{s}$ implies thumb along $d\vec{A}$. This is consistent with Lenz's law.

B. State that the induced electric field is like an electrostatic field in that it exerts a force on a charge but that it is unlike an electrostatic field in that it is not conservative. For an electrostatic field, the integral defining the emf vanishes. An electric potential cannot be associated with an induced electric field.

C. Consider a cylindrical region containing a uniform magnetic field along the axis. Assume a time dependence for $\vec{B}$ and derive expressions for the electric field inside the region and outside the region. See Sample Problem 30–4. Point out that the lines of $\vec{E}$ form closed circles concentric with the cylinder and that the magnitude of $\vec{E}$ is uniform around a circle.

VI. Definition of inductance.

A. Connect a light bulb and choke coil in parallel across a switchable dc supply. Close the switch and note that the lamp is initially brighter than when steady state is reached. Open the switch and note that the light brightens before going off. Remark that this behavior is due to the changing magnetic flux through the coil and that the flux is created by the current in the coil itself.

B. Point out that when current flows in a loop, it generates a magnetic field and the loop contains magnetic flux due to its own current. If the current changes, so does the flux and an emf is generated around the loop. The total emf, due to all sources, determines the current. Remark that the self-flux is proportional to the current and the induced emf is proportional to the rate of change of the current.

C. Define the inductance by $L = N\Phi_B/i$, where N is the number of turns, Φ_B is the magnetic flux through each turn, and i is the current in the circuit. Unit: 1 henry = 1 V·s/A.

D. Remark that Faraday's law yields $\mathcal{E} = -L\,di/dt$ for the induced emf.

E. Inductors are denoted by  in circuit diagrams. Point out that if the circuit element looks like 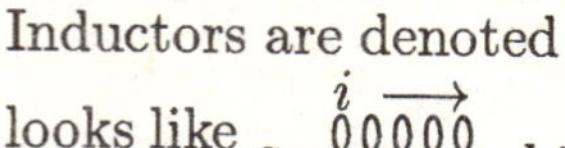, then $V_b - V_a = -L\,di/dt$ is algebraically correct. As an example, use $i(t) = i_m \sin(\omega t)$. Note that i is positive when it is directed from a to b and negative when it is directed from b to a. Compute $V_a - V_b = L i_m \omega \cos(\omega t)$. Graph i and the potential difference as functions of time to show the phase relationship. Remark that a real inductor can be regarded as a pure inductance in series with a pure resistance.

F. Show how to calculate the inductance of an ideal solenoid. Use the current to calculate the field, then the flux, and finally equate $N\Phi_B$ to Li and solve for L. Point out that L is independent of i but depends on geometric factors such as the cross-sectional area, length, and the number of turns per unit length.

G. Optional: Show how to calculate the inductance of a toroid.

VII. An LR circuit.

A. Derive the loop equation for a single loop containing a source of emf (an ideal battery), a resistor, and an inductor in series: $\mathcal{E} - iR - L\,\mathrm{d}i/\mathrm{d}t = 0$, where the current is positive if it leaves the positive terminal of the seat of emf. Use the prototypes developed earlier:

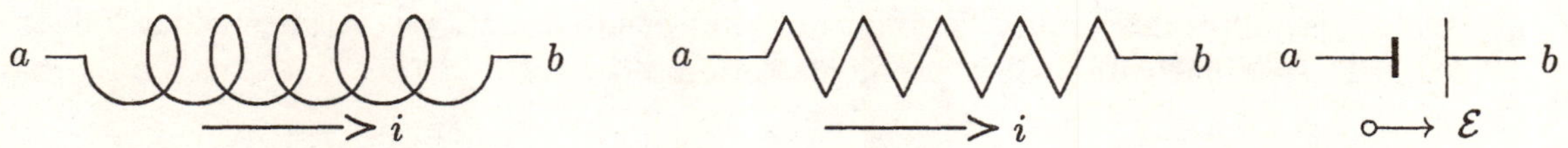

$V_b - V_a = -L\,\mathrm{d}i/\mathrm{d}t$ $\qquad$ $V_b - V_a = -iR$ $\qquad$ $V_b - V_a = \mathcal{E}$

Remark that these are correct no matter if the current is positive or negative or if it is increasing or decreasing. Write down the solution for the current as a function of time for the case $i(0) = 0$: $i = (\mathcal{E}/R)[1 - e^{-Rt/L}]$. Show that the expression satisfies the loop equation and meets the initial conditions. Show a graph of $i(t)$; point out the asymptotic limit $i = \mathcal{E}/R$ and the time constant $\tau_L = L/R$. Remark that if L/R is large, the current approaches its limit more slowly than if L/R is small.

B. Explain the qualitative physics involved. When the battery is turned on and the current increases, the emf of the coil opposes the increase and the current approaches its steady state value more slowly than if there were no inductance. At long times, the current is nearly constant so $\mathrm{d}i/\mathrm{d}t$ and the induced emf are small. The current is nearly the same as it would be in the absence of an inductor. Just after the battery is turned on, the potential difference across the resistor is zero and the potential difference across the inductor is $\mathcal{E}$. After a long time, the potential difference across the resistor is $\mathcal{E}$ and the potential difference across the inductor is zero.

C. Repeat the calculation for a circuit with an inductor and resistor but no battery. Take the initial current to be i_0 and show that $i(t) = i_0 e^{-t/\tau_L}$. Graph the solution and show the position of τ_L on the time axis. Point out that the emf of the coil opposes the decrease in current.

D. Demonstrate the two circuits by connecting a resistor and coil in series to a square-wave generator. Observe the current by placing oscilloscope leads across the resistor. Observe the potential difference across the coil. Vary the time constant by varying the resistance.

VIII. Energy considerations.

A. Consider a single loop circuit containing an ideal battery, a resistor, and an inductor. Assume the current is increasing. Write down the loop equation, multiply it by i, and identify the power supplied by the battery and the power lost in the resistor. Explain that the remaining term describes the power being stored by the inductor, in its magnetic field. Point out the similarity between $i\mathcal{E}$ and $-iL\,\mathrm{d}i/\mathrm{d}t$ for the rate at which work is being done by an ideal battery and by an inductor (with emf $-L\mathrm{d}i/\mathrm{d}t$).

B. Integrate $P = iL\,\mathrm{d}i/\mathrm{d}t$ to obtain $U_B = \frac{1}{2}Li^2$ for the energy stored in the magnetic field (relative to the energy for $i = 0$).

C. Consider the energy stored in a long current-carrying solenoid and show that the energy density is $u_B = B^2/2\mu_0$. Explain that this gives the energy density at a point in any magnetic field and that the energy required to establish a given magnetic field can be calculated by integrating the expression over the volume occupied by the field.

IX. Mutual induction.

A. Repeat the demonstration experiment discussed in note IC. Explain it in terms of the concept of mutual induction. Point out that the flux through the second coil is proportional to the current in the first. Define the mutual induction of the second coil with respect to the first by $M_{12} = N_2\Phi_{12}/i_1$. Show that $\mathcal{E}_2 = -M_{12}\,di_1/dt$ is the emf induced in the second coil when the current in the first changes. State without proof that $M_{12} = M_{21}$.

B. Example: derive the mutual inductance for a small coil placed at the center of a solenoid or for a small, tightly wound coil placed at the center of a larger coil.

C. Show that two inductors connected in series and well separated have an equivalent inductance of $L = L_1 + L_2$. Then, show that if their fluxes are linked, $L = L_1 + L_2 \pm 2M$, where the minus sign is used if the field lines have opposite directions. Also consider inductors in parallel. Alternatively, assign problems 45 and 46.

SUGGESTIONS

1. Assignments
   a. Question 2 deals with the magnitudes of induced emf and current. Questions 1 and 3 deal with Lenz's law. Use several as examples and several to test students.
   b. Assign problems 1, 2, and 3 to cover the emf's generated by various time dependent magnetic fields. Addition of emf's is covered in problem 9. This is a good problem to test for understanding of the sign of an induced emf.
   c. Motional emf is covered in problems 11, 15, 29, 30 and 32. If you use a flip coil in the lab, assign problem 13. Problems 31 and 33 deal with energy transfers.
   d. Assign problems 34 and 35 in connection with the discussion of induced electric fields.
   e. Assign problem 38 (coil) or 41 (two parallel wires) as an example of a typical inductance calculation.
   f. Use some of questions 7, 9, and 10 when discussing LR circuits. Assign problem 50. LR time constants are considered in problem 49.
   g. After discussing energy flow in a simple LR circuit with increasing current, assign problems 60 and 62.
   h. Problem 63 deals with energy storage and energy density in an inductor.
2. Demonstrations
   a. As a supplementary demonstration, take a large, long coil, mount it vertically, insert a solid, soft-iron rod with a foot or so sticking out, and connect the coil via a switch to a large dc power supply. Place a solid aluminum ring around the iron rod. The ring should fit closely but be free to move. Close the switch and the ring will jump up, then settle down. Repeat with a ring that has a gap in it. Finally, use an ac power supply. The effect can be enhanced by cooling the ring with liquid nitrogen.

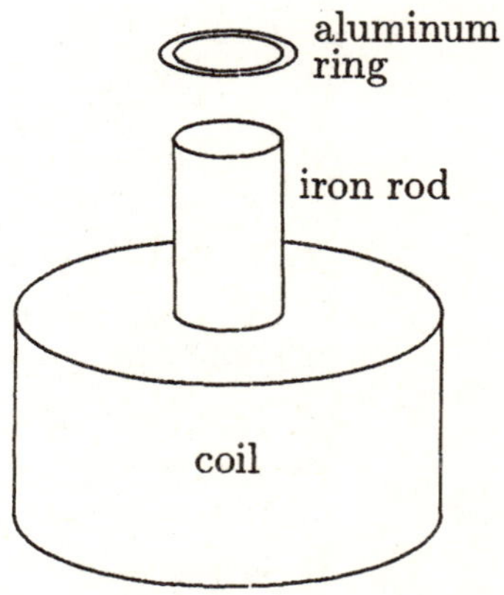

   b. Generation of induced currents: Freier and Anderson Ek1 — 6.
   c. Eddy currents: Freier and Anderson Ei1 — 6.
   d. Generators: Freier and Anderson: Eq4 — 7, Er1.
   e. Self-inductance: Freier and Anderson Eq1 — 3.
   f. LR circuit: Freier and Anderson Eo11, En5 — 7.
3. Audio/Visual
   a. *Electromagnetism*; from Cinema Classics DVD 4: Waves (II) & Electricity and Magnetism; available from Ztek Co. (PO Box 11768, Lexington, KY 40577–1768, www.ztek.com) and from the American Association of Physics Teachers (AAPT, One Physics Ellipse, College Park MD 20740–3845; www.aapt.org).

b. *Electricity and Magnetism*; *Electromagnetic Effects*; *Induction Application*; *Eddy Currents*; from Physics Demonstrations in Electricity and Magnetism, Part III; VHS video tape, DVD; ≈3 min each; Physics Curriculum & Instruction (22585 Woodhill Drive, Lakeville, MN 55044; www.PhysicsCurriculum.com).

4. Computer Software

   Electricity and Magnetism from Exploration of Physics Volume I; Windows and Macintosh; Physics Curriculum & Instruction (22585 Woodhill Drive, Lakeville, MN 55044; www.PhysicsCurriculum.com). Simulated experiments with analysis. Includes a section on Lenz's law.

5. Laboratory

   a. Probeware Activity 26: *Faraday's Law.* A voltage senor, connected to a computer, is used to measure the emf generated in a coil as a permanent magnet is pushed through it. The computer displays the voltage as a function of time.

   b. Bernard Experiment 35: *Electromagnetic Induction.* Students measure the magnitude and observe the direction of current induced by a changing magnetic flux in a simple galvanometer circuit. Changing flux is produced by moving permanent magnets, by moving current-carrying coils, and by changing current in a coil.

   c. Meiners Experiment 11–4: *The Magnetic Field of a Circular Coil.* The emf generated in a small search coil when a low frequency ac current flows in a given circuit (a circular coil in this case) is used to determine the magnetic field produced by the circuit. The field is investigated as a function of position, specified in spherical coordinates.

Chapter 31 ELECTROMAGNETIC OSCILLATIONS AND ALTERNATING CURRENT

BASIC TOPICS

I. LC oscillations.

   A. Draw a diagram of an LC series circuit and assume the capacitor is charged. Explain that as charge flows, energy is transferred from the electric field of the capacitor to the magnetic field of the inductor and back again. When the capacitor has maximum charge, the current (dq/dt) vanishes, so no energy is stored in the inductor. When the current is a maximum, the charge on the capacitor vanishes and no energy is stored in that element.

   B. Write down the loop equation, then convert it so the charge q on the capacitor is the dependent variable. If the direction of positive current is into the capacitor plate with positive charge q, then $i = dq/dt$. If it is out of that plate, then $i = -dq/dt$.

   C. Write down the solution: $q(t) = Q\cos(\omega t + \phi)$. Show by direct differentiation that this is a solution if $\omega^2 = 1/LC$. Show that ϕ is determined by the initial conditions and treat the special case for which $q = Q$, $i = 0$ at $t = 0$.

   D. Once the solution is found, derive expressions for the current, the energy stored in the capacitor, and the energy stored in the inductor, all as functions of time. Sketch graphs of these quantities. Show that the total energy is constant.

   E. Derive expressions for the potential differences across the capacitor and the inductor. Draw graphs of them as well. Mention that the charge on the capacitor is proportional to the potential difference across its plates and that the time rate of change of the current is proportional to the emf of the inductor.

   F. Note that the form of the differential equation for q is the same as that for the displacement x of a block oscillating on the end of a spring. Make the analogy concrete by explaining that if q is replaced by x, L is replaced by m, and C is replaced by $1/k$, the equation for q

becomes the equation for x. Also point out that the current corresponds to the velocity of the block, that the energy in the inductor corresponds to the kinetic energy of the block, and that the energy in the capacitor corresponds to the potential energy in the spring.

II. Damped oscillations.

A. Write down the loop equation for a single RLC loop, then convert it so q is the dependent variable. State that $q(t) = Q\,e^{-Rt/2L}\cos(\omega' t + \phi)$ satisfies the differential equation. Here ω' is somewhat less than $1/\sqrt{LC}$. If time permits, the expression for ω' can be found by substituting the assumed solution into the differential equation.

B. Draw a graph of $q(t)$ and point out that the envelope decreases exponentially. Each time the capacitor is maximally charged, the charge on the positive plate is less than the previous time. Explain that this does not violate the conservation of charge principle since the total of the charge on both plates of the capacitor is always zero. Energy is dissipated in the resistor.

C. To show the oscillations, wire a resistor, inductor, and capacitor in series with a square-wave generator and connect an oscilloscope across the capacitor. The scope shows a function proportional to the charge. Also connect the oscilloscope across the resistor to show a function that is proportional to the current. If you have a dual-trace scope, show the functions simultaneously. Show the effect of varying C (use a variable capacitor), R (use a decade box), and L (insert an iron rod into the coil). If time permits, show that oscillations occur only if $1/LC > (R/2L)^2$.

III. Elements of ac circuit analysis.

A. Consider a resistor connected to a sinusoidally oscillating emf. State that the potential difference across the resistor has the same angular frequency ω_d as the emf. State that the potential difference across the resistor is in phase with the current and that the amplitudes are related by $I_R = V_R/R$. Emphasize that $v_R(t)$ gives the potential of one end of the resistor relative to the other. Draw a phasor diagram: two arrows along the same line with length proportional to I_R and V_R, respectively. Both make the angle $\omega_d t$ with the horizontal axis and rotate in the counterclockwise direction. Point out that the vertical projections represent $i_R(t)$ and $v_R(t)$ and these vary in proportion to $\sin(\omega_d t)$ as the arrows rotate.

B. Consider a capacitor connected to a sinusoidally oscillating emf. Start with $i_C = dq/dt = C\,dv_C/dt$, substitute $v_C = V_C \sin(\omega_d t)$, and show that v_C lags i_C by 90° and that the amplitudes are related by $I_C = V_C/X_C$, where $X_C = 1/\omega_d C$ is the capacitive reactance. Draw a phasor diagram to show the relationship. Mention that the unit of reactance is the ohm.

C. Consider an inductor connected to a sinusoidally oscillating emf. Start with $v_L = L\,di_L/dt$, substitute $v_L = V_L \sin(\omega_d t)$, and show that v_L leads i_L by 90° and that the amplitudes are related by $I_L = V_L/X_L$, where $X_L = \omega_d L$ is the inductive reactance. Draw a phasor diagram to show the relationship.

D. Wire a small resistor in series with a capacitor and a signal generator. Use a dual trace oscilloscope with one set of leads across the resistor and the other set across the capacitor. Remind students that the potential difference across the resistor is proportional to the current, so the scope shows i_C and v_C. Point out the difference in phase. Repeat with an inductor in place of the capacitor.

IV. Forced oscillations of an RLC series circuit.

A. Draw the circuit. Assume the generator emf is given by $\mathcal{E}(t) = \mathcal{E}_m \sin(\omega_d t)$ and the current is given by $i(t) = I\sin(\omega_d t - \phi)$. Pick consistent directions for positive emf and positive current. Construct a phasor diagram step-by-step (see Fig. 31–13). First draw the current and resistor voltage phasors, in phase. Remind students that the current is the same in every element of the circuit so voltage phasors for the other elements can be drawn using the phase relations between voltage and current developed earlier. Draw the capacitor voltage phasor lagging by 90° and the inductor voltage phasor leading by 90°. Make $V_L > V_C$. Their lengths are IX_C and IX_L, respectively. Draw the projections of the phasors on the vertical axis and remark that the algebraic sum must be $\mathcal{E}(t)$.

B. Draw the impressed emf phasor. Remark that its projection on the inductance phasor must be $V_L - V_C$ and that its projection on the resistance phasor must be V_R. Make the analogy to a vector sum.

C. Use the phasor diagram to derive the expression for the current amplitude: $I = \mathcal{E}_m/Z$, where $Z = \sqrt{R^2 + (X_L - X_C)^2}$ is the impedance of the circuit. Show that the impedance is frequency dependent by substituting the expressions for the reactances.

D. Use the phasor diagram to derive the expression for the phase angle of i relative to $\mathcal{E}$: $\tan\phi = (X_L - X_C)/R$. Point out that $\mathcal{E}$ leads i if $X_L > X_C$, but $\mathcal{E}$ lags i if $X_L < X_C$. For later use, show that $\cos\phi = R/Z$.

V. Resonance

A. Sketch graphs of the current amplitude as a function of the generator frequency for several values of the resistance (see Fig. 33–13). Point out that the current amplitude is greatest when the generator frequency matches the natural frequency of the circuit and that the peak becomes larger as the resistance is reduced. Use the expression derived above for the current amplitude to show that I is greatest for $X_C = X_L$ and that this means $\omega_d = 1/\sqrt{LC}$. Remark that this is the resonance condition. Also show that the phase angle between the current and the generator emf vanishes at resonance.

B. Demonstrate resonance phenomena by wiring an RLC loop in series with a sinusoidal audio oscillator. Look at the current by putting the leads of an oscilloscope across the resistor. Use a decade box for the resistor and measure the current amplitude for various frequencies and for several resistance values. Be sure the amplitude of the oscillator output remains the same. Explain that similar circuits are used to tune radios and TV's.

C. Use a sweep generator to show the current amplitude. Set the oscilloscope sweep rate to accommodate that of the generator and put a small diode in series with the scope leads. Usually this will have enough capacitance that only the envelope will be displayed.

VI. Power considerations.

A. Discuss average values over a cycle. Show that the average of $\sin^2(\omega_d t + \phi)$ is $\frac{1}{2}$ and that the average of $\sin(\omega_d t)\cos(\omega_d t)$ is 0. Define the rms value of a sinusoidal quantity. Point out that ac meters are usually calibrated in terms of rms values.

B. Derive the expression for the power input of the ac source: $P = i\mathcal{E} = i_m\mathcal{E}_m \sin(\omega_d t + \phi)\sin(\omega_d t)$. Show that the average over a cycle is given by $\overline{P} = \mathcal{E}_{\rm rms} i_{\rm rms}\cos\phi$. Do the same for the power dissipated in the resistor. In particular, show that its average value can be written $i_{\rm rms}^2 R$ or $\mathcal{E}_{\rm rms} i_{\rm rms} R/Z$. Recall that $R/Z = \cos\phi$ and then use this relationship to show that the average power input equals the average power dissipated in the resistor.

C. Show that the average rate of energy flow into the inductor and capacitor are each zero.

D. Explain that $\cos\phi$ is called the power factor. If it is 1, the source delivers the greatest possible power for a fixed generator amplitude. Remark that the power factor is 1 at resonance.

SUPPLEMENTARY TOPIC

The transformer. Say that ac is in common use because it is efficient to transmit power at high potential and low current but safety considerations require low potential at the user and producer ends of a transmission line. Transformers can be used to change the potential. Use Faraday's law to show how the potential difference across the secondary is related to the potential difference across the primary. Explain what step-up and step-down transformers are. A dual trace oscilloscope can be used to demonstrate transformer voltages. Assume a purely resistive load and show how to find the primary and secondary currents. Show that, as far as the primary current is concerned, the transformer and secondary circuit can be replaced by a resistor with $R_{eq} = (N_p/N_s)^2 R$, where N_p is the number of turns in the primary coil, N_s is the number of turns in the secondary coil, and R is the load resistance. Explain impedance matching.

SUGGESTIONS

1. Assignments
   a. Questions 1 through 5 can be used to help students think about *LC* circuit relationships.
   b. If you compare an oscillating *LC* circuit to an oscillating mass on a spring, assign problem 6 or 7.
   c. Assign problems 1, 3, 4, and 11 to test for understanding of the fundamentals of *LC* oscillations. The frequency of oscillation is covered in problems 9, 10, and 15.
   d. When discussing solutions to the *RLC* loop equation, include questions 8, 9, and 11.
   e. Assign problem 33 in connection with discussions of the phase and amplitude of separate inductive and capacitive circuits.
   f. Resonance is covered in problems 45 and 49.
   g. Power in an *RLC* circuit is covered in problems 56 and 57 and the power factor in problem 55. Include question 12 in the discussion.
2. Demonstrations
   a. *LCR* series circuit: Freier and Anderson En12, Eo13.
   b. Measurements of reactance and impedance: Freier and Anderson Eo9.
   c. Transformers: Freier and Anderson Ek7, Em1, 2, 4, 5, 7, 8, 10.
3. Laboratory
   a. Meiners Experiment 10–11: *A.C. Series Circuits.* Students use an oscilloscope and ac meters to investigate voltage amplitudes, phases, and power in *RC* and *RLC* circuits. Voltage amplitudes and phases are plotted as functions of the driving frequency to show resonance. Reactances and impedances are calculated from the data.
   b. Bernard Experiment 37: *A Study of Alternating Current Circuits.* An ac voltmeter is used to investigate the voltages across circuit elements in *R*, *RC*, *RL*, and *RLC* circuits, all with 60 Hz sources. Reactances and impedances are computed. If possible, oscilloscopes should be used. A section labeled optional describes their use. This experiment is pedagogically similar to the text and can be used profitably to reenforce the ideas of the chapter. Warning: the lab book uses the word vector rather than phasor.

Chapter 32 MAXWELL'S EQUATIONS; MAGNETISM OF MATTER

BASIC TOPICS

I. Gauss' law for magnetism.
   A. Explain that a magnetic monopole is a particle that produces a magnetic field even while at rest, with magnetic field lines starting or stopping on it. Remark that no magnetic monopole has been observed yet but it is currently being sought. Write down Gauss' law for the magnetic field and state that magnetic field lines form closed contours so the flux

through any closed surface vanishes. If monopoles were found to exist, the law would be modified to include them. Compare with Gauss' law for the electric field.

B. To show that the ends of a magnet are not monopoles, magnetize a piece of hard iron wire. Use a compass to locate and mark the north and south poles. Break the wire into pieces and again use the compass to show that each piece has a north and a south pole. Repeat a few times using smaller pieces each time. Remark that the same results would be obtained if the breaking process were continued to the atomic level. Individual atoms and particles are magnetic dipoles, not monopoles.

II. The Maxwell induction law.

A. In the material discussed so far, note the absence of any counterpart to Faraday's law, i.e. the creation of magnetic fields by changing electric flux. Tell students it should be there and you will now discuss its form.

B. Consider the charging of a parallel-plate capacitor. Remind students that in Ampere's law $d\vec{s}$ and $d\vec{A}$ are related by a right-hand rule and the surface integral is over any surface bounded by the closed contour.

C. In the diagram, surfaces A, B, and C are all bounded by the contour that forms the left end of the figure. If we choose surface A or C, then Ampere's law as we have taken it gives $\oint \vec{B}\cdot d\vec{s} = \mu_0 i$, but if we choose surface B, it gives $\oint \vec{B}\cdot d\vec{s} = 0$. Since the integral on the left side is exactly the same in all cases, something is wrong.

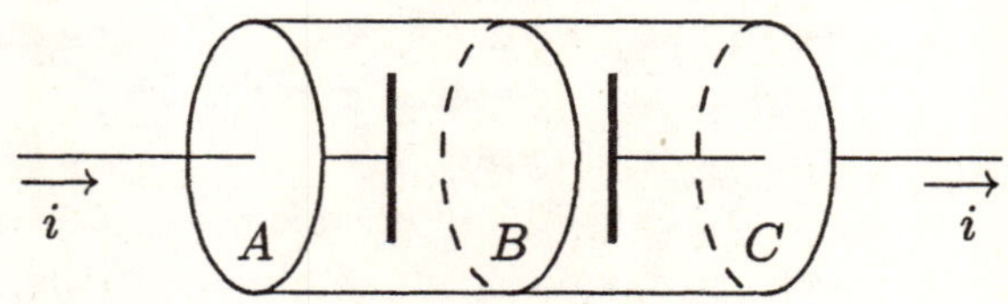

D. Note that the situation discussed and the lack of symmetry in the electromagnetic equations suggests that Ampere's law as used so far must be changed. Experiment confirms this conjecture.

E. Explain that if the electric flux through an open surface changes with time, then there is a magnetic field and the magnetic field has a tangential component at points on the boundary. Write down the Maxwell law of induction: $\oint 8\vec{B}\cdot d\vec{s} = \mu_0\epsilon_0\, d\Phi_E/dt$, where Φ_E is the electric flux through the surface. Compare to Faraday's law and point out the interchange of $\vec{B}$ and $\vec{E}$, the change in sign, and the appearance of the factor $\mu_0\epsilon_0$. State that this law can be combined with Ampere's law and write the complete Ampere-Maxwell law: $\oint \vec{B}\cdot d\vec{s} = \mu_0 i + \mu_0\epsilon_0\, d\Phi_E/dt$.

F. Give the right-hand rule that relates the normal to the surface used to calculate Φ_E and the direction of integration around its boundary. State that the surface may be a purely mathematical construction and that the law holds for any surface.

G. Consider a charging parallel-plate capacitor with circular plates and derive expressions in terms of dE/dt for the magnetic field on a plane between the plates, at points both inside and outside the capacitor. See Sample Problem 32–1.

III. Displacement current.

A. Define the displacement current: $i_d = \epsilon_0\, d\Phi_E/dt$. Explain that it does not represent the flow of charge and is not a true current, but that it enters the Ampere-Maxwell law in the same way as a true current. Discuss the direction of i_d. Consider a region in which the electric field is uniform and is changing. Find the direction for both an increasing and a decreasing field.

B. Refer to the Ampere-Maxwell law. Explain that there are no changing electric fields in the examples of previous chapters so only true currents were considered. Explain that there is no true current in the region between the plates of a charging capacitor, but there is a displacement current.

C. Consider a parallel-plate capacitor with circular plates, for which $\mathrm{d}E/\mathrm{d}t$ is given. Show that the total displacement current in the interior of the capacitor equals the true current into the capacitor. Explain that the sum of the true and displacement currents is continuous. Optional: discuss a leaky capacitor.

D. Derive expressions in terms of i_d for $\vec{B}$ at various points along the perpendicular bisector of the line joining the plate centers. Consider points between the plates and outside them.

E. Show that the total displacement current between the plates of a capacitor is the same as the true current into or out of the plates.

IV. Maxwell's equations.

A. Write down the four equations in integral form and review the physical processes that each describes. See Table 32–1.

B. Carefully distinguish between the line and surface integrals that appear in the equations and give the right-hand rules that relate the direction of integration for the contour integrals and the normal to the surface for the surface integrals.

C. Review typical problems: the electric field of a point charge, the magnetic field of a uniform current in a long straight wire, the magnetic field at points between the plates of a capacitor with circular plates, the electric field accompanying a changing uniform magnetic field with cylindrical symmetry.

D. State that in the absence of dielectric and magnetic materials these equations describe all electromagnetic phenomena to the atomic level and the natural generalizations of them provide valid descriptions of electromagnetic phenomena at the quantum level. They are consistent with modern relativity theory. Optional: for completeness you may want to rewrite the equations and include magnetization and electric polarization terms.

V. Magnetic dipoles in matter.

A. Explain that current loops and bar magnets produce magnetic fields which, for points far away, are dipole fields. Review the expressions for the magnetic field of a dipole and for the dipole moment of a loop in terms of the current and area. Place a bar magnet under a piece of plastic sheet on an overhead projector. Sprinkle iron filings on the sheet and show the field pattern of the magnet. Remind students that field lines emerge from the north pole and enter at the south pole.

B. Explain that the electron and many other fundamental particles have intrinsic dipole moments, which are related to their intrinsic spin angular momenta. Say that only one component, usually taken to be the z component, can be measured at a time. The z component of the spin angular momentum is $S_z = \pm h/4\pi = \pm 5.2729 \times 10^{-35}\,\mathrm{J \cdot s}$, where h is the Planck constant. The z component of the associated magnetic dipole moment is $\mu_{S,z} = -(e/m)S_z = \mp(eh/4\pi m) = \mp 9.27 \times 10^{-24}\,\mathrm{J/T}$, where m is the electron mass. Since an electron is negatively charged, the dipole moment and spin angular momentum are in opposite directions. Mention that particle and atomic magnetic moments are often measured in units of the **Bohr magneton** μ_B: $\mu_B = eh/4\pi m$. This is the magnitude of the electron spin dipole moment.

C. Explain that electrons in atoms create magnetic fields by virtue of their orbital motions. Derive Eq. 32–28, which gives the relationship between orbital angular momentum and dipole moment for a negative particle, such as an electron. Say that quantum mechanically the z component of the orbital angular momentum is given by $L_{\mathrm{orb},z} = m_\ell h/2\pi$, where $m_\ell = \pm 1, \pm 2, \ldots, \pm(\text{limit})$ and "limit" is the largest magnitude of m_ℓ. The z component of the dipole moment is $\mu_{\mathrm{orb},z} = -m_\ell(eh/4\pi) = -m_\ell \mu_B$.

D. Say that if an electron is placed in an external magnetic field, in the z direction, its magnetic energy is given by $U = -\mu_z B_{\text{ext}}$. Draw an energy level diagram to show the splitting of the levels in a magnetic field.

E. Remark that it is chiefly the orbital and spin dipole moments of electrons that are responsible for the magnetic properties of materials. Explain how to calculate the dipole moment of an atom: $\vec{\mu} = (-e/2m)\vec{L} + (-e/m)\vec{S}$, where $\vec{L}$ is the total orbital angular momentum and $\vec{S}$ is the total spin angular momentum of the electrons of the atom.

F. Explain that protons and neutrons also have intrinsic dipole moments, but that these are much smaller than the dipole moment of an electron because the masses are so much larger. Remark that nuclear magnetism has found medical applications.

VI. Magnetization.

A. Define magnetization as the dipole moment per unit volume. Although only uniformly magnetized objects are considered in the text, you may wish to state the definition as the limiting value as the volume shrinks to zero.

B. State that a magnetized object produces a magnetic field both in its exterior and interior and write $\vec{B} = \vec{B}_0 + \vec{B}_M$ for the total field. Here $\vec{B}_0$ is the applied field and $\vec{B}_M$ is the field due to dipoles in the material. Remark that for some materials $\vec{B}_M$ is in the same direction as $\vec{B}_0$, while for others it is in the opposite direction.

VII. Diamagnetism and paramagnetism.

A. Give a qualitative discussion of diamagnetism. Explain that an external field changes the electron orbits so there is a net dipole moment and that the induced moment is directed opposite to the field. This tends to make the total field weaker than the external field alone. Bismuth is an example of a diamagnetic substance.

B. Give a qualitative discussion of paramagnetism. Explain that paramagnetic substances are composed of atoms with net dipole moments and, in the absence of an external field, the moments have random orientations, so that they produce no net magnetic field. An external field tends to align the moments and the material produces its own field. Since the moments, on average, are aligned with the external field, the total field is stronger than the external field alone. Alignment is opposed by thermal agitation and both the net magnetic moment and magnetic field decrease as the temperature increases.

C. Remind students that the potential energy of a dipole $\vec{\mu}$ in a magnetic field $\vec{B}$ is given by $U = -\vec{\mu} \cdot \vec{B}$ and show that the energy required to turn a dipole end for end, starting with it aligned with the field, is $2\mu B$. Calculate U for $\mu = \mu_B$ and $B = 1\,\text{T}$. Calculate the mean translational kinetic energy for an ideal gas at room temperature ($\frac{3}{2}kT$) and remark that there is sufficient energy for collisions to reorient the dipoles. Calculate the temperature for which $2\mu B = \frac{3}{2}kT$.

D. Give the Curie law for small applied fields. Explain that for small applied fields M is proportional to B and inversely proportional to T. Draw the full graph of magnetization as a function of the applied field and point out the linear region. Describe saturation and explain that there is an upper limit to the magnetization. Point out this region on the graph. The limit occurs when all atomic dipoles are aligned. Use a teslameter or flip coil to measure the magnetic field just outside the end of a large, high-current coil. Put a large quantity of manganese in the coil and again measure the field.

E. Explain that diamagnetic effects are present in all materials but are overshadowed by paramagnetic or ferromagnetic effects if the atoms have dipole moments.

VIII. Ferromagnetism.

A. Explain that, for iron and other ferromagnetic substances (such as Co, Ni, Gd, and Dy), the atomic dipoles are aligned by an internal mechanism (exchange coupling) so the substance can produce a magnetic field spontaneously, in the absence of an external field. At temperatures above its Curie temperature, a ferromagnetic substance becomes paramagnetic. Gadolinium is ferromagnetic with a Curie temperature of about 20° C. Put a sample in a beaker of cold water ($T < 20°$ C) and use a weak magnet to pick it up from the bottom of the beaker but not out of the water. Add warm water to the beaker and the sample will drop from the magnet.

B. Describe ferromagnetic domains and explain that the dipoles are aligned within any domain but are oriented differently in neighboring domains. The magnetic fields produced by the various domains cancel for an unmagnetized sample. When the sample is placed in a magnetic field, domains with dipoles aligned with the field grow in size while others shrink. The dipoles in a domain may also be reoriented somewhat as a unit.

C. Define hysteresis (see Fig. 32–19) and explain that the growth and shrinkage of domains are not reversible processes. Domain size is dependent not only on the external field but also on the magnetic history of the sample. When the external field is turned off, the material remains magnetized. Draw a hysteresis curve and point out the approach to saturation and the residual field. Explain that to demagnetize an ferromagnet an external field must be applied in the direction opposite to the magnetization.

D. Explain the difference between soft and hard iron in terms of hysteresis. Use a large, high-current coil to magnetize a piece of hard iron and show that it remains magnetized when the current is turned off. Also magnetize a piece of soft iron and show it is magnetized only as long as the current remains on. When the current is turned off, very little permanent magnetization remains. Soft iron is used for transformer coils.

SUPPLEMENTARY TOPIC

Earth's magnetic field. Section 32–6 describes the magnetic field of Earth. The shape, cause, and some of the ramifications of Earth's field are important topics and should be covered if you have the time. If not, you might intersperse some of the information in your other lectures. Explain that the field can be approximated by a magnetic dipole field. Draw a sphere, label the north and south geographic poles, draw a dipole moment vector at the center (pointing roughly from north to south, about 10° away from the axis of rotation), and draw some magnetic field lines. Remark that the north pole of the dipole is near the south geographic pole. Define declination and inclination.

SUGGESTIONS

1. Assignments
   a. To test for understanding of Gauss' law for magnetism, assign problem 2 or 3.
   b. Ask students to think about a permanent bar magnet that pierces the surface of a sphere and explain why the net magnetic flux through the surface is zero. Also ask them about the electric flux as a single charge as it crosses the surface and the magnetic flux of a single magnetic monopole as it crosses the surface.
   c. To test for understanding of the direction of the magnetic field induced by a changing electric field, assign questions 1 and 3.
   d. Question 4 helps students think carefully about displacement current. Also assign problems 15, 16, and 19.
   e. Questions 5 and 6 deal with the energy of a magnetic dipole in an external magnetic field. They also deal with the intrinsic dipole moments of electrons. Assign problem 29 in connection with the dipole moments of electrons.
   f. Questions 7, 8, and 11 deal with diamagnetism and question 9 deals with paramagnetism. Magnetization in a paramagnetic substance is covered in problems 38 and 39.

g. The Curie temperature of a ferromagnet is covered in problem 42 and the magnetization of a ferromagnet is covered in problem 44. Use problem 43 to show that magnetic interactions are not responsible for ferromagnetism.

2. Demonstrations
   a. Field of a magnet: Freier and Anderson Er4.
   b. Gauss' law: Freier and Anderson Er12.
   c. Paramagnetism: Freier and Anderson Es3, 4.
   d. Ferromagnetism: Freier and Anderson Es1, 2, 6 — 10.
   e. Levitation: Freier and Anderson Er10, 11.

3. Books and Monographs

   Magnetic Monopoles; edited by Alfred S. Goldhaber and W. Peter Trower; available from the AAPT, One Physics Ellipse, College Park MD 20740–3845. Reprint collection, with a resource letter.

4. Audio/Visual
   a. *Ferromagnetic Domain Wall Motion*; *Paramagnetism of Liquid Oxygen*; from the AAPT Miller collection of single-concept films; DVD; available from Ztek Co. (PO Box 11768, Lexington, KY 40577–1768, www.ztek.com) and from the AAPT (see above for address).
   b. *Electromagnetism*; VHS video tape; Films for the Humanities and Sciences (PO Box 2053, Princeton, NJ 08543–2053; www.films.com). Contains sections on Earth's magnetic field, origins of magnetism, ferromagnetic domains, and electromagnetic induction.

5. Laboratory

   Meiners Experiment 11–6: *Magnetization and Hysteresis.* Faraday's law is used to measure the magnetic field inside an iron toroid for various applied fields. A plot of the field as a function of the applied field shows hysteresis. A method for obtaining the hysteresis curve as an oscilloscope trace is also given.

Chapter 33 ELECTROMAGNETIC WAVES

BASIC TOPICS

I. Qualitative features of electromagnetic waves.
   A. Explain that an electromagnetic wave is composed of electric and magnetic fields. The disturbance, analogous to the string shape that moves on a taut string, is made up of the fields themselves, moving through space or a material medium. Also explain that electromagnetic waves carry energy and momentum.
   B. State that the wave speed in a vacuum is given by $c = 1/\sqrt{\mu_0\epsilon_0}$ and is about 3.00×10^8 m/s. The existence of waves and this expression for the wave speed in vacuum are predicted by Maxwell's equations. Since the values of c and μ_0 are fixed, this fixes ϵ_0.
   C. Show the electromagnetic spectrum (Fig. 33–1 of the text) and point out the visible, ultraviolet, infrared, x-ray, microwave, and radio regions. Remark that all the waves are fundamentally the same, differing only in wavelength and frequency. Remind students that ϵ_0 and μ_0 enter electrostatics and magnetostatics, respectively, and were first encountered in situations that had nothing to do with wave propagation.
   D. Restate that the visible spectrum extends from just over 400 nm to just under 700 nm. Remark that while color is largely subjective, violet is at the short wavelength end while red is at the high wavelength end. Use a prism to display the spectrum. Show Fig. 33–2 of the text and remark that human eyes are most sensitive in the green–yellow portion of the spectrum and that sensitivity falls off rather rapidly on either side.

E. State that an accelerating charge creates electromagnetic radiation. Show a diagram of an oscillating electric dipole antenna and its fields (see Fig. 33–3. Point out that $\vec{E}$ and $\vec{B}$ are perpendicular to each other and to the direction of propagation and that they oscillate in phase with each other at any point. Explain the term polarization.

II. Traveling sinusoidal waves.

A. Take $E(x,t) = E_m \sin(kx - \omega t)$, along the y axis, and $B(x,t) = B_m \sin(kx - \omega t)$, along the z axis. Remark that both fields travel in the positive x direction and that they are in phase. Remind students that the minus sign in the argument becomes a plus sign for a wave traveling in the negative x direction.

B. Consider a rectangular area in the xy plane, with infinitesimal width dx and length h (along y). Evaluate $\oint \vec{E} \cdot d\vec{s}$ and Φ_B, then show that Faraday's law yields $\partial E/\partial x = -\partial B/\partial t$. Substitute the expressions for E and B to show that $E = cB$, where $c = \omega/k$. Stress that the magnitudes of $\vec{E}$ and $\vec{B}$ are related. Remark that $\vec{E}$ is different at different points because $\vec{B}$ changes with time.

C. Consider a rectangular area in the xz plane, with infinitesimal width dx and length h (along z). Evaluate $\oint \vec{B} \cdot d\vec{s}$ and Φ_E, then show that the Ampere-Maxwell law yields $-\partial B/\partial x = \mu_0\epsilon_0\, \partial E/\partial t$. Combine this with the result of part B to show that $c = 1/\sqrt{\mu_0\epsilon_0}$. Remark that $\vec{B}$ is different at different points because $\vec{E}$ changes with time. Emphasize the role played by the displacement current.

III. Energy and momentum transport.

A. Define the Poynting vector $\vec{S} = (1/\mu_0)\vec{E} \times \vec{B}$ and explain that it is in the direction of propagation and that its magnitude gives the electromagnetic energy per unit area that crosses an area perpendicular to the direction of propagation per unit time. Remark that for a plane wave, $S = EB/\mu_0 = E^2/\mu_0 c = cB^2/\mu_0$.

B. Consider the plane wave of Section II, propagating in the positive x direction. Consider a volume of width Δx and cross section A (in the yz plane) and show that the electric and magnetic energies in it are equal and that the total energy is $\Delta U = (EBA/\mu_0 c)\,\Delta x$, for small Δx. This energy passes through the area A in time $\Delta t = \Delta x/c$ so the rate of energy flow per unit area is EB/μ_0, as previously postulated.

C. Explain that most electromagnetic waves of interest oscillate rapidly and we are not normally interested in the instantaneous values of the energy or energy density. Explain how to find the average over a period of the square of a sinusoidal function. Define the intensity as the time average of the magnitude of the Poynting vector and write expressions for it in terms of the average energy density and in terms of the field amplitudes.

D. Explain that electromagnetic waves transport momentum and that S/c gives the momentum that crosses a unit area per unit time. The momentum is in the direction of $\vec{S}$. Also explain that if an object absorbs energy U, then it receives momentum U/c. If the object reflects energy U, then it receives momentum $2U/c$.

E. Show that if a wave, incident normal to a surface of area A, is completely absorbed, then the force on the surface is IA/c and the radiation pressure is I/c, where I is the intensity. Show that the force and radiation pressure have twice these values if the wave is completely reflected.

F. As an example of radiation pressure, you may wish to consider solar pressure. S can be determined from the solar constant $1.38\,\mathrm{kW/m^2}$ (valid just above Earth's atmosphere).

IV. Polarization.

A. Remind students that a linearly polarized electromagnetic wave is one for which the electric field is everywhere parallel to the same line. As the wave passes by any point, the field oscillates along the line of polarization.

B. Explain that a linearly polarized wave can be resolved into two other linearly polarized waves with mutually orthogonal polarization directions. Take the original polarization direction to be at the angle θ to one of the new directions and show that the amplitudes are given by $E_1 = E_m \cos\theta$ and $E_2 = E_m \sin\theta$, where E_m is the original amplitude.

C. Explain that the electric field associated with unpolarized light does not remain in the same direction for more than about 10^{-8} s and the new direction is unrelated to the old.

D. Shine unpolarized light through crossed Polaroid sheets and note the change in intensity as the second sheet is rotated. Show that the intensity does not change if the first sheet is rotated. Remark that for an ideal polarizing sheet the transmitted intensity is half the incident intensity.

E. Derive the law of Malus. Explain that the light emerging from the first Polaroid sheet is linearly polarized in a direction determined by the orientation of the sheet. Remark that this direction is called the polarizing direction of the sheet. Draw a diagram of the electric field amplitude as the light enters the second sheet, at an angle θ to the polarizing direction of the second sheet. Resolve the amplitude into components along the polarizing direction and perpendicular to it. Explain that the first component is transmitted through the sheet while the second is absorbed. The amplitude of the transmitted wave is proportional to $\cos\theta$ and the intensity is proportional to $\cos^2\theta$.

F. Shine unpolarized light onto two crossed Polaroid sheets and remark that no light is transmitted. Then, slide another sheet between the two and point out the change in transmitted intensity as you rotate the sheet in the middle. The sheets can be taped to ringstands to hold them. Explain the phenomenon by examining the polarization at each stage of the transmission.

V. Wave and geometrical optics.

A. Explain that optical phenomena outside the quantum realm can be understood in terms of Maxwell's equations and that the wave nature of electromagnetic radiation must be taken into account to explain many important phenomena. State that some of these will be discussed later.

B. Explain that if the wavelength of the light is much smaller than any obstacles it meets or any slits through which it passes, then the important property is the direction of motion, not details of the wave nature. This is the realm of geometrical optics.

C. Define a ray as a line that gives the direction of travel of a wave. It is perpendicular to the wave fronts (surfaces of constant phase). Explain that geometrical optics deals largely with tracing rays as light is reflected from surfaces or passes through materials.

VI. Reflection and refraction.

A. Explain that when light traveling in one medium strikes a boundary with another medium, some is reflected and some is transmitted into the second medium. Draw a plane boundary between two media and show an incident, a reflected, and a refracted ray. Label the angles these rays make with the normal to the surface. Use θ_1 to label the angle of incidence, θ_1' to label the angle of reflection, and θ_2 to label the angle of refraction. Emphasize that these angles are measured relative to the normal to the surface.

B. Tell students that the speed of light may be different in different materials and state that the speeds of light in the two media are crucial for determination of the amplitudes of the reflected and refracted light and for determination of the angle of refraction. Define the index of refraction of a medium as the ratio of the wave speed in vacuum to the wave speed in the medium and write $v = c/n$. Remark that the index of refraction is a property of the medium and depends on the wavelength. Point out Table 33–1, which gives the indices of refraction of various materials. Note that the index of refraction for a vacuum is 1 and is nearly 1 for air. State it is wavelength dependent and point out Fig. 33–19.

C. Consider a plane wave incident on a plane surface. Write down the law of reflection: $\theta_1 = \theta_1'$.
D. Write down the law of the law of refraction: $n_1 \sin\theta_1 = n_2 \sin\theta_2$.
E. Explain that light rays are bent toward the normal when light enters a more optically dense medium (higher index of refraction) and are bent away from the normal when it enters a less optically dense medium.
F. Consider light striking a water surface from air and trace a few rays. Consider light from an underwater source and trace a few rays as they enter the air. Consider a slab of glass with parallel sides and show that the emerging ray has the same direction as the entering ray but is displaced along the slab. Optional: derive the expression for the displacement.
G. Trace a ray through a prism and derive the expression for the angle of deviation: $\psi = \theta_1 + \theta_2 + \phi$, where θ_1 is the angle of incidence, θ_2 is the angle of emergence, and ϕ is the prism angle. Explain that ψ is different for different colors because n depends on wavelength.
H. Shine an intense, monochromatic, well-collimated beam on a prism and point out the reflected and refracted beams. A laser works reasonably well but it is difficult for the class to see the beam. Use smoke or chalk dust to make it visible. To avoid the mess, use an arc beam or the beam from a 35 mm projector, filtered by red glass. Make a $\frac{1}{2}$ in. hole in a 2 in. by 2 in. piece of aluminum and insert it in the film gate. Use white light from the projector and the prism to show that different wavelengths are refracted through different angles.
I. Explain total internal reflection. Show that no wave is transmitted when the angle of incidence is greater than the critical angle and derive the expression for the critical angle in terms of the indices of refraction. Stress that the index for the medium of incidence must be greater than the index for the medium of the refracted light. Total internal reflection can be demonstrated with some pieces of solid plastic tubing having a diameter larger than that used for fiber optics. The beam inside is quite visible. If time permits, discuss fiber optics and some of its applications.

VII. Polarization by reflection.
A. Reflect a well collimated beam of unpolarized light from a plane glass surface. A slide projector beam does nicely. Darken the room and obtain a reflection spot on the ceiling. Place a Polaroid sheet in the reflected beam and note the change in intensity of the spot as you rotate it. Remark that the reflected light is partially polarized.
B. Orient the incident beam so the angle of incidence is Brewster's angle and use the Polaroid sheet to show the reflected light is now entirely polarized.
C. Discuss Brewster's law. Explain that unpolarized light incident on a boundary is partially or completely polarized on reflection. When the angle of incidence and the angle of refraction sum to 90°, the reflected light is completely polarized, with $\vec{E}$ perpendicular to the plane of the incident and reflected rays. Show that the angle of incidence θ_B for completely polarized reflected light is given by $\tan\theta_B = n_2/n_1$, where medium 1 is the medium of the reflected ray.

SUGGESTIONS

1. Assignments
   a. Relationships among frequency, wavelength, and speed are explored in problems 2, 3, and 4. These also give some examples of high and low frequency electromagnetic radiation and ask students to interpret Fig. 33–2, which graphs the sensitivity of the human eye as a function of wavelength.

b. To stress the relationship between $\vec{E}$ and $\vec{B}$ in an electromagnetic wave, assign problem 8. Use questions 1 and 2 in the discussion of the relationships among the directions of the electric field, the magnetic field, and the direction of propagation.
c. To emphasize the magnitude of the energy and momentum carried by an electromagnetic wave, assign problems 13 and 23. Also consider some problems that deal with point sources: 15 and 17, for example.
d. Use questions 3, 4, and 6 to test for understanding of polarization. The fundamentals of polarizing sheets are covered in problems 35 and 37. In the first, the incident light is unpolarized while in the second, it is polarized. Also consider problem 43.
e. Problem 45 covers the law of refraction. Also consider questions 7, 9, and 10.
f. Assign problems 54 and 57 in connection with total internal reflection. Assign either problem 65 or 87 in connection with polarization by reflection.

2. Demonstrations
   Radiation: Freier and Anderson Ep4, 5.

3. Books and Monographs
   a. *Resource Letters, Book Four* and *Resource Letters, Book Five*; American Association of Physics Teachers (AAPT, One Physics Ellipse, College Park MD 20740–3845; www.aapt .org). Contains resource letters on light.
   b. *Connecting Time and Space.* See Chapter 2 SUGGESTIONS.

4. Audio/Visual
   a. *Color, Scattering, Polarization*; from Cinema Classics DVD 4: Waves (II) & Electricity and Magnetism; available from Ztek Co. (PO Box 11768, Lexington, KY 40577–1768, www.ztek.com) and from the AAPT (see above for address).
   b. *Light*; VHS video tape, DVD; Films for the Humanities and Sciences (PO Box 2053, Princeton, NJ 08543–2053; www.films.com).
   c. *The Determination of the Velocity of Light*; VHS video tape, DVD; Films for the Humanities and Sciences (see above for address).
   d. *Reflection*, *Refraction*, and *Dispersion*; from Cinema Classics DVD 3: Waves (I); available from Ztek Co. (see above for address) and from the AAPT (see above for address).
   e. *Propagation of Light*; *Visible and Infrared Spectrum*; from Physics Demonstrations in Light, Part I; VHS video tape, DVD; ≈3 min each; Physics Curriculum & Instruction (22585 Woodhill Drive, Lakeville, MN 55044; www.PhysicsCurriculum.com).

5. Computer Software
   a. *Optics* from Exploration of Physics Volume I; Windows and Macintosh; Physics Curriculum & Instruction (Physics Curriculum & Instruction (see above for address). Simulated experiments with analysis. Includes a section on Snell's law.
   b. *Light and Optics* from Exploration of Physics Volume II; Windows and Macintosh; Physics Curriculum & Instruction (see above for address). Includes sections on measurements of the speed of light, polarization, and refraction.
   c. *Physics Simulation Programs*. See Chapter 16 SUGGESTIONS.

6. Laboratory
   a. Probeware Activity 27: *Polarization.* A light intensity probe, connected to a computer, is used to investigate the intensity of light transmitted through two or more polarizing sheets. A rotary motion sensor automates the procedure so a plot of the intensity as a function of the relative orientation of the sheets is displayed on the monitor. Students verify the cosine squared law.
   b. Meiners Experiment 13–7; *Polarization of Light.* Polaroid sheets are first investigated and the law of Malus is verified. Then, a Polaroid sheet is used to investigate polarization

by reflection, by refraction, and by scattering. Brewster's angle is found. Rotation of the direction of polarization by a sugar solution is also studied and crossed polarizer are used to check various objects for stresses. This is essentially a series of demonstrations performed by students.

c. Bernard Experiment 46; *Polarized Light.* Similar to Meiners Experiment 13–7 except light transmitted by a calcite crystal is also investigated. A photodetector is used to obtain quantitative data.

d. Meiners Experiment 13-3: *Prism Spectrometer.* Helium lines are used to determine the index of refraction as a function of wavelength for a glass prism. A good example of dispersion and excellent practice in carrying out a rather complicated derivation involving Snell's law. Also see Bernard Experiment 43: *Index of Refraction with the Prism Spectrometer* and Bernard Experiment 44: *The Wavelength of Light.* In the second of these experiments, students use a prism spectrometer to determine the wavelength of lines from a sodium source.

Chapter 34 IMAGES

BASIC TOPICS

I. Plane mirrors.

A. Consider a plane wave incident on a plane mirror. Remind students of the law of reflection: $\theta_1 = \theta_1'$.

B. Consider a point source in front of a plane mirror. Draw both incident and reflected rays and show that the reflected rays appear to come from a point behind the mirror. Show that the object and image lie on the same normal to the mirror and that they are the same distance from the mirror. Remark that no light comes from the image and that the image is said to be virtual.

C. Define the object distance p and image distance i and explain that the latter is taken to be negative for virtual images. The law of equal distance is written $p = -i$.

D. Give the condition for being able to see an image. Draw a mirror, an eye, a source, and its image. Draw the line from the image to the eye and state that the image can be seen if this line intersects the mirror. Show that length of a wall mirror with its top edge at eye level need reach only halfway to the floor for a person to see his feet. Demonstrate with a mirror resting on the floor and half-covered with a cloth. Have a student stand in front of the mirror. Start with the cloth about shoulder height and lower it until the student can see his feet.

II. Spherical mirrors.

A. Consider a point source in front of a concave spherical mirror. Draw a diagram that shows the central axis, the center of curvature, and the source on the axis, outside the focal point. Show that small-angle rays form an image and that object and image distances are related by $1/p + 1/i = 2/r$. To emphasize the small-angle approximation, consider the case $p = 2r$ and use a full hemispherical concave surface. The small-angle formula predicts all rays cross the axis at $i = (2/3)r$, but the ray that strikes the edge of the mirror crosses at the vertex.

B. Explain that the mirror equation is also valid for convex mirrors and for any position of the object, even virtual objects for which incoming rays converge toward a point behind the mirror. Give the sign convention: p and i are positive for real objects and images (in front of the mirror) and are negative for virtual objects and images (behind the mirror); r is positive for concave mirrors (center of curvature in front of the mirror) and negative for convex mirrors (center of curvature behind the mirror). Remark that a surface is concave or convex according to its shape as seen from a point on the incident ray.

C. Define the focal point as the image point when the incident light is parallel to the axis. By considering a source far away, show that $f = r/2$. Consider a concave mirror and show that for $p > f$, the image is real; for $p < f$, the image is virtual. Also show that for $p = f$, parallel rays emerge after reflection.

D. Describe a geometric construction for finding the image of an extended source. Trace rays from an off-axis point: one through the center of curvature, one through the focal point, and one parallel to the axis. Use both concave and convex mirrors as examples. Explain that the geometric construction gives the same result as the small-angle approximation if reflection is assumed to take place at a plane through the mirror vertex and perpendicular to the optic axis. The law of reflection cannot be applied at this plane, of course.

E. Define lateral magnification and show that $m = -i/p$. Explain the sign: m is positive for erect images and negative for inverted images. Virtual images of real objects are erect and real images of real objects are inverted.

F. Take the limit $r \to \infty$ and show that the mirror equation makes sense for a plane mirror.

III. Spherical refracting surfaces.

A. Draw a convex spherical boundary between two media, use the law of refraction to trace a small-angle ray from a source on the central axis, and show that $n_1/p + n_2/i = (n_2 - n_1)/r$, where n_1 is the index of refraction for the region of incident light and n_2 is the index of refraction for the region of refracted light. You can demonstrate the bending of the light using a laser and a round-bottom flask. Use a little smoke or chalk dust to make the beam visible in air and a pinch of powdered milk in the water to make it visible inside the flask.

B. Explain the sign convention. Point out that real images are on the opposite side of the boundary from the incident light and virtual images are on the same side. Explain that p and i are positive for real objects and images, negative for virtual objects and images. r is positive for convex surfaces, negative for concave. With this sign convention, the equation holds for concave or convex surfaces and for $n_2 > n_1$ or $n_1 > n_2$.

C. Consider the limit $r \to \infty$, which yields $i = -pn_2/n_1$. This is the solution to the apparent depth problem. For water four inches deep, a ball on the bottom appears to be at a depth of about three inches. Use an aquarium filled with water and a golf ball to make a hallway display.

IV. Thin lenses.

A. Explain that a lens consists of two refracting surfaces close together in vacuum. State or derive the thin lens equation: $1/p + 1/i = (n - 1)(1/r_1 - 1/r_2)$, where n is the index of refraction for the lens material. Stress that the equation holds for small-angle rays. State that it also holds to a good approximation for a lens in air. r_1 is the radius of the first surface struck by the light and r_2 is the radius of the second. They are positive or negative according to whether the surfaces are convex or concave when viewed from a point on the incident ray. You may wish to generalize the equation by retaining the indices of refraction. The result is $1/p + 1/i = (n_2/n_1 - 1)(1/r_1 - 1/r_2)$. This allows you to consider a thin glass or air lens in water.

B. By considering $p \to \infty$, show that the focal length is given by $1/f = (n - 1)(1/r_1 - 1/r_2)$ or more generally by $1/f = (n_2/n_1 - 1)(1/r_1 - 1/r_2)$. Show that the same value, including sign, is obtained no matter which surface is struck first by light. Then show that $1/p + 1/i = 1/f$. Point out that there are two focal points, the same distance from the lens but on opposite sides. For a converging lens, rays from a point source at f on one side are parallel on the other side; incident parallel rays converge to f on the other side. For a diverging lens, rays that converge toward f on the other side emerge parallel; rays that are parallel emerge as diverging from f on the incident side.

C. Show how to locate the image of an extended object by tracing a ray parallel to the axis, a ray through the lens center, and a ray along a line through the first focal point (on the incident side for a converging lens and on the other side for a diverging lens).

D. Define lateral magnification and show that $m = -i/p$. Explain that the sign tells whether the image is erect or inverted.

E. Consider all possible situations: converging lens with $p > f$, $p < f$, and $p = f$; diverging lens with $p > f$, $p < f$, and $p = f$. In each case, show whether the image is real or virtual, erect or inverted, and find its position relative to the focal point.

F. Note that most optical instruments are constructed from a combination of two or more lenses. Point out that to analyze them, one considers one lens at a time, with the image of the previous lens as the object of the lens being considered. This sometimes leads to virtual objects. Note that the overall magnification is given by $m = m_1 m_2 m_3 \ldots$ and that the sign of m tells whether the image is erect or inverted. If the image lies on the opposite side of the system from the object and is outside the system, then it is real; otherwise it is virtual.

SUPPLEMENTARY TOPIC

Optical instruments. This section may be studied in the laboratory. Ask students to experiment with the image forming properties of positive and negative lenses, then construct one or more optical instruments. Display several instruments in the lab.

SUGGESTIONS

1. Assignments
   a. Interesting applications of plane mirrors are covered in problems 4 (can observer see an image?) and 120 (rotation of mirror). Problem 102 and question 3 deal with images in multiple mirrors.
   b. Use questions 3 and 5 to discuss images in spherical mirrors. Problems 19 through 21 cover nearly all possibilities. Lateral magnification is covered in problem 7.
   c. Assign problem 32 in connection with spherical refracting surfaces. Problems 34 through 40 cover all possibilities.
   d. Use question 6 to discuss images formed by thin lenses. For comprehensive coverage of nearly all relationships, assign problems 66, 67, 68 and 74 through 77. Problems 43 and 45 test understanding of the lensmaker's equation. Also assign problem 125, which deals with a compound system and includes a ray tracing exercise.
   e. Consider expanding the course a little by including problem 91, which deals with the human eye.
2. Demonstrations
   a. Plane mirrors: Freier and Anderson Ob1 - 6, Ob8.
   b. Refraction at a plane surface: Freier and Anderson Od1 - 7.
   c. Prisms: Freier and Anderson Of1 - 4.
   d. Total internal reflection: Freier and Anderson Oe1 - 7.
3. Books and Monographs
   a. *Resource Letters, Book Four* and *Resource Letters, Book Five*; American Association of Physics Teachers (AAPT, One Physics Ellipse, College Park MD 20740–3845; www.aapt .org). Contains a list of journal articles on optics.
   b. *The Camera*; by Bill G. Aldridge, Gary S. Waldman, and John Yoder III.; available from the AAPT (see above for address). Concepts important for understanding cameras.

4. Audio/Visual

   Optics; VHS video tape, DVD; Films for the Humanities and Sciences (PO Box 2053, Princeton, NJ 08543–2053; www.films.com).

5. Computer Software

   a. *Ray*; Miky Ronen; Macintosh; available from Physics Academic Software (Centennial Campus, 940 Main Campus Drive, Suite 210, Raleigh, NC 27606–5212; www.aip.org/pas). A ray-tracing program. The user can place reflecting surfaces, refracting surfaces, mirrors, lenses, and prisms on the screen and control their orientation. Rays are traced using either the paraxial approximation or the actual path.

   b. *Optics* from Exploration of Physics Volume I; Windows and Macintosh; Physics Curriculum & Instruction (22585 Woodhill Drive, Lakeville, MN 55044; www.PhysicsCurriculum.com). Simulated experiments with analysis. Includes a section on lenses.

   c. *Optics Phenomena*; Helmut F. Mikelskis; Windows; available from Physics Academic Software (see above for address). Interactive modules dealing mostly with geometrical optics.

   d. *Light and Optics* from Exploration of Physics Volume II; Windows and Macintosh; Physics Curriculum & Instruction (see above for address). Includes sections on formation of real images and ray tracing.

6. Laboratory

   a. Meiners Experiment 13–1: *Laser Ray Tracing.* A laser beam is used to investigate the laws of reflection and refraction and to observe total internal reflection and the formation of images by spherical mirrors. Measurements are used to calculate the index of refraction of several materials, including liquids, and the focal length of mirrors. Tracing is done by arranging the apparatus so the laser beam grazes a piece of white paper on the lab table. Much the same set of activities are described in Bernard Experiment 38: *Reflection and Refraction of Light*, but pins are used as objects rather than a laser source and rays are traced by positioning other pins along them. The technique can be used if you do not have sufficient lasers for the class.

   b. Bernard Experiment 39: *The Focal Length of a Concave Mirror.* Several methods are described, including a technique that involves finding the radius of curvature. Others involve finding the image when the object distance is extremely long, when it is somewhat greater than $2f$, and when it is somewhat less than $2f$. Then, the mirror equation is used to solve for f.

   c. Meiners Experiment 13–2: *Lenses.* A light source and screen on an optical bench are used to find the focal lengths and magnifications of both convex and concave lenses. Chromatic and spherical aberrations are also studied. Also see Bernard Experiment 40: *Properties of Converging and Diverging Lenses*, a compendium of techniques for finding focal lengths.

   d. Bernard Experiment 41: *Optical Instruments Employing Two Lenses.* Students construct simple two-lens telescopes and microscopes on optical benches, then investigate their magnifying powers. By trying various lens combinations, they learn the purposes of the objective and eyepiece lenses.

Chapter 35 INTERFERENCE

BASIC TOPICS

I. Huygens' principle.

   A. Shine monochromatic light through a double slit and project the pattern on the wall. Either use a laser or place a single slit between the source and the double slit. Use a diagram to explain the setup. Point out the appearance of light in the geometric shadow and the occurrence of dark and bright bands. You can make acceptable double slits by coating a microscope slide with lamp black or even black paint. Tape a pair of razor blades together and draw them across the slide. By inserting various thicknesses of paper or shim stock between the blades, you can obtain various slit spacings.

B. Explain that Huygens' principle will be used to understand the pattern, then state the principle. Describe plane wave propagation in terms of Huygen wavelets: draw a plane wave front, construct spherical wave fronts of the same radius centered at several points along the plane wave front, then draw the plane tangent to these.

C. Use Huygens' principle to derive the law of refraction. Assume different wave speeds in the two media and show that the wavelengths are different. Consider wavefronts one wavelength apart and show that $\sin\theta_1/\sin\theta_2 = v_1/v_2$. Explain that $n = c/v$ and obtain the law of refraction.

D. Go back to the double-slit pattern and explain that those parts of an incident wave front that are within the slit produce spherical wavelets that travel to the screen while wavelets from other parts are blocked. Some wavelets reach the geometric shadow. The spreading of the pattern beyond the shadow is called diffraction and will be studied in the next chapter. Wavelets from different slits arrive at the same point on the screen and interfere to produce the bands. This phenomena will be studied in this chapter.

II. Two-slit interference patterns.

A. Draw a diagram of a plane wave incident normally on a two-slit system and draw a ray from each slit to a screen far away. Remark that the waves are in phase at the slits but they travel different distances to get to the same point on the screen and may have different phases there. The electric fields sum to the total electric field. At some points, the two fields cancel, at other points they reinforce each other. Remind students that the intensity is proportional to the square of the total field, not to the sum of the squares of the individual fields.

B. Point out that if the screen is far away, the two rays are nearly parallel, then show that the difference in distance traveled is $d\sin\theta$, where d is the slit separation and θ is the angle the rays make with the forward direction. Explain the condition $d\sin\theta = m\lambda$ for a maximum of intensity and the condition $d\sin\theta = (m+\frac{1}{2})\lambda$ for a minimum.

C. Show that a lens can be used to obtain the same pattern, even if the screen is not far away.

III. The intensity.

A. Take the two fields to be $E_1 = E_0\sin(\omega t)$ and $E_2 = E_0\sin(\omega t+\phi)$, where $\phi = (2\pi/\lambda)d\sin\theta$. This is easily shown by remarking that $\phi = k\Delta d$, where $k = 2\pi/\lambda$ and $\Delta d = d\sin\theta$ (derived earlier).

B. Explain how the fields can be represented on a phasor diagram. Explain that a phasor has a length proportional to the amplitude and makes the angle ωt or $\omega t+\phi$ with the horizontal axis. Its projection on the vertical axis is proportional to the field. Sum the phasors to obtain the total field. Show that the amplitude E_θ of the total field is $2E_0\cos(\phi/2)$. Plot the intensity $4E_0^2\cos^2(\phi/2)$ as a function of ϕ. Point out that $\phi = 0$ produces a maximum, that maxima occur at regular intervals, and that the minima are halfway between adjacent maxima.

C. Show that the intensity at a maximum is four times the intensity due to one source alone. Remark that no energy is gained or lost. All energy through the slits arrives at the screen. The presence of the slitted barrier, however, redistributes the energy.

D. Note the half-width of each maximum, at half the peak, is given by $\sin\theta = \lambda/4d$. The smaller λ/d, the sharper the maximum. Near the central maximum, where $\sin\theta \approx \tan\theta \approx \theta$, the linear spread on the screen is $y \approx (\lambda/2d)D$, where D is the distance from the slits to the screen.

E. It is also worth noting that since $\sin\theta = m\lambda/d \leq 1$ for a maximum, the smaller λ/d, the more maxima occur.

F. For completeness, you might mention the amplitude of the wavelets fall off as $1/r$ and are not quite the same at the screen. Show this is a negligible effect for the patterns considered here.

IV. Coherence.

A. Explain that two waves are coherent if their relative phase does not change with time.

B. Explain that the two interfering waves must be coherent to obtain an interference pattern. The phase difference at the observation point must be constant over the observation time. Explain why two incandescent lamps, for example, do not produce a stable interference pattern. The light is from many atoms and the emission time for a single atom is about 10^{-8} s. The phase difference changes in a random way over times that are short compared to the observation time. State that in this case the intensities add.

C. Explain that an extended source can be used to obtain an interference pattern. Light from each atom goes through both slits and forms a pattern, but the patterns of different atoms are displaced from each other, according to the separation of the atoms in the source. No pattern is seen unless the incident light comes only from a small region of the source. If you did not use a laser in the demonstration, explain the role of the single slit in front of the double slit.

D. Explain that a laser produces coherent light even though many atoms are emitting simultaneously. Because emission is stimulated, light from any atom is in phase with light from all other atoms. A laser can be used to form an interference pattern without restricting the incident beam.

V. Thin-film interference.

A. Cut a 1 to 2 mm slit in a 2" square piece of aluminum and insert it in the film gate of a 35 mm projector. Let the beam impinge on a soap bubble to show the effect.

B. Consider normal incidence on a thin film of index n_1 in a medium of index n_2 and suppose the medium behind the film has index n_3. Explain that a wave reflected at the interface with a medium of higher index undergoes a phase change of π. If $n_1 < n_2 < n_3$, waves reflected at both surfaces undergo phase changes of π. Consider all other possibilities and then specialize to a thin film of index n in air. Give the conditions for maxima and minima for both the reflected light and the transmitted light, assuming near normal incidence. Note that the wavelength in the medium must be used to calculate the phase change on traveling through the medium. Define optical path length and point out its importance for thin-film interference.

C. Broaden the discussion qualitatively by including non-normal incidence. Note that for some angles, conditions are right for destructive interference of a particular color while at other angles, conditions are right for constructive interference of the same color. Also note that these angles depend on λ. Hence the soap bubble colors.

D. If time permits, discuss Newton's rings. Use a plano-convex lens and a plane sheet of glass together with a laser. Use a diverging lens to spread the beam.

SUPPLEMENTARY TOPIC

The Michelson interferometer. This is an excellent example of an application of interference effects. Set up a hallway demonstration and give a brief explanation.

SUGGESTIONS

1. Assignments
   a. In the discussion of coherence, give a more detailed explanation of the single slit placed between the source and the double slit.
   b. Questions 1, 2, 3, and 4 and problems 1 through 13 are good tests of the fundamentals. Some of the problems deal with the speed of light in material media while others deal with the calculation of phase differences. Use them in the discussion or ask students to answer a few of them for homework.
   c. Problems 14 through 17 and 19, 20, and 23 deal with the basics of interference. Assign one or two. Use problem 27 or 117 to test for understanding of the derivation of the double-slit equation.
   d. Assign problem 20, 21, or 22 in connection with the double-slit interference pattern. Also consider questions 6, 7, and 8.
   e. Use questions 10, 11, and 12 and a few of problems 41 through 52 to help with the discussion of thin films. Problems 57 through 68 deal with interference on transmission through a thin film. Some problems in these two groups ask for the wavelength and some ask for the film thickness. Assign one of each.
   f. Problems 79, 80, and 81 illustrate some applications of a Michelson interferometer.
2. Demonstrations
   a. Double-slit interference: Freier and Anderson Ol4, 5, 9.
   b. Thin-film interference: Freier and Anderson Ol15 — 18.
   c. Michelson interferometer: Freier and Anderson Ol19.
3. Audio/Visual
   a. *Michelson Interferometer*; from the AAPT Miller collection of single-concept films; DVD; available from Ztek Co. (PO Box 11768, Lexington, KY 40577–1768, www.ztek.com) and from the American Association of Physics Teachers (AAPT, One Physics Ellipse, College Park MD 20740–3845; www.aapt.org).
   b. *Interference*; from Cinema Classics DVD 4: Waves (II) & Electricity and Magnetism; available from Ztek Co. and from the AAPT (see above for addresses).
   c. *Interference*; *Thin Film Interference*; from Physics Demonstrations in Light, Part II; VHS video tape, DVD; $\approx$3 min each; Physics Curriculum & Instruction (22585 Woodhill Drive, Lakeville, MN 55044; www.PhysicsCurriculum.com).
4. Computer Software
   a. *Wave Interference*; Mike Moloney; DOS; available from Physics Academic Software (Centennial Campus, 940 Main Campus Drive, Suite 210, Raleigh, NC 27606–5212; www.aip.org /pas). Uses phasors to obtain intensity patterns.
   b. *Light and Optics* from Exploration of Physics Volume II; Windows and Macintosh; Physics Curriculum & Instruction (see above for address). Includes a section on double-slit interference.
5. Laboratory
   a. Probeware Activity 28: *Diffraction of Light.* A light intensity detector is moved along the interference pattern produced as laser light passes through double slits. The pattern is shown on the computer monitor and is used to compute the wavelength of the light. A single-slit diffraction pattern is also obtained and analyzed.
   b. Meiners Experiment 13–4: *Interference and Diffraction.* Students observe double-slit patterns of water waves in a ripple tank, sound waves, microwaves, and visible light. In each case except water waves, they measure and plot the intensity as a function of angle, then

use the data to calculate the wavelength. A microcomputer can be used to take data and plot the intensity of a visible light pattern.

c. Meiners Experiment 13–6: *The Michelson Interferometer.* An interferometer is used to measure the wavelengths of light from mercury and a laser and to find the index of refraction of a glass pane and air. Good practical applications.

Chapter 36 DIFFRACTION

BASIC TOPICS

I. Qualitative discussion of single-slit diffraction.

A. Shine coherent monochromatic light on a single slit and project the pattern on the wall. Point out the broad central bright region and the narrower, less bright regions on either side, with dark regions between. Also point out that light is diffracted into the geometric shadow.

B. Remark that diffraction can be discussed in terms of Huygens wavelets emanating from points in the slit. Explain that they not only spread into the shadow region but that they arrive at any selected point with a distribution of phases and interfere to produce the pattern. Explain that for quantitative work, this chapter deals with Fraunhofer diffraction, with the screen far from the slit.

C. Draw a single slit with a plane wave incident normal to it. Also draw parallel rays from equally spaced points within the slit, all making the same angle θ with the forward direction. Point out that all wavelets are in phase at the slit. The first minimum can be located by selecting θ so that, at the observation point, the ray from the top of the slit is 180° out of phase with the ray from the middle of the slit. All wavelets then cancel in pairs. Show that this leads to $a \sin\theta = \lambda$, where a is the slit width. Point out that this value of θ determines the width of the central bright region and that this region gets wider as the slit width narrows. Use $\sin\theta \approx \tan\theta \approx \theta$ (in radians) to show that the linear width of the central region on a screen a distance D away is $2D\lambda/a$. Use a variable width slit or a series of slits to demonstrate the effect.

D. By dividing the slit into fourths, eighths, etc. and showing that in each case the wavelets cancel in pairs if θ is properly selected, find the locations of other minima. Show that $a \sin\theta = m\lambda$ for a minimum.

E. Explain that for $a < \lambda$, the central maximum covers the whole forward direction. No point of zero intensity can be observed. Also remark that the intensity becomes more uniform as a decreases from λ. This was the assumption made in the last chapter when the interference of only one wavelet from each slit was considered.

F. Qualitatively discuss the intensity. Draw a phasor diagram showing ten or so phasors representing wavelets from equally spaced points in the slit. Show that each wavelet at the observation point is out of phase with its neighbor by the same amount. First, show the phasors with zero phase difference ($\theta = 0$), then show them for a larger value of θ. Show that they approximate a circle at the first minimum and then, as θ increases, they wrap around to form another maximum, with less intensity than the central maximum. Point out that as θ increases, the pattern has successive maxima and minima and that the maxima become successively less intense.

II. The intensity.

A. Draw a diagram showing ten or so phasors along the arc of a circle and let ϕ be the phase difference between the first and last. See Fig. 36–8. Explain that you will take the limit as the number of wavelets increases without bound and draw the phasor addition diagram as an arc. Use geometry to show that $E_\theta = E_m(\sin\alpha)/\alpha$, where $\alpha = \phi/2$. Point out that the intensity can be written $I_\theta = I_m(\sin^2\alpha)/\alpha^2$, where I_m is the intensity for $\theta = 0$. By examining the path difference for the rays from the top and bottom of the slit, show that $\alpha = (\pi a/\lambda)\sin\theta$. Explain that these expressions give the intensity as a function of the angle θ.

B. Sketch the intensity as a function of θ (see Fig. 36–7) and show mathematically that the expression just derived predicts the positions of the minima as found earlier.

C. (Optional) Set the derivative of $(\sin\alpha)/\alpha$ equal to 0 and show that $\tan\alpha = \alpha$ at an intensity maximum. State that the first two solutions are $\alpha = 4.493$ rad and 7.725 rad. Use these results to show that the intensity at the first two secondary maxima are 4.72×10^{-2} and 1.65×10^{-2}, relative to the intensity for $\theta = 0$. You might also want to pick a wavelength and slit width, then find the angular positions of the first two secondary maxima. Remark that they are close to but not precisely at midpoints between zeros of intensity.

III. Double-slit diffraction.

A. Consider the double-slit arrangement discussed in the previous chapter. Point out that the electric field for the light from each of the slits obeys the equation developed for single-slit diffraction and these two fields are superposed. They have the same amplitude, $E_m(\sin\alpha)/\alpha$, and differ in phase by $(2\pi d/\lambda)\sin\theta$, where d is the center-to-center slit separation. The result for the intensity is $I_\theta = I_m(\cos^2\beta)(\sin^2\alpha)/\alpha^2$, the product of the single-slit diffraction equation and the double-slit interference equation. Here $\beta = (\pi d/\lambda)\sin\theta$.

B. Sketch I_θ versus θ for a double slit and point out that the single-slit pattern forms an envelope for the double-slit interference pattern. Remark that this is so because d must be greater than a. See Fig. 36–14.

C. Show how to calculate the number of interference fringes within the central diffraction maximum and remark that the result depends on the ratio d/a but not on the wavelength.

D. Discuss missing maxima. Point out that the first diffraction minimum on either side of the central single-slit diffraction maximum might coincide with a double-slit interference maximum, in which case the maximum would not be seen. Show that the maximum of order m is missing if $d/a = m$.

IV. Diffraction gratings.

A. Make or purchase a set of multiple-slit barriers with 3, 4, and 5 slits, all with the same slit width and spacing. Multiple slits can be made using razor blades and a lamp blackened microscope slide. Use a laser to show the patterns in order of increasing number of slits. Finish with a commercial grating.

B. Qualitatively describe the pattern produced as the number of slits is increased. Point out the principle maxima and, if possible, the secondary maxima. Remark that the principle maxima narrow and that the number of secondary maxima increases as the number of slits increases. Remark that for gratings with a large number of rulings, the principal maxima are called lines. For each barrier, sketch a graph of the intensity as a function of angle. Explain that the single-slit diffraction pattern forms an envelope for the pattern.

C. Remark that you will assume the slits are so narrow that the patterns you will consider lie well within the central maximum of the single-slit diffraction pattern and you need to consider only one wave from each slit. Explain that lines occur whenever the path difference for rays from two adjacent slits is an integer multiple of the wavelength: $d\sin\theta = m\lambda$. Remark that m is called the order of the line. Also remark that the angular positions of the lines depend only on the ratio d/λ and not on the number of slits or their width.

D. Consider N phasors of equal magnitude that form a regular polygon and remark this is the configuration for an interference minimum adjacent to a principal maximum. Show that for one of these minima the phase difference for waves from adjacent slits is $2\pi(m + 1/N)$ and the path difference is $d \sin\theta = \lambda(m + 1/N)$. Replace θ with $\theta + \delta\theta$, where $d \sin\theta = m\lambda$, to derive the expression $\delta\theta = \lambda/Nd\cos\theta$ for the angular half-width of the principal maximum at angle θ. Explain that this predicts narrowing of the principal maxima as the number of slits is increased. Also explain that principal maxima at large angles are wider than those at small angles.

E. Show a commercial transmission grating and tell students a typical grating consists of tens of thousands of lines ruled over a few centimeters. Explain that light is transmitted through both the rulings and the regions between but since these represent different thickness of material, the phase of the waves leaving the rulings is different from that of waves leaving the regions between. As a result, the diffraction pattern is the same as that of a multiple-slit barrier. Say that a diffraction pattern is also produced by lines ruled on a reflecting surface.

F. Put a grating in front of a white-light source and point out the spectrum. Put a grating in front of a discharge tube to display the emission spectrum of hydrogen or mercury. Note the separation of the lines corresponding to the same principal maximum produced by different frequency light. Explain that atoms produce light with certain discrete frequencies and that these are separated by the grating. Remark that measurements of the angles can be used to compute the wavelengths present if the ruling separation is known. Point out the colors of a compact disk or CD ROM.

V. X-ray diffraction (optional).

A. Explain that x rays are electromagnetic radiation with wavelength on the order of 10^{-10} m (1 Å). Point out that crystals are regular arrays of atoms with spacings on that order and so can be used to diffract x rays.

B. Consider a set of parallel crystalline planes and explain that reflection of the incident beam occurs at each plane, with the angle of reflection equal to the angle of incidence. Draw a diagram like Fig. 37–28 and state that x-ray diffraction is conventionally described in terms of the angle between the ray and the plane, rather than the normal to the plane. Show that waves reflected from the planes interfere constructively if $2d \sin\theta = m\lambda$.

C. Explain that for a given set of planes intense diffracted waves are produced only if waves are incident at an angle θ that satisfies the Bragg condition, given above. Measurements of these angles can be used to investigate the crystal structure. Show how to calculate the distance between planes, given the wavelength and the scattering angle. Explain that a crystal with a known structure can be used as a filter to obtain x rays of a given wavelength from a source with a broad range of wavelengths.

SUPPLEMENTARY TOPICS

1. Diffraction from a circular aperture. This topic is important for its application to diffraction patterns of lenses and the diffraction limit to the resolution of objects by a lens system. Show a diagram or picture (like Fig. 36–9) and point out the bright central disk and the secondary rings. Tell students that the angular position of the smallest ring of zero intensity occurs for $\theta = 1.22\lambda/d$, where d is the diameter of the aperture. If you intend to discuss the resolving power of a grating, the Rayleigh criterion for a circular aperture should be covered first since it is easier to present and understand. You can demonstrate the Rayleigh criterion by drilling two small holes, closely spaced, in the bottom of a tin can. Place the can over a light bulb and let students view it from various distances. See problem 36–17. Also use red and blue filters to show the dependence on wavelength.

2. Dispersion and resolving power of a grating. Define the dispersion of a grating and show it is $m/d\cos\theta$ for a line of order m occurring at angle θ. Note that dispersion can be increased by decreasing the ruling separation but dispersion does not depend on the number of rulings. If

you have gratings with different ruling separations, use them to show the hydrogen spectrum and point out the difference. Define the resolving power of a grating and show it is Nm for the line of order m. Remark that the resolving power does depend on the number of rulings and that the greater this number, the greater the resolving power. Show the sodium spectrum with a grating for which the two D lines cannot be resolved, then show it with one for which they can. Explain that dispersion and resolving power measure different aspects of the pattern produced by a grating. The lines produced by two different wavelengths may be fairly well separated in angle (large dispersion) but cannot be resolved because the principal maxima are so wide (small resolving power).

SUGGESTIONS

1. Assignments
   a. Basic relationships for the single-slit pattern are explored in problems 1, 2, 3, 6, and 7. Assign a few of these. To test for qualitative understanding, consider questions 1, 2, and 4.
   b. Following the discussion of the equation for the double-slit pattern, ask question 8. Characteristics of the pattern are explored in problems 29, 30, and 33.
   c. Diffraction from a circular aperture with application to the Rayleigh criterion for resolution is covered in problems 16 through 28. Assign one or two if you cover this topic.
   d. After discussing diffraction patterns of multiple slits, ask question 11. Problems 37 and 39 cover the fundamental equation for an intensity maximum. Questions 9 and 10 and problem 42 deal with line width.
   e. Ask question 12 in connection with the dispersion and resolving power of a grating. Assign problems 49 and 50 if you cover this topic.
   f. After discussing x-ray diffraction by crystals, assign problems 55 and 58. Problems 60 and 63 are a little more challenging. Problem 61 deals with the geometry of a square lattice.
2. Demonstrations
   a. Single-slit diffraction: Freier and Anderson Ol2, 3, 6, 7.
   b. Multiple-slit diffraction: Freier and Anderson Ol10, 13.
   c. Diffraction by circular and other objects: Freier and Anderson Ol21 — 23.
   d. Diffraction by crystals: Freier and Anderson Ol14.
3. Computer Software
   a. *Light and Optics.* See Chapter 35 SUGGESTIONS.
   b. *EM Field*. See Chapter 22 SUGGESTIONS.
4. Computer Projects
   Have students use a computer to plot the intensity pattern for various situations including the case when the screen is not far from the sources.
5. Laboratory
   a. Probeware Activity 28: *Diffraction of Light.* See Chapter 35.
   b. Meiners Experiment 13–4: *Interference and Diffraction.* See Chapter 40 notes.
   c. Probeware Activity 29: *Spectral Lines.* A diffraction grating, along with the equipment of Activity 28, is used to measure the wavelengths of the emission line of mercury vapor.
   d. Meiners Experiment 13–5; *Diffraction Gratings.* Wavelengths of the helium spectrum are found using a grating spectrometer and the influence of the number of grating rulings is investigated.
   e. Bernard Experiment 44; *The Wavelength of Light.* Wavelengths of the sodium spectrum are found using a grating spectrometer. The wavelength of a laser is also found.
   f. Bernard Experiment 45; *A Study of Spectra with the Grating Spectrometer.* Sources used are a sodium lamp, an incandescent bulb, a mercury lamp, and a lamp containing an

unknown element. The limits of the visible spectrum are determined and the unknown element is identified.

Chapter 37 RELATIVITY

BASIC TOPICS

I. Introduction.

A. Consider a wave on a string and remind students that its speed relative to the string is given by $v_w = \sqrt{\tau/\mu}$, where τ is the tension and μ is the linear mass density. Explain that, according to non-relativistic mechanics, an observer running with speed v_0 with the wave measures a wave speed of $v_w - v_o$ and an observer running against the wave measures a wave speed of $v_w + v_o$. Remark that these results are *not* valid for light (or fast moving waves and particles). The speed of light in a vacuum is found to be the same, regardless of the speed of the observer (or the speed of the source).

B. Remark that this fact has caused us to revise drastically our idea of time. If, for example, two observers moving at high speed with respect to each other both time the interval between two events, they obtain different results.

C. Explain that special relativity is a theory that relates measurements taken by two observers who are moving with respect to each other. Although it sometimes seems to contradict everyday experience, it is extremely well-supported by experiment.

D. State the postulates: the laws of physics are the same for observers in all inertial frames; the speed of light in a vacuum is the same for all directions and in all inertial frames. Remind students what an inertial frame is. Explain that the laws of physics are relationships between measured quantities, not the quantities themselves. Newton's laws and Maxwell's equations are examples. State that relativity has forced us to revise Newton's second law but not Maxwell's equations.

II. Time measurements.

A. Explain the term *event* and note that three space coordinates and one time coordinate are associated with each event. Explain that each observer may think of a coordinate system with clocks at all places where events of interest occur and that the clocks are synchronized. Outline the synchronization process involving light. State that the coordinate system and clock used by an observer are at rest with respect to the observer and may be moving from the viewpoint of another observer.

B. State that two observers in relative motion cannot both claim that two events at different places are simultaneous if their motion is not perpendicular to the line joining the coordinates of the events. To illustrate, show Fig. 37–4 and explain that the events are simultaneous in Sam's frame but the Red event occurs before the Blue event in Sally's frame. Show that signals from the events meet at the mid-point of Sam's spaceship but the signal from the Red event gets to the mid-point of Sally's spaceship before the signal from the Blue event. Stress the importance of the second postulate for reaching these conclusions.

C. Explain the light flasher used to measure time, in principle. Consider a flasher at rest in one frame, take two events to be a flash and the subsequent reception of reflected light back at the instrument, then remark that the time interval is $\Delta t_0 = 2D/c$, where D is the separation of the mirror from the flash bulb. Consider the events as viewed in another frame, moving with speed v perpendicularly to the light ray, and show the interval is $\Delta t = 2D/c\sqrt{1-v^2/c^2} = \Delta t_0/\sqrt{1-v^2/c^2}$. This is also written $\Delta t = \gamma\Delta t_0$, where γ $(= 1/\sqrt{1-v^2/c^2})$ is called the Lorentz factor. State that $v/c < 1$ and $\gamma > 1$.

D. Remark that Δt_0 is the *proper time interval* and that both events occur at the same coordinate in the frame in which it is measured. Point out that Δt is larger than Δt_0. Explain that the same result is obtained no matter what clocks are used for the measurement (as long as they are accurate and each is at rest in the appropriate frame). Ask students to identify a frame to estimate the proper time interval for a ball thrown from third to first base. Note that $\Delta t \approx \Delta t_0$ if $v \ll c$.

E. State that time dilation has been observed by comparing clocks carried on airplanes to clocks remaining behind and by comparing the average decay time of fast moving fundamental particles to their decay time when at rest. You might want to discuss the twin paradox here.

III. Length measurements.

A. Point out the problem with measuring the length of an object that is moving relative to the meter stick: the position of both ends must be marked *simultaneously* (in the rest frame of the meter stick) on the meter stick. If the speed v of the object is known, another method can be used to measure its length: put a mark on a coordinate axis along the line of motion of the object, then measure the time Δt_0 taken by the object to pass the mark. The length is given by $L = v\Delta t_0$. Note that Δt_0 is a proper time interval but L is not the proper length.

B. Explain that the length of the object, as measured in its rest frame, is $L_0 = v\Delta t$, where Δt is the time interval measured in that frame. Substitution of $\Delta t = \gamma\Delta t_0$ leads to $L = L_0/\gamma$. State that L_0, the length as measured in the rest frame of the object, is called the *proper length.* Since $\gamma > 1$, all observers moving with respect to the object measure a length that is less than the rest length. The same result is obtained no matter what method is used to measure length. Note that $L \approx L_0$ if $v \ll c$.

IV. The Lorentz transformation.

A. Consider two reference frames: S' moving with speed v in the positive x direction relative to S. Remark that the coordinates of an event as measured in S are written x, y, z, t while the coordinates as measured in S' are written x', y', z', t'. Write down the Lorentz transformation for the coordinate differences of two events: $\Delta x' = \gamma(\Delta x - v\Delta t)$, $\Delta y' = \Delta y$, $\Delta z' = \Delta z$, $\Delta t' = \gamma(\Delta t - v\Delta x/c^2)$. Remark that these equations reduce to the Galilean transformation if $v \ll c$: $\Delta x' = \Delta x - v\Delta t$, $\Delta y' = \Delta y$, $\Delta z' = \Delta z$, $\Delta t' = \Delta t$.

B. Explain that the transformation equations can be solved for Δx and Δt, with the result $\Delta x = \gamma(\Delta x' + v\Delta t')$, $\Delta t = \gamma(\Delta t' + v\Delta x'/c^2)$. From the viewpoint of an observer in S', S is moving in the negative x' direction, so the two sets of equations are obtained from each other when v is replaced by $-v$ and the primed and unprimed symbols are interchanged.

C. Discuss some consequences of the Lorentz transformation equations:

1. Simultaneity. Take $\Delta t = 0$, $\Delta x \neq 0$ and show that $\Delta t' = -\gamma v\Delta x/c^2$ ($\neq 0$). If two events are simultaneous and occur at different places in S, then they are not simultaneous in S'. Point out that $\Delta t'$ is positive for Δx negative and is negative for Δx positive. Similarly, take $\Delta t' = 0$, $\Delta x' \neq 0$ and show $\Delta t = \gamma v\Delta x'/c^2$ ($\neq 0$).
2. Time dilation. Consider two events that occur at the same place in S and show that $\Delta t' = \gamma\Delta t$. Point out that Δt is the proper time interval. Also show that the events do not occur at the same place in S': $\Delta x' = -\gamma v\Delta t$. Work the same problem for two events that occur at the same place in S'.
3. Length measurement. Suppose the object is at rest in S' and the meter stick is at rest in S. Marks are made simultaneously in S on the meter stick at the ends of the object. Thus, $\Delta t = 0$. Show that $\Delta x' = \gamma\Delta x$ and point out that $\Delta x'$ is the rest length. Work the same problem with the object at rest in S and the meter stick at rest in S'.
4. Causality. Consider two events, the first of which influences the second. For example,

a particle is given an initial velocity along the x axis and collides with another particle. Remark that t_2 (the time of the collision) must be greater than t_1 (the time of firing). Take $\Delta t = t_2 - t_1$ and $\Delta x > 0$, then show that the Lorentz transformation predicts $\Delta t'$ is positive for every frame for which $v < c$. The collision cannot happen before the firing in any frame moving at less than the speed of light.

5. Velocity transformation. Tell students that v represents the velocity of frame S' relative to S and that $\vec{u}$ and $\vec{u}'$ represent the velocity of a particle, as measured in S and S', respectively. Now take u and u' to be the x components of the particle velocity. Divide the Lorentz equation for Δx by the Lorentz equation for Δt to show that the x component of the particle velocity in S is $u = (u' + v)/(1 + vu'/c^2)$. Show this reduces to the Galilean transformation $u = u' + v$ for $v \ll c$. Take $u' = c$ and show that $u = c$. If $u' < c$, then $u < c$ for all frames moving at less than the speed of light.

V. Relativistic momentum and energy.

A. Explain that the non-relativistic definition of momentum must be generalized if momentum is to be conserved in collisions involving particles moving at high speeds. State that the proper generalization is $\vec{p} = m\vec{v}/\sqrt{1 - v^2/c^2}$. Remark that $\vec{p}$ is unbounded as the particle speed approaches the speed of light. In this text, m is used for the rest mass and is called simply the mass. The concept of relativistic mass is not used.

B. Remark that the definition of energy must be changed if the work-energy theorem is to hold for particles at high speeds. State that the relativistically correct expression for the energy of a free particle is $E = mc^2/\sqrt{1 - v^2/c^2}$. Take the limit as v/c becomes small and show that E can then be approximated by $mc^2 + \frac{1}{2}mv^2$. Thus, the correct relativistic definition of the kinetic energy is $K = E - mc^2$. Point out that the particle has energy mc^2 when it is at rest and remark that mc^2 is called the rest energy.

C. Explain that mass and rest energy are not conserved in many interactions involving fundamental particles but that total energy E is; rest energy can be converted to kinetic energy and vice versa.

D. Derive $E^2 = (pc)^2 + (mc^2)^2$ and explain that this expression replaces $E = p^2/2m$ ($= mv^2/2$). Remark that $E = pc$ for a massless particle, such as a photon.

SUPPLEMENTARY TOPIC

The Doppler effect for light. The expression for the frequency transformation can be derived easily by considering the measurement of the period in two frames. Suppose an observer in S obtains T for the interval between successive maxima at the same place. This is a proper time interval and the interval in another frame S' is γT. If S' is moving parallel to the wave, however, the two events do not occur at the same place in S' and γT is not the period in that frame. An observer in S' must wait for a time $|\Delta x'|/c$ longer before the next maximum is reached at the place of the first. Thus, $T' = \gamma T + |\Delta x'|/c$ or since $\Delta x' = -\gamma T v$, $T' = \gamma T(1 + v/c) = T\sqrt{(1+\beta)/(1-\beta)}$. Thus, $f' = f\sqrt{(1-\beta)/(1+\beta)}$. If S' is moving perpendicularly to the wave, the two events occur at the same place in both frames and $T' = \gamma T$, so $f' = f/\gamma$.

SUGGESTIONS

1. Assignments
   a. Simultaneity and time measurements are the issues in questions 2, 3, 5, and 7. Ask some of them to test for understanding. Also assign problems 5 and 6.
   b. When length contraction is covered, assign problems 9, 13, and 14.
   c. Assign problems 16, 17, and 25 in support of the discussion of the Lorentz transformation.
   d. Assign problems 27 and 30 in connection with the relativistic velocity transformation.

e. Assign problems 29 and 33 in connection with the relativistic Doppler effect.

f. Use question 10 to broaden the discussion of mass and rest energy. Assign problems 42, 44, 47, and 54 in connection with relativistic energy and momentum. If you covered cyclotrons in Chapter 28, assign problem 50.

2. Computer Software

a. *RelLab*; Paul Horwitz, Edwin F. Taylor, and Kerry Shetline; Macintosh; available from Physics Academic Software (Centennial Campus, 940 Main Campus Drive, Suite 210, Raleigh, NC 27606–5212; www.aip.org/pas). Shows the coordinates and times of events, as measured in user-selected reference frames. Presents some relativity paradoxes that can be understood with the aid of the program.

b. *Spacetime*; Edwin F. Taylor; Windows, Macintosh; available from Physics Academic Software (see above for address). Shows a "spacetime highway", on which objects in different lanes move with different speeds. Shows the corresponding spacetime diagram, on which events are identified. All the clocks and rulers are also shown so the user can compare readings in different frames.

c. *Relativity* from Exploration of Physics Volume II; Windows and Macintosh; Physics Curriculum & Instruction (22585 Woodhill Drive, Lakeville, MN 55044; www.PhysicsCurriculum.com). Includes sections on the Michelson-Morley experiment, simultaneity, length contraction, and time dilation.

d. *Relativistic Collision*; Edwin F. Taylor; DOS, Macintosh; available from Physics Academic Software (see above for address). Data is entered and the computer calculates the energy, momentum, mass, and geometry.

3. Computer Project

Have students write a computer program or design a spreadsheet to evaluate the Lorentz transformation equations. Then have them use it to investigate simultaneity, length contraction, and time dilation.

Chapter 38 PHOTONS AND MATTER WAVES

BASIC TOPICS

I. Introduction.

A. Explain that this chapter deals with some of the fundamental results of quantum physics. The first few sections describe experimental results that can be understood only if light is regarded as made up of particles. Remark that interference and diffraction phenomena require waves for their explanation. Reconciliation of these opposing views will be discussed later.

B. Explain that the energy of a photon is related to the frequency of the wave through $E = hf$ and the momentum of a photon is related to the wavelength of the wave through $p = h/\lambda$. Show these equations predict $p = E/c$, the classical relationship. Also explain that the energy density is nhf, where n is the photon concentration, and that the intensity is Rhf, where R is the rate per unit area with which photons cross a plane perpendicular to their direction of motion. Recall the discussion of the Poynting vector in Chapter 33. Explain that the Planck constant is a constant of nature and pervades quantum physics. Give its value (6.63×10^{-34} J·s) and calculate the photon energy and momentum for visible light, radio waves, and x rays.

C. Point out that classically monochromatic electromagnetic radiation can have any value of energy. Quantum mechanically, this is not true, but since h is so small, the discreteness of the energy values is important only at the atomic level.

II. The photoelectric effect.

A. Sketch a schematic of the experimental setup. Explain that monochromatic light is incident on a sample. It is absorbed and part of the energy goes to electrons, some of which are emitted. The energy of the most energetic electron is found by measuring the stopping potential V_0.

B. Point out that the stopping potential is independent of the light intensity. As the intensity is increased, more electrons are emitted but they are not more energetic. Show a plot of the stopping potential as a function of frequency and point out that the relationship is linear and that as the frequency is increased the electrons emitted are more energetic. Also state that electrons are emitted promptly when the light is turned on. If the radiation energy were distributed throughout the region of a wave, it would take a noticeable amount of time for an electron to accumulate sufficient energy to be emitted, since an electron has a small surface area. This argument can be made quantitative (see Sample Problem 38–2).

C. Give the Einstein theory. Electromagnetic radiation is concentrated in photons, with each photon having energy hf. The most energetic electrons after emission are those with the greatest energy while in the material and, in the interaction with a photon, receive energy hf. If the light intensity is increased without changing the frequency, there are more photons and, hence, more electrons emitted, but no single electron can receive more energy. Furthermore, the electron receives energy immediately and need not wait to absorb the proper amount.

D. Show that this analysis leads to $hf = \Phi + K_m$, where Φ is the work function, the energy needed to remove the most energetic electron from the material. It is characteristic of the material. Remark that $K_m = eV_0$ and that the Einstein theory predicts a linear relationship between V_0 and f and predicts a minimum frequency for emission: $hf = \Phi$. Remark that the emitted electrons have a distribution of speeds if $hf > \Phi$ because they come from states with different energies.

III. The Compton effect.

A. Note that in the explanation of the photoelectric effect, a photon is assumed to give up all its energy to an individual electron. The photon then ceases to exist. Explain that a photon might transfer only part of its original energy in an interaction with an electron. Since a lower energy means a lower frequency, the scattered light has a longer wavelength than the incident light. State that a photon also carries momentum and part of it is transferred to a target electron.

B. Discuss the experiment. Light is scattered from electrons in matter and the intensity of the scattered light is measured as a function of wavelength for various scattering angles. Show Figs. 38–3 and 38–4. Stress that the experimental data can be explained by considering the interaction to be a collision between two particles, with energy and momentum conserved. Relativistic expressions, however, must be used for energy and momentum.

C. Remark that the situation is exactly like a two-dimensional collision between two particles. Write down the relativistic expressions for the momentum and energy of a particle with mass (the electron) and remind students of the rest energy. Assume the electron is initially at rest and that the photon is scattered through the angle ϕ. The electron leaves the interaction at an angle θ to the direction of the incident photon. Write down the equations for the conservation of energy and the conservation of momentum in two dimensions. Write down the momentum and energy of the photon in terms of the wavelength and solve for the change on scattering of the wavelength: $\Delta\lambda = (h/mc)(1 - \cos\phi)$. Emphasize that agreement with experiment strongly supports the conclusion that the momentum of a photon is $p = h/\lambda$.

D. Note that the change in wavelength is independent of wavelength and that the change is significant only for short wavelength light, in the x-ray and gamma ray regions. Also state that the theoretical results successfully predict experimental data. The widths of the curves are due chiefly to moving electrons, for which $\Delta\lambda$ is slightly different, and the peak near $\Delta\lambda = 0$ is due to scattering from more massive particles (atoms as a whole). Stress that the particle picture of light accounts for experimental data.

IV. Matter waves.

A. Explain that electrons and all other particles have waves associated with them, just as photons have electromagnetic waves associated with them. State that the waves exhibit interference and diffraction effects. Draw a diagram of a single-slit barrier with a beam of monoenergetic electrons incident on it and a fluorescent screen or other mechanism for detecting electrons behind it. Explain that an intense central maximum is obtained and that many electrons arrive in this region. Secondary maxima are also obtained.

B. State that the width of the central maximum depends on the speed of the electrons and narrows if the speed is increased. The maximum also narrows if more massive particles are used at the same speed. Remind students that when they studied the single-slit diffraction of electromagnetic waves, they found the width of the central maximum narrowed as the wavelength decreased. Conclude that the momentum of the particle is related to the wavelength of the wave and that one is proportional to the reciprocal of the other.

C. State that the particle energy and the wave frequency are related by $E = hf$ and that the particle momentum and the wavelength are related by $p = h/\lambda$, just as for photons. Calculate the wavelengths of a 1-eV electron and a 35-m/s baseball.

D. By way of example, state that crystals diffract electrons of appropriate wavelength ($\approx 10^{-10}$ m) and the angular positions of the scattering maxima can be found using Bragg's law, suitably modified to account for changes in the propagation direction that occur when matter waves enter the crystal.

E. Explain that, at the atomic and particle level, physics deals with probabilities. What can be analyzed is the probability for finding a particle, not its certain position. State that a one-dimensional matter wave is denoted by $\psi(x)$ and that $|\psi|^2$ gives the probability density for finding the particle near x. That is, the probability that the particle is in the region between x and $x + dx$ is given by $|\psi(x)|^2\,dx$. Similarly, if E is the electric field amplitude for an electromagnetic wave, then E^2 is proportional to the probability density for finding a photon. In the limit of a large number of particles, $|\psi|^2$ is proportional to the particle concentration.

F. State that space-dependent part of a particle wave function obeys the Schrödinger equation:

$$\frac{d^2\psi}{dx^2} + \frac{8\pi^2 m}{h^2}\left[E - U(x)\right]\psi = 0\,,$$

where E is the energy of the particle and $U(x)$ is its potential energy function. Explain that, for a free particle, we may take $U(x) = 0$ and write

$$\frac{d^2\psi}{dx^2} + k^2\psi = 0\,,$$

where $k^2 = (8\pi^2 m/h^2)E$. The most general solution is

$$\psi(x) = Ae^{ikx} + Be^{-ikx}\,,$$

where A and B are arbitrary constants. The first term represents a particle moving in the positive x direction with momentum $hk/2\pi$ and the second represents a particle moving in the negative x direction with the same magnitude momentum.

G. You may want to review some properties of complex numbers. Explain that a complex number can be written $\psi = \psi_R + i\psi_I$, where ψ_R is the real part and ψ_I is the imaginary part. Say that $i = \sqrt{-1}$. Define the complex conjugate: state that $|\psi|^2 = \psi^*\psi$, where ψ^* is the complex conjugate of ψ, and show that $|\psi|^2 = \psi_R^2 + \psi_I^2$. Also show that $|e^{ikx}|^2 = 1$.

V. The uncertainty principle.

A. Because a different answer might result each time the position of the electron is measured, there is an uncertainty in the position. It can be defined similarly to the standard deviation of a large collection of experimental results. Similar statements can be made about momentum measurements. Explain that the uncertainties in position and momentum are both determined by the particle wave function. Explain that if the electron is placed in a state for which the uncertainty in position is small then the uncertainty in momentum is large and vice versa.

B. Give the Heisenberg uncertainty relations: $\Delta x \cdot \Delta p_x \geq \hbar$, $\Delta y \cdot \Delta p_y \geq \hbar$, and $\Delta z \cdot \Delta p_z \geq \hbar$ and state that $\hbar$ is the Planck constant divided by 2π. Note that it is impossible to reduce both the uncertainty in position and the uncertainty in momentum simultaneously to zero. Compare this conclusion with the classical result by setting $\hbar = 0$.

VI. Barrier tunneling.

A. Show Figs. 38–15 and 38–16 and explain that the wave function penetrates a finite barrier. It is oscillatory (in position) outside the barrier, where $E > U_0$, and exponential inside, where $E < U_0$. The figure shows the probability density.

B. Explain that the particle has a probability of being found on either side of the barrier. Contrast to the behavior of a classical particle.

C. Write down Eqs. 38–21 and 38–22 for the transmission coefficient and explain that this measures the probability of transmission through the barrier. Remark that transmission is small for high, wide barriers and becomes larger as the barrier height decreases and as the barrier width narrows. Also define the reflection coefficient R by $R = 1 - T$.

SUGGESTIONS

1. Assignments
   a. Ask questions 3, 9, and 10 and assign one or two of problems 4 and 7 as part of a discussion of photon properties. Emphasize that the energy in a light beam is the product of the number of photons and the energy of each photon. Assign some of problems 3, 4, 8, 9, 10, 11, and 12.
   b. After discussing the photoelectric effect, ask some of questions 1, 2, 4, 5, and 6. Assign problems 22 and 24.
   c. After discussing the Compton effect, ask questions 6, 7, and 11 and assign problem 28. Also consider problem 34.
   d. In the discussion of the properties of matter waves, include questions 12 through 15 and 18. Assign problems 43 and 50.
   e. Following the discussion of the uncertainty principle, assign problems 59 and 60.
   f. Ask some of questions 16, 17, 19, and 20 in connection with tunneling. Also assign problems 62 and 63.

2. Demonstrations

   Photoelectric effect: Freier and Anderson MPb1

3. Audio/Visual

   Photons and X-rays, *Electrons*, and *Particles and Waves*; from Cinema Classics DVD 6: Angular Momentum and Modern Physics; available from Ztek Co. (PO Box 11768, Lexington, KY 40577–1768, www.ztek.com) and from the American Association of Physics Teachers (AAPT, One Physics Ellipse, College Park MD 20740–3845; www.aapt.org).

   a. *The Quantum Idea* from the *Wave-Particle Duality* series; VHS video tape; Films for the Humanities and Sciences (PO Box 2053, Princeton, NJ 08543–2053; www.films.com).
   b. *Photons* from the *Wave-Particle Duality* series; VHS video tape; Films for the Humanities and Sciences (see above for adress).
   c. *Matter Waves* from the *Wave-Particle Duality* series; VHS video tape; Films for the Humanities and Sciences (see above for address).
   d. *Electron Diffraction*; VHS video tape, DVD; Films for the Humanities and Sciences (see above for address).
   e. *Understanding Uncertainty*; VHS video tape, DVD; Films for the Humanities and Sciences (see above for address).
   f. *The Particle Model* from the *Wave-Particle Duality* series; VHS video tape; Films for the Humanities and Sciences (see above for address).
   g. *The Wave Model* from the *Wave-Particle Duality* series; VHS video tape; Films for the Humanities and Sciences (see above for address).
   h. *The Electromagnetic Model* from the *Wave-Particle Duality* series; VHS video tape; Films for the Humanities and Sciences (see above for address).

4. Computer Software

   a. *Modern Physics* from Exploration of Physics Volume II; Windows and Macintosh; Physics Curriculum & Instruction (22585 Woodhill Drive, Lakeville, MN 55044; www.PhysicsCurriculum.com). Includes sections on the photoelectric effect and double-slit interference of electron waves.
   b. *Photoelectric Tutor*; Graham Oberon and Richard Steinberg; DOS, Macintosh; available from Physics Academic Software (Centennial Campus, 940 Main Campus Drive, Suite 210, Raleigh, NC 27606–5212; www.aip.org/pas). A tutorial on the photoelectric effect.

5. Computer Project

   A commercial math program or a student-generated root-finding program can be used to solve the equations for the photoelectric and Compton effects. Students may be interested, for example, in seeing how the Compton lines broaden when the electrons are not initially at rest. Assign some exercises as homework or set aside some laboratory time for a more detailed investigation.

6. Laboratory

   a. Probeware Activity 30: Photoelectric Effect — Planck's Constant. The stopping potential is measured as a function of the frequency of the incident light and Planck's constant is computed.
   b. Meiners Experiment 14–2: *The Photoelectric Effect.* Students investigate the characteristics of various photocells, then use a plot of stopping potential versus. frequency to determine the Planck constant. A mercury source and optical filters are used to obtain monochromatic light of various frequencies.
   c. Meiners Experiment 14–5: *Electron Diffraction.* The Sargent-Welch electron diffraction apparatus is used to investigate the diffraction of electrons by aluminum and graphite. Since powder patterns (rings) are obtained, you will need to explain their origin.

Chapter 39 MORE ABOUT MATTER WAVES

BASIC TOPICS

I. One-dimensional particle traps

A. Explain that, for a particle confined by infinite potential energy barriers to the region between 0 and L on the x axis, possible wave functions are given by $\psi_n(x) = A\sin(n\pi x/L)$, where $n = 1, 2, \ldots$. Show that these satisfy the Schrödinger equation

$$\frac{d^2\psi}{dx^2} + \frac{8\pi^2 m}{h^2}E\psi = 0$$

inside the trap and that ψ goes to zero at the boundaries. Explain that a condition for the given function to be a solution is that the energy of the particle must be $E_n = n^2h^2/8mL^2$. You might want to include the time dependence by writing $\Psi = A\sin(n\pi x/L)f_n(t)$ and explaining that $f_n(t)$ is a function of time with magnitude 1.

B. Explain that confinement of the particle leads to energy quantization and that energy is quantized for any bound particle. Plot the allowed values of the energy, as in Fig. 39–3. Point out that the particle has kinetic energy even in the ground state and mention that this energy is called its zero-point energy.

C. Explain that the particle can certainly be found between $x = 0$ and $x = L$, so $\int_0^L |\psi_n|^2\,dx = 1$. The wave function is said to be *normalized* if it obeys this condition. Show that the normalization condition leads to $A = \sqrt{2/L}$.

D. Use the particle confined to a one-dimensional trap as an example and explain that $\psi_n^2\,dx = (2/L)\sin^2(n\pi x/L)\,dx$ gives the probability that the particle can be found between x and $x+dx$ when it is in the state with the given wave function. Sketch several of the probability density functions and point out that there are several places where the probability density vanishes. See Fig. 39–6.

E. Explain that experimentally the probability can be found, in principle, by performing a large number of position measurements and calculating the fraction for which the particle is found in the designated segment of the x axis. Since a position measurement changes the state of the particle, it must be restarted in the same state each time.

F. Explain that a particle may jump from some initial state to a lower energy state with the emission of a photon and that the photon energy is equal to the change in the particle energy. Write $hf = \Delta E$, where f is the frequency of the electromagnetic wave associated with the photon. Remind students that the frequency and wavelength are related by $\lambda f = c$. Use an energy level diagram to show that only a set of discrete wavelengths occur. Explain that the particle may also absorb a photon and jump to a higher energy state but only if the photon energy equals the difference between two allowed energy values for the particle.

G. Draw a diagram of a one-dimensional trap with finite potential energy barriers at the ends and state that the particle wave function now extends into the barriers, although it decreases exponentially there. Show Fig. 39–8. Mention that the allowed values of the energy are different from those for infinite barriers, but that the energy is still quantized. Also mention that the particle might absorb a photon of any energy that makes the final particle energy greater than the barriers. Particle energy above the barriers is not quantized.

II. Two- and three-dimensional particle traps

A. Describe a two-dimensional rectangular trap with sides of length L_x and L_y, such that the particle has infinite potential energy at the boundaries and zero potential energy within.

Give the expression for the energies:

$$E_{n_x,n_y} = \frac{h^2}{8m}\left[\frac{n_x^2}{L_x^2} + \frac{n_y^2}{L_y^2}\right],$$

where n_x and n_y are integers. Explain that neither n_x or n_y can be zero since either of those values would make the wave function zero everywhere.

B. Repeat the discussion for a three-dimensional trap in the form of a rectangular solid with sides of lengths L_x, L_y, and L_z. Show that the energies are given by

$$E_{n_x,n_y,n_z} = \frac{h^2}{8m}\left[\frac{n_x^2}{L_x^2} + \frac{n_y^2}{L_y^2} + \frac{n_z^2}{L - z^2}\right],$$

C. Mention the possibility of degeneracy. In some cases two or more states (with different quantum numbers) may have the same energy.

III. The hydrogen atom and line spectra.

A. Use a commercial hydrogen tube to show the visible hydrogen spectrum. Since the intensity is low, you will not be able to project this but you can purchase inexpensive 8″ × 10″ sheets of plastic grating material, which can be cut into pieces and passed out to the students. Point out Fig. 39–19.

B. Give the expression for the hydrogen energy levels in terms of the principal quantum number: $E_n = -(me^4/8\epsilon_0^2h^2)(1/n^2)$. State that quantum physics predicts these allowed values. Say that a photon is emitted when a hydrogen atom changes state and derive $f = (me^4/8\epsilon_0^2h^3)[(1/n_2^2) - (1/n_1^2)]$ for the frequency of the emitted electromagnetic wave.

C. Explain that the Schrödinger equation is a differential equation for the wave function of a particle and that the main ingredient that causes two identical particles to have different wave functions is their potential energy function. For an electron in a hydrogen atom, the potential energy function is $U(r) = -e^2/4\pi\epsilon_0 r$, where r is the distance from the proton to the electron. Mention that when this potential energy function is used in the Schrödinger equation and the reasonable condition that the wave functions remain finite everywhere is applied, then the allowed energy values are predicted. Draw a graph of $U(r)$ and draw lines across it to indicate the values of the first few energy levels.

D. Explain that states for hydrogen are classified using three quantum numbers:

1. The principal quantum number n, which determines the energy.
2. The orbital quantum number ℓ, which determines the magnitude of the orbital angular momentum.
3. The orbital magnetic quantum number m_ℓ, which determines the z component of the orbital angular momentum.

E. Explain that traditionally each value of n is said to label a shell. Remark that a shell may consist of many states, but each is associated with the same value of the energy. Tell students that for a given shell, ℓ may take on the values 0, 1, 2, ..., $n - 1$. There are n different values in all. Explain that all the states with given values of n and ℓ are said to form a subshell. Say that for a given value of ℓ, m_ℓ may take on any integer value from $-\ell$ to $+\ell$ and there are $2\ell + 1$ values in all. As examples, list all the states for $n = 1$, 2, and 3. Group them according to n and remark that all states with the same n have the same energy, all states with the same ℓ have the same magnitude of orbital angular momentum, and all states with the same m_ℓ have the same z component of orbital angular momentum. Remark that states with different values of n, ℓ, or m_ℓ have different wave functions.

F. Give the ground state wave function and obtain the expression for the probability density. Define the Bohr radius ($a = h^2\epsilon_0/\pi me^2 = 52.9\,\text{pm}$). Remark that ψ has spherical symmetry and explain that this is true of all $\ell = 0$ wave functions. Remind students that the volume of a spherical shell with thickness dr is $4\pi r^2\,dr$ and define the radial probability density as $P(r) = 4\pi r^2|\psi(r)|^2$. Sketch $P(r)$ for the hydrogen atom ground state (Fig. 39–20) and point out there is a range of radial distances at which the electron might be found. Locate the most probable radius and the average radius.

G. Show a dot plots (Figs. 39–22 and 39–24) for the $n = 2$ states and write the expressions for the wave functions (see problems 43 and 48). Remark that the individual probability densities are not spherically symmetric but their sum is.

SUGGESTIONS

1. Assignments
   a. Use questions 1 through 7 in your discussion of a particle in a one-dimensional infinite well. Assign problems 5, 6, and 49.
   b. Ask questions 8, 10, 13, and 14 in connection with a particle trapped in a one-dimensional finite well. Consider problems 15, 52, and 53.
   c. Assign problems 20 and 21 in connection with two-dimensional traps and problems 22 and 23 in connection with three dimensional traps.
   d. After discussing the hydrogen spectrum, ask questions 9, 16, and 17. Also assign problems 25, 27, 29, and 31. If you have discussed the terms *binding energy* and *excitation energy*, assign problem 38.
   e. When you discuss the enumeration of hydrogen atom states, assign problem 51.
   f. Discuss problems 39 and 42 in connection with the ground state of a hydrogen atom. You might also assign problems 37 and 41 if you did not show the given ground state wave function is a solution to the Schrödinger equation and is normalized. Give problem 47 to students who are math oriented.
   g. The $n = 2$ hydrogen wave functions are covered in problems 43 and 48.

2. Demonstrations

   Thompson and Bohr models of the atom.

3. Audio/Visual
   a. *Absorption Spectra*; from the AAPT Miller collection of single-concept films; DVD; available from Ztek Co. (PO Box 11768, Lexington, KY 40577–1768, www.ztek.com) and from the American Association of Physics Teachers (AAPT, One Physics Ellipse, College Park MD 20740–3845; www.aapt.org).
   b. *Electron Distribution in the Hydrogen Atom*; A.F. Burr and Robert Fisher; slide set; AAPT (see above for address). Probability distributions for n =1 to n = 6.
   c. *Atoms, Molecules, and Models*; from Cinema Classics DVD 6: Angular Momentum and Modern Physics; available from Ztek Co. and from the AAPT (see above for addresses).

4. Computer Software
   a. *Bellbox*; Darrel J. Conway; available from Physics Academic Software (Centennial Campus, 940 Main Campus Drive, Suite 210, Raleigh, NC 27606–5212; www.aip.org/pas). This simulation program allows students to experiment with the Einstein-Podolsky-Rosen paradox.

5. Laboratory

   Meiners Experiment 14–3: *Analysis of Spectra.* A spectroscope is used to obtain the wavelengths of hydrogen and helium lines. Hydrogen lines are compared with predictions of the Balmer equation.

Chapter 40 ALL ABOUT ATOMS

BASIC TOPICS

I. Orbital and spin angular momentum.
   A. Remark that orbital angular momentum is quantized and that the allowed values of its magnitude are given by $L = \sqrt{\ell(\ell+1)}\hbar$, where $\hbar = h/2\pi$. The orbital quantum number ℓ takes on positive integer values.
   B. State that the z component of the angular momentum is given by $L_z = m_\ell \hbar$, where $m_\ell = 0$, ± 1, ± 2, ..., $\pm \ell$. m_ℓ is called the orbital magnetic quantum number. The z axis can be in any direction, perhaps defined by an external magnetic field. Point out that the angle θ between the angular momentum vector and the z axis is given by $\cos\theta = m_\ell/\sqrt{\ell(\ell+1)}$. The smallest value of θ occurs when $m_\ell = \ell$ and it is not zero. Explain that the angles $\vec{L}$ makes with the x and y axes cannot be known if the angle between $\vec{L}$ and the z axis is known. Discuss this in terms of the precession of $\vec{L}$ about the z axis.
   C. Explain that the electron and some other particles have intrinsic angular momentum, as if they were spinning. The magnitude of the electron spin angular momentum is $S = \sqrt{s(s+1)}\hbar = \sqrt{3/4}\hbar$ and the z component is either $m_s = -\frac{1}{2}\hbar$ or $+\frac{1}{2}\hbar$ (there are two possible states), where m_s is called the spin magnetic quantum number. You might want to remark that spin is not predicted by the Schrödinger equation but that it is predicted by relativistic modifications to quantum physics.
   D. Say that the total angular momentum of an atom is the vector sum of the orbital angular momenta and the spin angular momenta of its electrons.

II. Magnetic dipole moments.
   A. Explain that the electron has a magnetic dipole moment because of its orbital motion and write $\vec{\mu}_{\text{orb}} = -(e/2m)\vec{L}$ and $\mu_{\text{orb},z} = -(e/2m)L_z = -(e\hbar/2m)m_\ell$. Give the value of the Bohr magneton ($\mu_B = e\hbar/2m = 9.28 \times 10^{-24}$ J/T). Remind students that because of its motion, the electron experiences a torque in an external magnetic field and produces its own magnetic field (provided $\vec{\mu}_{\text{orb}} \neq 0$).
   B. State that the spin magnetic moment is $\mu_{sz} = -2m_s\mu_B$. Stress the appearance of the factor 2. The electron produces a magnetic field and experiences a torque in a magnetic field because of this moment.
   C. Remark that the energy of an electron is changed by$-\mu_z B$ when an external field $\vec{B}$ is applied in the positive z direction. Thus states with the same n but different m_ℓ have different energies in a magnetic field. This is called the Zeeman effect. Photons with an energy equal to the energy difference of the two spin states cause the spin to flip. The phenomenon can be detected by measuring the absorption of the beam.
   D. Briefly describe the Stern-Gerlach experiment. Explain that a magnetic dipole in a *non-uniform* magnetic field experiences a force and that $F_z = \mu_z\, dB/dz$ for a field in the z direction that varies along the z axis. Atoms with different values of m_ℓ experience different forces and arrive at different places on a screen. That discrete regions of the screen receive atoms is experimental evidence for the quantization of the z component of angular momentum.
   E. To emphasize the practical, qualitatively explain NMR and its use in diagnostic medicine. You might also explain how local magnetic fields in solids, for example, can be measured using magnetic resonance techniques.

IV. Pauli exclusion principle.
   A. State the principle. For any two electrons in the same trap at least one of their quantum numbers must be different. State that this is a principle that holds for electrons, protons,

neutrons, and many other particles. Also state that it does not hold for all particles and give the photon as an example of a particle for which it does not hold.

B. As an example consider a small group of electrons in a square trap. State that the single-particle energy levels are given by $(h^2/8mL^2)(n_x^2 + n_y^2)$, where L is the length of an edge of the trap and n_x and n_y are quantum numbers that may take on any integer value greater than zero. For each of the (n_x, n_y) pairs of values $(1,1)$, $(1,2)$, $(2,1)$, $(1,3)$, $(3,1)$, $(2,3)$, and $3,1)$ calculate the energy in units of $h^2/8mL^2$, order the states according to energy, and point out the degeneracies, including spin. Note how many electrons can have each value of the energy. Assume there are five electrons, give the ground state configuration, and calculate the ground state energy of the system in units of $h^2/8mL^2$. Emphasize the role of the Pauli exclusion principle.

C. Repeat for the first excited state of the system.

IV. Atomic states.

A. Explain that quantum mechanical states for an electron in an atom are classified using four quantum numbers:

1. The principal quantum number n, which determines the energy.
2. The orbital quantum number ℓ, which determines the magnitude of the orbital angular momentum and, to a lesser extent, the energy.
3. The orbital magnetic quantum number m_ℓ, which determines the z component of the orbital angular momentum.
4. The spin magnetic quantum number m_s, which determines the z component of the spin angular momentum.

B. Explain that traditionally each value of n is said to label a shell and the shells are named $K, L, M, N, \ldots$, in order of increasing n. Remark that a shell may consist of many states.

C. Remind students that for a given shell, ℓ may take on the values 0, 1, 2, ..., $n-1$. There are n different values in all. Explain that all the states with given values of n and ℓ are said to form a subshell. Remind students that for a given value of ℓ, m_ℓ may take on any integer value from $-\ell$ to $+\ell$. Since m_s can have either of two values, a subshell consists of $2(2\ell+1)$ states. Either state or prove that the shell with principal quantum number n has $2n^2$ states.

D. Give the spectroscopic notation: s labels an $\ell = 0$ subshell, p labels an $\ell = 1$ subshell, d labels an $\ell = 2$ subshell, etc. Explain that the value of n is placed in front of the letter and the number of electrons in the subshell is given as a superscript: $3d^2$ indicates two electrons in the $n = 3$, $\ell = 2$ subshell.

E. As you may have done for the last chapter, list all the states for $n = 1$, 2, and 3. Group them according to n and remark that all states with the same ℓ have the same magnitude of orbital angular momentum and all states with the same m_ℓ have the same z component of orbital angular momentum. Remark that states with different values of n, ℓ, and m_ℓ have different wave functions.

V. Atom building and the periodic table.

A. Give the "rules" for atom building:

1. The four quantum numbers n, ℓ, m_ℓ, m_s can be used to label states. Remark that wave functions and energies are different for electrons with the same quantum numbers in different atoms.
2. The electrons in an atom obey the Pauli exclusion principle: No more than one electron can have any given set of quantum numbers.

B. Explain that as more protons are added to the nucleus, the electron wave functions pull in toward regions of low potential energy. This and the dependence of the energy on ℓ means that states associated with one principal quantum number may not be filled before states associated with the next principal quantum number are started. For example, a $5s$ state is lower in energy than a $4d$ state, in different atoms. It also accounts for the fact that all atoms are nearly the same size.

C. Show a periodic table. Point out the inert gas atoms and explain they all have filled s and p subshells. Point out the alkali metal and alkaline earth atoms and state they have one and two electrons, respectively, outside closed shells. Remark that electrons in partially filled shells are chiefly responsible for chemical activity. Point out the atoms in which d and f states are being filled and finally those in which p states are being filled.

VI. X rays and the numbering of the elements.

A. Explain that x rays are produced by firing energetic electrons into a solid target. Show Fig. 40–14 and point out the continuous part of the spectrum and the peaks. Also point out that there is a sharply defined minimum wavelength to the x-ray spectrum. Explain that the continuous spectrum results because the electrons lose some or all of their kinetic energy in close (decelerating) encounters with nuclei. This energy appears as photons and $\Delta K = hf$. Explain that a photon of minimum wavelength is produced when an electron loses all its kinetic energy in a single encounter. Derive the expression for the minimum wavelength in terms of the original accelerating potential and point out it is independent of the target material.

B. Explain that the line spectrum in Fig. 40–14 appears because incident electrons interact with atomic electrons and knock some of the deep-lying electrons out of the atoms. Electrons in higher levels drop to fill the holes, emitting photons with energy equal to the difference in energy of the initial and final atomic levels. The K_α line is produced when electrons drop from the L $(n = 2)$ shell to the K $(n = 1)$ shell and the K_β line is produced when electrons drop from the M $(n = 3)$ shell to the K shell. Explain Fig. 40–16.

C. Show Fig. 40–17 and state that when the square root of the frequency for any given line is plotted as a function of the atomic number of the target atom, the result is nearly a straight line. Argue that the innermost electrons have an energy level scheme close to that of hydrogen but with an effective nuclear charge of $(Z - 1)e$, where the 1 accounts for screening by electrons close to the nucleus. Z is the number of protons in the nucleus, the atomic number. Use the expression for hydrogen energy levels. For K_α, put $n = 2$ for the initial state and $n = 1$ for the final state, then show that $\sqrt{f}$ is proportional to $(Z - 1)$.

D. Remark that this relationship was used to position the chemical elements in the periodic table independently of their chemical properties. This technique was particularly important for elements in the long rows of the periodic table, which contain many elements with similar chemical properties. Today the technique is used to identify trace amounts of impurities in materials.

VII. The laser.

A. List the characteristics of laser light: monochromatic, coherent, directional, can be sharply focused. See the text for quantitative comparisons with light from other sources.

B. Explain the mechanism of light absorption: an incident photon is absorbed if hf corresponds to the energy difference of two electron states of the material and the upper state is initially empty. An electron jumps from the lower to the upper state. Explain spontaneous emission: an electron spontaneously (without the aid of external radiation) makes the transition from one state to a lower state (if that state is empty) and a photon with hf equal to the energy difference is emitted. Emphasize that in most cases the electron remains in the upper state for a time on the order of 10^{-9} s but that there are metastable states in which the electron remains for a longer time ($\approx 10^{-3}$ s). Explain stimulated emission: with the electron in an upper state, an incident photon with the proper energy can cause it to make the jump to a lower state. The result is two photons of the same energy, moving in the same direction, with waves having the same phase and polarization. Remark that laser light is produced by a large number of such events, each triggered by a photon from a previous event. Hence, all laser photons are identical. Explain that metastable states are important since the electron must remain in the upper state until its transition is induced. Compare with light produced by random spontaneous transitions.

C. Explain that, in thermodynamic equilibrium, upper levels are extremely sparsely populated compared to the ground state. To obtain laser light, the population of an upper level must be increased; otherwise absorption events would equal or exceed stimulated emission events. A laser must be pumped. Write down the expression for the thermal equilibrium number of atoms in the state with energy E: $n(E) = Ce^{-E/kT}$. Explain that C is independent of energy but depends on the number of atoms present. State that the temperature T is on the Kelvin scale.

D. Discuss the helium-neon laser, paying particular attention to the role played by helium atoms in maintaining population inversion in the neon atoms. Also explain the roles played by the walls and mirror ends. Go over the four characteristics of laser light discussed earlier and tell how each is achieved.

SUGGESTIONS

1. Assignments
   a. To test for understanding of the angular momentum quantum numbers, go over questions 1 through 5 and assign problems 5 and 7. To stress the connection between angular momentum and magnetic dipole moment, assign problem 9.
   b. To discuss the Stern-Gerlach experiment in more detail, include question 10. Also assign problems 12 and 14.
   c. Use problems 19 and 21 (two-dimensional trap) or problems 23 and 24 (three-dimensional trap) to test for understanding of the Pauli exclusion principle. To emphasize the role played by spin in the building of the periodic table, ask problem 17. To help in the discussion of the periodic table, assign problems 25 and 27.
   d. The existence of a minimum wavelength in the continuous x-ray spectrum provides an argument for the particle nature of light. Either discuss this or see if the students can devise the argument. Assign problems 31 and 32. After discussing characteristic x-ray lines and Moseley plots, ask questions 11 and 12. Assign problems 33, 37, and 40.
   e. Ask questions 13 and 14 to see if students understand how lasers work. Also assign problems 45, 51, and 52. Populations of states are covered in problems 46, 48, and 55.

2. Demonstrations

   Zeeman effect: Freier and Anderson MPc1.

3. Books and Monographs

   Resource Letters, Book Five; American Association of Physics Teachers (AAPT, One Physics Ellipse, College Park MD 20740–3845; www.aapt.org). Contains an exhaustive list of journal articles on atomic physics.

4. Audio/Visual
   a. *Structure of the Atom*; VHS video tape ; Films for the Humanities and Sciences (PO Box 2053, Princeton, NJ 08543–2053; www.films.com).
   b. *Atoms and Molecules*; VHS video tape, DVD; Films for the Humanities and Sciences (PO Box 2053, Princeton, NJ 08543–2053; www.films.com).

5. Computer Software
   a. *Modern Physics* from Exploration of Physics Volume II; Windows and Macintosh; Physics Curriculum & Instruction Physics Curriculum & Instruction (22585 Woodhill Drive, Lake-

ville, MN 55044; www.PhysicsCurriculum.com). Includes sections on models of the atom, the atomic nucleus and Rutherford's experiment.

b. *Atomic Orbitals*; interactive CD-ROM; Films for the Humanities and Sciences (see above for address).

Chapter 41 THE CONDUCTION OF ELECTRICITY IN SOLIDS

BASIC TOPICS

I. Electron energy bands.

A. Explain that a crystalline solid is a periodic arrangement of atoms and show some ball and stick models or Fig. 41–1.

B. Explain that energy levels for electrons in crystalline solids are grouped into bands with the levels in any band being nearly continuous and with gaps of unallowed energies between. Remark that bands are produced when atoms are brought close together. Wave functions for outer electrons then overlap and extend throughout the solid. Show Fig. 41–3 and remark that low energy bands are narrow since the wave functions are highly localized around nuclei and overlap is small. High energy bands are wide because overlap is large. When the atoms are close together, outer-shell electrons are influenced by many atoms rather than just one.

C. Remind students that since the Pauli exclusion principle holds, the lowest total energy is achieved when electrons fill the lowest states with one electron in each state. Thus, at $T = 0\,\mathrm{K}$, all states are filled up to a maximum energy.

D. Remark that for a metal at $T = 0\,\mathrm{K}$, the highest occupied state is near the middle of a band, while for an insulator or semiconductor, it is at the top of a band.

E. Write down the Fermi-Dirac occupancy probability $P(E)$, given by Eq. 41–6, and state that it gives the thermodynamic probability that a state with energy E is occupied. State that E_F is a parameter, called the Fermi energy, that is different for different materials. Show that for $T = 0\,\mathrm{K}$, $P(E) = 1$ for $E < E_F$ and $P(E) = 0$ for $E > E_F$. To give a numerical example, calculate the probabilities of occupation for states 0.1 and 1 eV above the Fermi energy, then 0.1 and 1 eV below, at room temperature. Graph $P(E)$ versus. E for $T = 0$ and for $T > 0$. See Fig. 41–6. Also show the graph for a still higher temperature and point out that the central region (from $P = 0.9$ to $P = 0.1$, say) widens. This quantitatively describes the thermal excitation of electrons to higher energy states. Remark that the Fermi-Dirac occupancy probability is valid for any large collection of electrons, including the collections in metals, insulators, and semiconductors.

II. Metallic conduction.

A. Write down Eq. 26–25 for the resistivity and remark that n is the concentration of conduction electrons and τ is the mean time between collisions of electrons with atoms. Ask students to review Section 26–6. Remark that a low resistivity results if the electron concentration is large or the mean free time is long. In a rough way, if there are few collisions per unit time, then the mean free time is long and the electrons are accelerated by the electric field for a long time before colliding, so the drift velocity is large. Remark that quantum physics must be used to determine n and τ.

B. Explain that for metals, the energies of conduction electrons (those in partially filled bands) are primarily kinetic and to a first approximation, we may take the electrons to be trapped in a box the size of the sample. The so-called free electron model of a metal takes the potential energy to be zero in the box.

C. Define the density of states function $N(E)$ and the density of occupied states function $N_o(E)$. Explain that $N_o(E) = N(E)P(E)$ and that the total electron concentration in a metal is given by $n = \int N(E)P(E)\,\mathrm{d}E$. In principle, this equation can be solved for the Fermi energy as a function of temperature. State that for nearly free electrons in a metal, $N(E)$ is given by Eq. 41–5, and that the Fermi energy is given by Eq. 41–9. Evaluate the expression for copper and show that E_F is about 7 eV above the lowest free electron energy. Strictly, this is the result for $T = 0$ but the variation of E_F and n with temperature is not important in a first approximation for metals.

D. Explain that the electric current is zero when no electric field is present because states for which the velocities are $+\vec{v}$ and $-\vec{v}$, for example, have the same energy. If one is filled, then so is the other. Thus the average velocity of the electrons vanishes. A current arises in an electric field because the electrons accelerate: they tend to make transitions within their band to other states such that the changes in their velocities are opposite to the field.

E. Explain that the acceleration caused by an electric field does not continue indefinitely because the electrons are scattered by atoms of the solid. As a result, the electron distribution distorts only slightly. Some states with energy slightly greater than E_F and velocity opposite the field become occupied while some states with energy slightly less than E_F and velocity in the direction of the field become vacant. Electrons with energy E_F have speeds v_F given by $E_F = \frac{1}{2}mv_F^2$ but the average speed (the drift speed) is considerably less because most electrons can be paired with others moving with the same speed in the opposite direction.

F. Explain that a steady state is reached and that the drift velocity is then proportional to the applied electric field. Only electrons near the Fermi energy suffer collisions and the additional velocities they obtain from the field between collisions are insignificant compared to their velocities in the absence of the field. Thus the mean free time is essentially independent of the field and Ohm's law is valid.

G. State that electrons in a perfectly periodic lattice do not suffer collisions, a result that is predicted by quantum physics. Collisions with the atoms occur because they are vibrating. Collisions also occur if the solid contains impurities or other imperfections. As the temperature increases, vibrational amplitudes of the atoms increase and so does the number of collisions per unit time. As a result, the mean free time becomes smaller. This explains the increase with temperature in the resistivity of a metal.

III. Insulators and semiconductors.

A. Explain that a filled band cannot contribute to an electric current because the average electron velocity is always zero, even in an electric field. State that insulators and semiconductors have just the right number of electrons to completely fill an integer number of bands and that, in the lowest energy state, all bands are either completely filled or completely empty. For metals, on the other hand, the highest occupied state is near the middle of a band. Metals always have partially filled bands. Show Fig. 41–4 and identify the valence and conduction bands for an insulator.

B. Explain that as the temperature is raised from $T = 0\,\mathrm{K}$, a small fraction of the electrons in the valence band of an insulator or semiconductor are thermally excited across the gap into the conduction band. For a semiconductor, the gap is small (about 1 eV) and at room temperature, both bands can contribute to the current. The conductivity, however, is still small compared to that of a metal. For an insulator, the gap is large (more than 5 eV), so the number of promoted electrons is extremely small and the current is insignificant for laboratory fields. Explain that silicon and germanium are the only elemental semiconductors although there are many semiconducting compounds. Carbon is a prototype insulator, with a gap of 5.5 eV. Compare with silicon, which has a gap of 1.1 eV. Resistivities of metals and semiconductors are compared in Table 41–1.

C. When electrons are promoted across the gap, they contribute to the current in an electric field. The valence band becomes partially filled and electrons there also contribute. It is usually convenient to think about the few empty states in this band rather than the large number of electrons there. That is, the electrons in a nearly filled band are replaced by a collection of fictitious particles, called holes, so that the properties of the hole system are identical to the properties of the electron system they replace. Holes behave as if they were positive charges. In contrast to electrons, holes drift in the direction of the electric field. Compare the carrier concentrations of metals and semiconductors at room temperature. See Table 41–1.

D. Explain the different signs for the temperature coefficients of resistivity, also given in Table 41–1. Explain that for both metals and semiconductors near room temperature, the mean free time decreases with increasing temperature. For metals, the electron concentration is essentially constant but for semiconductors, n increases dramatically with temperature as electrons are thermally promoted across the gap. This effect dominates and the resistivity of an intrinsic semiconductor decreases with increasing temperature.

E. Explain that the proper kind of replacement atoms (donors) can increase the number of electrons in the conduction band and another kind (acceptors) can increase the number of holes in the valence band. They produce n and p type semiconductors, respectively. By considering the number of electrons in their outer shells, explain why phosphorus is a donor and aluminum is an acceptor. Point out that wave functions for impurity states are highly localized around the impurity and so do not contribute to the conductivity. Go over Sample Problem 41–6, which shows that only a relatively small dopant concentration can increase the carrier concentration enormously. Doped semiconductors are used in nearly all semiconducting devices.

IV. Semiconducting devices.

A. Show a commercial junction diode and draw a graph of current vs. potential difference (Fig. 41–12). Include both forward and back bias. Explain that it is a rectifier, with high resistance for current in one direction and low resistance for current in the other direction. Demonstrate the i-V characteristics by placing a diode across a variable power supply and measuring the current for various values of the potential. Reverse the potential to show the rectification.

B. Describe a p-n junction and remark that the diffusion of carriers leaves a small depletion region, nearly devoid of carriers, straddling the metallurgical junction. Explain the origin of the electric field in the depletion region and the origin of the contact potential. Stress that the field is due to uncovered replacement atoms, positive donors on the n side and negative acceptors on the p side.

C. Describe a diffusion current as one that arises because particles diffuse from regions of high concentration toward regions of low concentration. Explain that this motion results from the random motion of the particles. More particles leave a high concentration region simply because there are more particles there, not because they are driven by any applied force. State that the diffusion current for both electrons and holes in an unbiased p-n junction is from the p to the n side, against the contact electric field. Point out that the drift current is from the n toward the p side and that the diffusion and drift currents cancel when no external field is applied. Point out the depletion zone and the currents on Fig. 41–14.

D. Draw a circuit with a battery across a p-n junction, the positive terminal attached to the n side. Explain that this is a back bias. The internal electric field is now larger, the barrier to diffusion is higher, and the reverse current is extremely small. Also explain that the width of the depletion zone is increased by application of a reverse bias.

E. Draw the circuit for forward bias. The internal electric field is now smaller, the barrier to diffusion is lower, and the current increases dramatically. The depletion zone narrows.

F. Explain how diodes are used for rectification and how light-emitting diodes work.

G. Optional. Explain how a field-effect transistor works. Explain the mechanism by which the gate voltage of a MOSFET controls current through the channel. Remove the covers from a few chips and pass them around with magnifying glasses for student inspection.

SUGGESTIONS

1. Assignments
   a. Questions 1 and 2 deal with crystal structure. Ask them if you include more than a passing mention of this topic. Also consider problem 43.
   b. Questions 3 and 4 deal with electrons in solids and questions 5 and 6 deal with energy bands.
   c. Assign problems 2 and 7 in connection with the Fermi energy of a metal. Assign problem 6 in connection with the density of states for a metal. Assign some of problems 8, 11, and 16 in connection with the occupancy probability. Also assign problem 17 and either problem 19 or 22. The justification for the free electron model of a metal is covered in problem 12.
   d. Problem 30 should be assigned or covered in class when you discuss intrinsic semiconductors.
   e. Doped semiconductors are considered in questions 9, 10, and 14. Discuss them and then assign problems 31 and 32. Also consider problem 33.
   f. *p-n* junctions are the subject of questions 12, 13, and 15. Be sure to assign or discuss problem 36. If you include LED's, assign either problem 37 or 38 and if you discuss field-effect transistors, assign problem 40.
2. Audio/Visual
   Condensed Matter; from Cinema Classics DVD 6: Angular Momentum and Modern Physics; available from Ztek Co. (PO Box 11768, Lexington, KY 40577–1768, www.ztek.com).
3. Computer Project
   Ask students to use a root finding program to carry out calculations of the electron concentration in the conduction band and hole concentration in the valence band of both intrinsic and doped semiconductors. Then ask them to calculate the contact potential for a *p-n* junction with given dopant concentrations.

Chapter 42 NUCLEAR PHYSICS

BASIC TOPICS

I. Nuclear properties.

A. Explain that the nucleus of an atom consists of a collection of tightly bound neutrons, which are neutral, and protons, which are positively charged. A proton has the same magnitude charge as an electron. Define the term nucleon and state that the number of nucleons is called the mass number and is denoted by A, the number of protons is called the atomic number and is denoted by Z, and the number of neutrons is denoted by N. Point out that $A = Z + N$. Remark that nuclei with the same Z but different N are called *isotopes.* The atoms have the same chemical properties and the same chemical symbol. Show a wall chart of the nuclides. Refer to Table 42–1 when discussing properties of nuclides.

B. Explain that one nucleon attracts another by means of the strong nuclear force and that this force is different from the electromagnetic force. It does not depend on electrical charge and is apparently the same for all pairs of nucleons. It is basically attractive; at short distances (a few fm), it is much stronger than the electrostatic force between protons, but it becomes very weak at larger distances. Two protons exert attractive strong forces on each other only at small separations but they exert repulsive electric forces at all separations. Because of the short range, a nucleon interacts only with its nearest neighbors via the strong force. Because the nucleus is small, the much stronger nuclear force dominates and both protons and neutrons can be bound in stable nuclei. Explain that the force is thought to be a manifestation of the strong force that binds quarks together to form nucleons.

C. Show Fig. 42–4 and point out the $Z = N$ line and the stability zone. Explain why heavy nuclei have more neutrons than protons. Also explain that unstable nuclei are said to be radioactive and convert to more stable ones with the emission of one or more particles. Show Fig. 42–12 and point out the stable and unstable nuclei.

D. Explain that although the surface of a nucleus is not sharply defined, nuclei can be characterized by their mean radii and these are given by $r = r_0 A^{1/3}$, where $r_0 \approx 1.2\,\text{fm}$ ($1\,\text{fm} = 10^{-15}\,\text{m}$). Stress how small this is compared to atomic radii. Show that this relationship between r and A leads to the conclusion that the mass densities of all nuclei are nearly the same. Show that the density of nuclear matter is about $2 \times 10^{17}\,\text{kg/m}^3$.

E. Explain that the mass of a nucleus is less than the sum of the masses of its constituent nucleons, well separated. The difference in mass is accounted for by the binding energy through $\Delta E_{\text{be}} = \Delta m\, c^2$, where Δm is the magnitude of the mass difference. The binding energy is the energy that must be supplied to separate the nucleus into well separated particles, at rest. Generalize this equation to the case of a nucleus with Z protons and N neutrons: $\Delta E_{\text{be}} = Zm_pc^2 + Nm_nc^2 - mc^2$. Also define the binding energy per nucleon ΔE_{ben}. Show Fig. 42–6 and point out that there is a region of greatest stability, near iron. For heavier nuclei, the binding energy per nucleon falls slowly but nevertheless does fall. For lighter nuclei, the binding energy per nucleon rises rapidly with increasing mass number. Explain the terms fission and fusion, then remark that the high mass number region is important for fission processes, the low mass number region is important for fusion processes.

F. State that nuclear masses are difficult to measure with precision, so masses are usually expressed in atomic mass units: $1\,\text{u} = 1.6605 \times 10^{-27}\,\text{kg}$. Also state that tables usually give atomic rather than nuclear masses and so include the mass of the atomic electrons. Show that the electron masses cancel in the expression for the binding energy. Give the mass-energy conversion factor: $931.5\,\text{MeV/u}$.

G. Explain that nuclei have discrete energy levels, with separations on the order of MeV. An excited nucleus can make a transition to a lower energy state with the emission of a photon, typically in the gamma ray region of the spectrum. Explain that a nucleus may have intrinsic angular momentum and a magnetic moment. Spins are on the order of $\hbar$, like atomic electrons, but moments are much less than electron moments because the mass of a nucleon is much greater than the mass of an electron.

II. Radioactive decay.

A. Explain that nuclei may be either stable or unstable and those that are unstable ultimately decay to stable nuclei. Decay occurs by spontaneous emission of an electron (e^-), a positron (e^+), a helium nucleus (α), or larger fragments. The resulting nucleus has a different complement of neutrons and protons than the original nucleus.

B. Explain that decay is energetically favorable if the total mass of the products is less than the original mass. Define a decay symbolically as $X \to Y + b$, where X is the original nucleus, Y is the daughter nucleus, and b is everything else. Point out that charge, number of nucleons, and energy are all conserved. Define the disintegration energy by $Q = (m_X - m_Y - m_b)c^2$. Note that an appropriate number of electron rest energies must be added or subtracted so that atomic masses may be used. Note also that Q must be positive for spontaneous decays and Q appears as the kinetic energy of the decay products or as an excitation energy if the daughter nucleus is left in an excited state.

C. Explain that each radioactive nucleus in a sample has the same chance of decaying and that the decay rate or activity ($R = -dN/dt$) is proportional to the number of undecayed nuclei present at time t: $-dN/dt = \lambda N$. This has the solution $N = N_0 \exp(-\lambda t)$, so the decay rate is given by $R = R_0 \exp(-\lambda t)$. Define the term half-life and show that it is related to λ by $T_{1/2} = (\ln 2)/\lambda$. Go over Sample Problems 42–4 and 42–5, show Fig. 42–8, and point out the half-life. Emphasize that R decreases by a factor of two in every half-life interval. Define the becquerel unit.

D. Discuss α decay. Write down Eq. 42–22 and explain that the daughter nucleus has two fewer neutrons and two fewer protons than the parent. Go over Sample Problem 42–6 to show that α decay is energetically favorable for ^{238}U. Show Fig. 42–9 and explain that the deep potential well is due to the strong attraction of the residual nucleus for the nucleons in the α particle, while the positive potential is due to Coulomb repulsion. The two forces form a barrier to decay. Explain that the α particle can tunnel through the barrier. Its wave function does not go to zero at the inside edge, but rather has a finite amplitude in the barrier and on the outside. There is a non-zero probability of finding the α particle on the outside. High, wide barriers produce a small probability of tunneling and a long half-life while low, narrow barriers produce the opposite effect. Note the wide range of half-lives that occur in nature (Table 42–2).

E. Discuss β decay. Explain that a neutron can transform into a proton with the emission of an electron and a neutrino (strictly, an antineutrino) and that a proton can transform into a neutron with the emission of a positron and a neutrino. Mention the properties of a neutrino: massless, neutral, weakly interacting. Only protons bound in nuclei can undergo β decay but both free and bound neutrons can decay. These transformations lead to decays such as the ones given in Eqs. 42–24 and 42–25. Explain that the energy is shared by the decay products and that the electrons or positrons show a continuous spectrum of energy up to some maximum amount (see Fig. 42–10). Explain that neutron rich nuclides generally undergo β^- decay while proton rich nuclides generally undergo β^+ decay. This is a mechanism for bringing the nucleus closer to stability. Carefully discuss the inclusion of electron rest energies in the equation for Q so that atomic masses can be used. In particular, show that in β^- decay there is no excess electron mass but in β^+ decay there is an excess of two electron masses.

F. Define the units used to describe radiation dosage: grey and sievert.

SUPPLEMENTARY TOPICS

1. Radioactive dating. If time permits, cover this topic as an application of radioactive decay processes.
2. Nuclear models. This topic adds a little breadth to the nuclear physics section and helps students understand nuclear processes a little better.

SUGGESTIONS

1. Assignments
   a. Nuclear constitution is covered in problems 4, 83, and 84. Nuclear radius and density are covered in problems 7, 9, 87.
   b. Include questions 2, 3, 6, 7, and 8 in the discussion of nuclear stability and nuclear binding. These ideas are illustrated in problems 6, 8, 14, 15, and 17. Be sure to include problem 6 if you intend to discuss fission (Chapter 42).
   c. Questions 9 through 14 deal with the decay law, activity, and half-life. Discuss a few. Problems 22 and 24 cover basic half-life calculations. Problems 26, 27, 67, and 69 involve half-life calculations drawn from many interesting applications. Assign some of them.

d. Following the discussion of α decay, students should be able to answer question 15. The disintegration energy and barrier height are covered in problems 44 and 81. Problem 43 asks students to take into account the recoil of the residual nucleus. Problem 40 shows why alphas are emitted rather than well-separated nucleons.

e. After discussing β decay, assign one or more of problems 45, 47, and 77. Problem 46 shows that β particles do not exist inside nuclei before decay occurs. The β decay discussion can be broadened somewhat by including the recoil of the nucleus. See problem 50.

2. Demonstrations

Geiger counter: Freier and Anderson MPa2.

3. Books and Monographs

Resource Letters, Book Four; American Association of Physics Teachers (AAPT, One Physics Ellipse, College Park MD 20740–3845; www.aapt.org). Contains an exhaustive list of journal articles on nuclear physics.

4. Audio/Visual

a. *Radioactive Decay*; *Scintillation Spectrometry*; from the AAPT Miller collection of single-concept films; DVD; available from Ztek Co. (PO Box 11768, Lexington, KY 40577–1768, www.ztek.com) and from the AAPT (see above for address).

b. *Nuclear Physics*; from Cinema Classics DVD 6: Angular Momentum and Modern Physics; available from Ztek Co. and from the AAPT (see above for addresses).

c. *Rutherford Scattering*, *Thomson Model of the Atom*; from the AAPT collection 1 of single-concept films; DVD; available from Ztek Co. (see above for address).

d. *Nuclear Physics*; VHS video tape; Films for the Humanities and Sciences (PO Box 2053, Princeton, NJ 08543–2053; www.films.com). Contains sections on radioactivity and nuclear energy.

e. *The Rutherford Model*; VHS video tape (10 min); Films for the Humanities & Sciences (see above for address).

f. *Nuclear Science Wall Chart*; available from the AAPT (see above for address). A poster describing basic nuclear processes, with applications to cosmology.

5. Computer Software

a. *Chart of the Nuclides: A Tutorial*; Philip DiLavore; available from Physics Academic Software (Centennial Campus, 940 Main Campus Drive, Suite 210, Raleigh, NC 27606–5212; www.aip.org/pas). The chart is shown and a click on any nuclide produces information about that nuclide.

6. Laboratory

Many of the following experiments make use of a Geiger tube and scalar.

a. Bernard Experiment 48: *The Characteristics of a Geiger Tube* describes how students can systematically investigate the plateau and resolving time of a Geiger tube. They also learn how to operate a scalar. Consider prefacing the other experiments either with this experiment or with a demonstration of the same material.

b. Meiners Experiment 14–7: *Half-Life of Radioactive Sources.* A Geiger counter and scalar are used to measure the decay rate as a function of time for indium, cesium 137, and barium 137. For the first and last, the data is used to compute the half-life. Other sections explain how to use a microcomputer to collect data and make the calculation and how to use a emanation electroscope to collect data. A neutron howitzer or minigenerator is required to produce radioactive sources.

c. Bernard Experiment 52: *Measurement of Radioactive Half-Life.* Nearly the same as Meiners 14–7. The generation of sources with short half-lives is discussed.

d. Meiners Experiment 14–6: *Absorption of Gamma and Beta Rays.* The particles are incident on sheets of aluminum and the number that pass through per unit time is counted. Students make a logarithmic plot of the counting rate as a function of the thickness of the aluminum and determine the range of the particles.

Chapter 43 ENERGY FROM THE NUCLEUS

BASIC TOPICS

I. The fission process.

A. Refer back to the binding energy per nucleon vs. A curve (Fig. 42–6). It suggests that a massive nucleus might split into two or more fragments nearer to iron, thereby increasing the total binding energy. Each fragment is more stable than the original nucleus. This is the fission process.

B. Remark that many massive nuclei can be rendered fissionable by the absorption of a thermal neutron. Such nuclei are called fissile. Give the example $^{235}\text{U} + \text{n} \rightarrow {}^{236}\text{U} \rightarrow \text{X} + \text{Y} + b\text{n}$. Explain that a thermal neutron ($\approx 0.04\,\text{eV}$) is absorbed by a ^{235}U nucleus and together they form the intermediate fissionable ^{236}U nucleus. This nucleus splits into two fragments (X and Y) and several neutrons. The sequence of events is illustrated in Fig. 43–2. Point out ^{236}U on Fig. 42–6. The disintegration energy for one possible fission event is calculated in Sample Problem 43–1.

C. Explain that different fission events, starting with the same nucleus, might produce different fragments. The fraction of events that produce a fragment of a given mass number A is graphed in Fig. 43–1. Point out that fragments of equal mass occur only rarely. Explain that the parent nucleus is neutron rich, the initial fragments are neutron rich, and that the initial fragments expel neutrons to produce the fragments X and Y. These generally decay further by β emission and some may emit delayed neutrons following β decay.

D. Show Fig. 43–3 and explain that the parent nucleus starts in the energy well near $r = 0$. The incoming neutron must supply energy to start the fission process. The required energy is slightly less than E_b since tunneling can occur. Point out the energy Q released by the process. Point out Table 43–2 and explain that E_n is the actual energy supplied by an incoming thermal neutron. Point out nuclides in the table for which fission does not occur.

E. Write out several fission modes for ^{235}U and note that on average more than one neutron is emitted. Explain that some neutrons come promptly while others come from later decays (the delayed neutrons). Point out that the average mode yields $Q \approx 200\,\text{MeV}$, of which 190 MeV or so appears as the kinetic energy of the fission fragments and 10 MeV goes to the neutrons.

II. Fission reactors.

A. Note that to have a practical reactor, the fission process must be self sustaining, once started. Also, there must be a way to control the rate of the process and to stop it, if desired.

B. To be self-sustaining, a chain reaction must occur: neutrons from one fission event trigger another. The neutrons emitted from a typical fission event share about 5 to 10 MeV energy and they must be slowed to thermal speeds to be useful. Some sort of moderator, often water, is used.

C. Explain that on average about 2.5 neutrons are produced per fission event. Describe in detail what happens to them. Some leak out of the system, some of the slowed neutrons are captured by ^{238}U, some are captured by fission fragments, and the rest start fission in ^{235}U. Fig. 43–4 gives some typical numbers.

D. Explain the terms critical, subcritical, and supercritical. Note that the control rods, which absorb slow neutrons, are used to achieve criticality. Point out that without the delayed neutrons, control would not be possible since time is needed to move the rods into or out of the reactor.

E. Define the multiplication factor k as the ratio of the number of neutrons present at one time that participate in fission to the number present in the previous generation. Remark that $k = 1$ for critical operation, $k < 1$ for subcritical operation, and $k > 1$ for supercritical operation. Explain that k is determined by the positions of the control rods. The rods are pulled out to increase k and thereby increase power output. They are pushed in to decrease k and thereby decrease power output. When the desired power level is obtained, the rods are positioned so $k = 1$.

F. Use Fig. 43–5 to describe the essential features of a nuclear power plant. Apart from the fact that the fission process is used to heat water or generate steam, this schematic could apply to any power plant. Remark on the special problems attendant on nuclear plants.

III. Fusion.

A. Return to Fig. 42–6 and remark that if two low-mass nuclei are combined to form a nucleus with greater mass, the binding energy is increased considerably. The energy is transformed to the kinetic energy of the resulting nucleus and any particles emitted. In order to carry out the fusion process, the nuclei must be given sufficient energy to overcome the electrostatic repulsion of their protons. They can then approach each other closely enough for the attraction of the strong force to bind them. For ^{3}He, the height of the barrier is about 1 MeV. Since tunneling is possible, fusion can occur at slightly smaller energies.

B. To achieve a large number of fusion events, hydrogen or helium gases must be raised to high temperatures. Even at the temperature of the Sun, only a small fraction of the nuclei have sufficient energy to overcome the Coulomb barrier. Go over Fig. 43–10.

C. Discuss fusion in the Sun. Remark that the core of the Sun is 35% hydrogen and 65% helium by mass. Outline the principal proton-proton cycle: two protons fuse to form a deuteron, a positron, and a neutrino. A deuteron fuses with a proton to form ^{3}He and two ^{3}He nuclei fuse to form ^{4}He and two protons. Remark that six protons are consumed and two are produced for a net loss of four. The two positrons are annihilated with electrons to produce photons. Note that the process can be simplified to $4\mathrm{p} + 2\mathrm{e}^- \to \alpha + 2\nu + 6\gamma$ and the Q value is computed from the mass difference between the alpha particle and the four protons.

D. Calculate the energy released. Show that $Q = 26.7\,\mathrm{MeV}$ and note that the neutrinos take about 0.5 MeV with them when they leave the Sun. Point out that the fusion process produces about 20 million times as much energy per kg of fuel as the burning of coal.

E. If time permits, discuss helium burning. Use the solar constant to calculate the rate at which the Sun converts mass to energy. Speculate on the future of the Sun. Also mention the carbon cycle, which is essentially the same as the proton-proton cycle. Carbon acts as a catalyst.

F. Discuss controlled thermonuclear fusion. Explain that deuteron-deuteron and deuteron-triton fusion events are being studied. Point out that high particle concentrations at high temperatures must be maintained for sufficiently long times in order to make the process work. Discuss some means for doing this: the tokamak for plasma confinement by magnetic fields, inertial confinement, and laser fusion. State that the right combination has not yet been achieved but work continues.

SUGGESTIONS

1. Assignments
   a. After explaining the basic fission process, test for understanding with questions 1, 3, 4, and 6. Also assign problems 2, 11, and 12.
   b. Following the discussion of the fission reactor, ask questions 7 and 8. To help students understand the role of a moderator, assign problem 21. To illustrate the role of the control rods, assign problems 18 and 23.
   c. Following the discussion of the basic fusion process, assign problems 28 and 29. Also ask question 12.
   d. To help students understand the fusion process as an energy source, assign problems 35 and 36. The carbon cycle is covered in problem 40.
2. Demonstrations
   Chain reaction: Freier and Anderson MPa1.
3. Books and Monographs
   a. *Fission Reactors*; edited by Melvin M. Levine.; available from the American Association of Physics Teachers (AAPT, One Physics Ellipse, College Park MD 20740–3845; www.aapt.org). Covers both physics and engineering aspects.
   b. *Introduction to Nuclear Fusion Power and the Design of Fusion Reactors*; edited by J.A. Fillo and P. Lindenfeld; available from the AAPT (see above for address). Covers both physics and engineering aspects.

Chapter 44 QUARKS, LEPTONS, AND THE BIG BANG

BASIC TOPICS

I. The particle "zoo".
   A. Show a list of particles already familiar to students. Include the electron, proton, neutron, and neutrino, then add the muon and pion. Explain that many other particles have been discovered in cosmic ray and accelerator experiments. To impress students with the vast array of particles and the enormous collection of data, make available to them a Review of Particle Properties paper, published roughly every two years in Reviews of Modern Physics.
   B. Explain that many new particles are discovered by bombarding protons or neutrons with electrons or protons and show a picture of a detector or a bubble chamber picture, such as Fig. 44–3. State that the picture shows tracks of charged particles in a strong magnetic field, hence the curvature. Remind students that the radius of curvature can be used to find the momentum of a particle if the charge is known. Indicate the collision point and emphasize that the new particles were not present before the collision: the original particles disappear and new particles appear. In most cases, the total rest energy after the collision is much greater than the total rest energy before. Kinetic energy was converted to rest energy.
   C. Mention that a few particles seem to be stable (electron, proton, neutrino) but most decay spontaneously to other particles. Point out decays on a bubble chamber picture. Explain the statistical nature of decays and remind students of the meaning of half-life. Examples: $\mathrm{n} \rightarrow \mathrm{p} + \mathrm{e}^- + \nu$, $\pi^+ \rightarrow \mu^+ + \nu$.
   D. Explain that for each particle there is an antiparticle with the same mass. A charged particle and its antiparticle have charge of the same magnitude but opposite sign. Their magnetic moments are also opposite, relative to their angular momenta. A particle and its antiparticle can annihilate each other, the energy (including rest energy) being carried by photons or other particles produced in the annihilation. Example: $\mathrm{e}^+ + \mathrm{e}^- \rightarrow \gamma + \gamma$.

Antiparticles (except the positron) are denoted by a bar over the particle symbol. Some uncharged particles (such as the photon and π^0) are their own antiparticles. The universe seems to be made of particles, not antiparticles.

II. Particle properties.

A. Spin angular momentum. Remind students that many particles have intrinsic angular momentum. Explain that the magnitude is always an integer or half integer times $\hbar$. Remark that particles with half integer spins are called fermions while particles with integer spins are called bosons. Remind students of the Pauli exclusion principle and its significance, then state that fermions obey the principle while bosons do not. Give examples: electrons, protons, neutrons, and neutrinos are fermions; photons, pions, and muons are bosons. Remark that spin angular momentum is conserved in particle decays and interactions. An odd number of fermions, for example, cannot interact to yield bosons only.

B. Charge. Remind students of charge quantization and charge conservation. Even if the character and number of particles change in an interaction, the total charge before is the same as the total charge after. Example: $\mathrm{n} \rightarrow \mathrm{p} + \mathrm{e}^- + \nu$.

C. Momentum and energy. Explain that energy and momentum are conserved in decays and interactions. Give masses and rest energies for the particles in the list of part I. Give the expressions for relativistic energy and momentum in terms of particle velocity.

D. Forces. Remark that all particles interact via the force of gravity and all charged particles interact via the electromagnetic force. The force of gravity is too weak to have observable influence at energies presently of interest. Remark that there are two additional forces, called strong and weak, respectively. Remind students of the role played by the strong force in holding a nucleus together and the role played by the weak force in beta decay. These topics were covered in Chapter 42. Note that lifetimes for strong decays are about 10^{-23} s, lifetimes for electromagnetic decays are about 10^{-14} to 10^{-20} s, and lifetimes for weak decays are about 10^{-8} to 10^{-13} s.

E. Leptons and hadrons. State that particles that interact via the strong force (as well as the weak) are called hadrons and that particles that interact via the weak force but not the strong are called leptons. List the leptons (electron, muon, tauon, and their neutrinos) and explain that a different neutrino is associated with each of the leptons. Remark that the neutrino that appears following muon decay is not the same as the neutrino that appears following beta decay. Neutrinos are labeled with subscripts giving the associated lepton: ν_e, ν_μ, and ν_τ.

F. Lepton numbers. State that a lepton number is associated with each lepton family, with particles in the family having a lepton number of +1, antiparticles in the family having a lepton number of −1, and all other particles having a lepton number of 0. Explicitly give the electron lepton numbers and the muon lepton numbers for members of the electron and muon families. Explain that each lepton number is conserved in all decays and interactions. Give some beta and muon decay examples.

G. Baryons and mesons. Remark that some strongly interacting particles (proton, neutron) are fermions and are called baryons while others (pion, kaon) are bosons and are called mesons. Explain that a baryon number of +1 is assigned to each baryon particle, a baryon number of −1 is assigned to each baryon antiparticle, and a baryon number of 0 is assigned to each meson. Then baryon number is conserved in exactly the same way charge is conserved: the total baryon number before a collision or decay is the same as the total baryon number after. This conservation law (and conservation of energy) accounts for the stability of the proton, the baryon with the smallest mass. There is some speculation that baryon number is not strictly conserved and that protons may decay to other particles, but the half-life is much longer than the age of the universe. Some physicists are trying to observe proton decay.

H. Strangeness. Explain that another quantity, called strangeness, is conserved in strong interactions. Neutrons and protons have $S = 0$, K^- and Σ^+ have $S = -1$. A particle and its antiparticle have strangeness of opposite sign. Conservation of strangeness allows $\pi^+ + p \rightarrow K^+ + \Sigma^+$ but prohibits $\pi^+ + p \rightarrow \pi^+ + \Sigma^+$, for example.

III. Quarks and the eight-fold way.

A. Show the eight-fold way patterns (Figs. 44–3 and 44–4) and point out the oblique axes. Remark that these patterns are to fundamental particles as the periodic table of chemistry is to atoms and that they have provided clues to the existence of particles not previously observed.

B. Remark that the properties of strongly interacting particles can be explained if we assume they are made up of more fundamental particles (called quarks). State that there are six quarks, not including the antiquarks, and list them and their properties (Table 44–5). Particularly note the fractional charge and baryon number. Baryons are constructed of three quarks, antibaryons of three antiquarks, and mesons of a quark and antiquark. Show that uud has the charge, spin, and baryon number of a proton and udd has the charge, spin, and baryon number of a neutron. Give the quark content of the spin 1/2 baryons (Fig. 44–5*a*) and the quark content of the spin 0 mesons. Point out that the strange quark accounts for the strangeness quantum number. Mention the charm, bottom, and top quarks and point out they lead to other particles.

C. Explain that the existence of internal structure allows for excited states: there are other particles with exactly the same quark content as those mentioned in III B but they are different particles because the quarks have different motions. The additional energy results in greater mass. Contrast this with the leptons, which have no internal structure. Quarks and leptons are believed to be truly fundamental.

D. Messenger particles. Explain that particles interact by exchanging other particles. Electromagnetic interactions proceed by exchange of photons, for example. Also explain that energy may not be conserved over short periods of time but this is consistent with the uncertainty principle. State that the strong interaction proceeds by the exchange of gluons by quarks and the weak interaction proceeds by the exchange of Z and W particles by quarks and leptons. The interaction that binds nucleons in a nucleus is the same as the interaction that binds quarks in a baryon or meson. In the former case, gluons are exchanged between quarks of different nucleons; in the latter, they are exchanged between quarks of the same baryon or meson.

E. Explain that quarks are conserved in strong interactions. Either the original quarks are rearranged to form new particles or quark-antiquark pairs are created, then both the original and the new quarks are rearranged. This accounts for conservation of strangeness. Example: $K^+ \rightarrow K^0 + \pi^+$ ($u\bar{s} \rightarrow d\bar{s} + u\bar{d}$). A $d\bar{d}$ pair is formed. The d quark couples to the $\bar{s}$ quark to form a K^0 and the $\bar{d}$ quark couples to the u quark to form a π^+. Contrast this with the weak interaction, which can change one type quark into another. Illustrate with beta decay, in which a d quark is converted to a u quark.

F. Explain that quarks have another property, called color. Color produces the gluon field, much as charge produces the electromagnetic field: baryons interact via the strong interaction because quarks have color. Be sure students understand that "color" in this context has nothing to do with the frequency of light. Mention that gluons carry color. The emission or absorption of a gluon changes the color of a quark. Contrast this with the electromagnetic interaction: a photon does not carry charge.

IV. The Big Bang and cosmology.

A. Remind students of the doppler shift for light and state that spectroscopic evidence convinces us that on a large scale, matter in the universe is receding from us and we are led to conclude that the universe is expanding. Write down Hubble's law and give the approximate value of the Hubble parameter: $63.0\,\text{km}/(\text{s}\cdot\text{Mpc}$ ($= 19.3\,\text{mm}/\text{s}\cdot\text{ly}$). Define the parsec ($3.084 \times 10^{13}\,\text{km}$). Show that this implies a minimum age for the universe of about $1.5 \times 10^{10}\,\text{y}$.

B. State that the future expansion (or contraction) of the universe depends on its mass density and that the density of matter that radiates is too small to prevent expansion forever. Explain that there is evidence for the existence of matter that does not radiate (dark matter). Explain how the rotational period of a star in a galaxy, as a function of its distance from the galactic center, provides such evidence. The nature of the dark matter is not presently known.

C. Discuss the microwave background radiation and state that physicists believe it was generated about 300, 000 years after the big bang, when the universe became tenuous enough to allow photons to exist without being quickly absorbed.

D. Remark that in the early universe the temperature was sufficiently high that the exotic particles now being discovered (and others) existed naturally. We need the results of high energy physics to understand the early universe.

E. Go over the chronological record given at the end of Section 44–14.

SUGGESTIONS

1. Assignments
   a. To test for understanding of the conservation laws and the stability of particles, ask questions 4, 6, and 7. Problems 1, 7 (or 9), 11, 12, 13, 15, and 17 each deal with one or more of the conservation laws. Assign several.
   b. To help clarify particle properties and classifications, ask questions 8, 11, 12, and 13.
   c. Problems 20, 21, 24, 39, and 28 provide excellent illustrations of the quark model.
   d. Assign problem 26 in connection with Hubble's law. Assign problem 27 or 28 in connection with the red shift. If you discussed the relativistic Doppler shift in connection with Chapter 37, assign problem 29. Problem 31 deals with the cosmic background radiation. Dark matter and the future of the universe are the subjects of problem 34.

2. Demonstrations
   Show nuclear emulsion plates, available from Brookhaven National Laboratory, Fermilab, and other high energy laboratories.

3. Books and Monographs
   a. *Resource Letters, Book Four* and *Resource Letters, Book Five*; American Association of Physics Teachers (AAPT, One Physics Ellipse, College Park MD 20740–3845; www.aapt .org). Contains lists of journal articles on high energy, particle physics, and cosmology.
   b. *Quarks*; edited by O.W. Greenberg; available from AAPT (see above for address). Reprints covering important aspects of the quark model.
   c. *Quarks, Quasars, and Quandaries*; edited by Gordon Aubrecht; available from AAPT (see above for address). Summaries of particle physics and cosmology.
   d. *Cosmology and Particle Physics*; edited by David Lindley, Edward W. Kolb, and David N. Schramm. Reprint collection dealing the evolution of the universe from the big bang.
   e. *Black Holes*; edited by Steven Detweiler; available from AAPT (see above for address). Reprints dealing with structure and dynamics of black holes.

4. Audio/Visual
   a. *Black Holes, Dark Matter* from *The Complete Cosmos* series; VHS video tape; Films for the Humanities and Sciences (PO Box 2053, Princeton, NJ 08543–2053; www.films.com).

b. *The Expanding Universe: From Big Bang to Big Crunch?*; VHS video tape; Films for the Humanities and Sciences (see above for address).

5. Laboratory
Meiners Experiment 14–8: *Nuclear and High Energy Particles.* A dry ice and alcohol cloud chamber is used to observe the tracks of alpha and beta particles as well as the tracks produced by cosmic rays. A magnet is used to make circular tracks.

SECTION FOUR
ANSWERS TO CHECKPOINTS

The following are the answers to the Checkpoints that appear throughout the text.

Chapter 2

1. b and c
2. zero (zero displacement for the entire trip)
3. (check the derivative dx/dt) (a) 1 and 4; (b) 2 and 3
4. (see Tactic 5) (a) plus; (b) minus; (c) minus; (d) plus
5. 1 and 4 ($a = d^2x/dt^2$ must be a constant
6. (a) plus (upward displacement on y axis); (b) minus (downward displacement on y axis); (c) $a = -g = -9.8\,\text{m/s}^2$

Chapter 3

1. (a) 7 m ($\vec{a}$ and $\vec{b}$ are in the same direction); (b) 1 m ($\vec{a}$ and $\vec{b}$ are in opposite directions)
2. c, d, f (components must be head-to-tail; $\vec{a}$ must extend from the tail of one component to the head of the other))
3. (a) +, +; (b) +, −; (c) +, + (draw vector from the tail of $\vec{d}_1$ to the head of $\vec{d}_2$)
4. (a) 90°; (b) 0 (vectors are parallel and in the same direction); (c) 180° (vectors are in opposite directions)
5. (a) 0 or 180°; (b) 90°

Chapter 4

1. (a) $(8\,\text{m})\hat{\text{i}} - (6\,\text{m})\hat{\text{j}}$; (b) yes, the xy plane (no z component)
2. (a) first; (b) third
3. (1) and (3) a_x and a_y are both constant, so $\vec{a}$ is constant; (2) and (4) a_y is constant but a_x is not, so $\vec{a}$ is not
4. $4\,\text{m/s}^2$, $-2\,\text{m/s}$, $3\,\text{m}$
5. (a) v_x constant; (b) v_y initially positive, decreases to zero, and then becomes progressively more negative; (c) $a_x = 0$ throughout; (d) $a_y = -g$ throughout
6. (a) $-(4\,\text{m/s})\hat{\text{i}}$; (b) $-(8\,\text{m/s}^2)\hat{\text{j}}$
7. (a) 0, distance not changing; (b) +70 km/h, distance increasing; (c) +80 km/h, distance decreasing
8. (a)–(c) increase

Chapter 5

1. c, d, and e
2. (a) and (b) a N, leftward
3. (a) and (b) 1, 4, 3, 2
4. (a) equal; (b) greater
5. (a) equal; (b) greater; (c) less
6. (a) increase; (b) yes; (c) same; (d) yes
7. (a) $F \sin\theta$; (b) increase
8. 0

Chapter 6

1. (a) zero; (b) 5 N; (c) n0; (d) yes; (e) 8 N
2. (a) same; (b) decreases; (c) decreases
3. greater
4. (a) $\vec{a}$ is downward, $\vec{N}$ is upward; (b) $\vec{a}$ and $\vec{N}$ are both upward
5. (a) same; (b) increases; (c) increases
6. (a) $4R_1$; (b) $4R_1$

Chapter 7

1. (a) decrease; (b) same; (c) negative, zero
2. d, c, b, a
3. (a) same; (b) smaller
4. (a) positive; (b) negative; (c) zero
5. zero

Chapter 8

1. no (consider round trip on the small loop)
2. 3, 1, 2
3. (a) all tie; (b) all tie

4. (a) CD, AB, BC (zero); (b) positive x direction
5. all tie

Chapter 9

1. (a) origin; (b) fourth quadrant; (c) on y axis below origin; (d) origin; (e) third quadrant; (f) origin
2. (a) to (c) at the center of mass, still at the origin
3. (a) 1, 3, then 2 and 4 tie (zero force); (b) 3
4. (a) unchanged; (b) unchanged; (c) decrease
5. (a) zero; (b) positive; (c) positive y direction
6. (a) zero; (b) no (c) negative x direction
7. (a) 500 km/h; (b) 2600 km/h; (c) 1600 km/h
8. (a) yes; (b) no
9. (a) 10 kg·m/s; (b) 14 kg·m/s; (c) 6 kg·m/s
10. (a) 4 kg · m/s; (b) 8 kg · m/s; (c) 3 J
11. (a) 2 kg · m/s; (b) 3 kg · m/s

Chapter 10

1. (b and (c)
2. (a) and (d)
3. (a) yes; (b) no; (c) yes; (d) yes
4. all tie
5. 1, 2, 4, 3
6. 1 and 3 tie, then 2 and 5 tie (zero)
7. (a) downward in the figure; (b) less

Chapter 11

1. (a) same; (b) less
2. less
3. (a) $\pm z$; (b) $+y$; (c) $-x$
4. (a) 1 and 3 tie, then 2 and 4 tie, then 5 (zero); (b) 2 and 3
5. (a) 3, 1, then 2 and 4 tie (zero)
6. (a) all tie; (b) sphere, disk, hoop
7. (a) decreases; (b) same; (c) increases

Chapter 12

1. c, e, f
2. directly below the rod (torque due to $\vec{F}_g$ on the apple, about the suspension point, is zero)
3. (a) no; (b) at site of $\vec{F}_1$, perpendicular to plane of figure; (c) 45 N
4. (a) at C (to eliminate forces there from a torque equation); (b) plus; (c) minus; (d) equal
5. d
6. (a) equal; (b) B; (c) B

Chapter 13

1. all tie
2. (a) 1, tie of 2 and 4, then 3; (b) line d
3. negative y direction
4. (a) increase; (b) negative
5. (a) 2; (b) 1
6. (a) path 1 [decreased E (more negative) gives decreased a]; (b) less than (decreased a gives decreased T)

Chapter 14

1. all tie
2. (a) all tie (gravitational force on the penguin is the same); (b) $0.95\rho_0$, ρ_0, $1.1\rho_0$
3. 13 cm^3/s, outward
4. (a) all tie; (b) 1, then 2 and 3 tie, 4 (wider means slower); (c) 4, 3, 2, 1 (wider and lower mean more pressure)

Chapter 15

1. (sketch x versus t) (a) $-x_m$; (b) $+x_m$; (c) 0
2. a (F must have the form of Eq. 6–10)
3. (a) 5 J; (b) 2 J; (c) 5 J
4. all tie (in Eq. 15–29, m is included in I)
5. 1, 2, 3 (the ratio m/b matters; k does not)

Chapter 16

1. a, 2; b, 3; c, 1 (compare with phase in Eq. 16–2, then see Eq. 16–5)
2. (a) 2, 3, 1 (see Eq. 16–13); (b) 3, then 1 and 2 tie (find amplitude of dy/dt)
3. same (independent of f); (b) decrease ($\lambda = v/f$); (c) increase; (d) increase
4. (a) increase; (b) increase; (c) increase
5. 0.20 and 0.80 tie, then 0.60, 0.45
6. (a) 1; (b) 3; (c) 2
7. (a) 75 Hz; (b) 525 Hz

Chapter 17

1. beginning to decrease (example: mentally move the curves of Fig. 17–7 rightward past the point at $x = 42$ m)
2. (a) 0, fully constructive; (b) 4λ, fully constructive
3. (a) 1 and 2 tie, then 3 (see Eq. 17–28); (b) 3, then 1 and 2 tie (see Eq. 17–26)
4. second (see Eqs. 17–39 and 17–41)
5. (a) greater; (b) less; (c) can't tell; (d) can't tell; (e) greater; (f) less
6. (measure speeds relative to the air) (a) 222 m/s; (b) 222 m/s

Chapter 18

1. (a) all tie; (b) 50° X, 50° Y, 50°W
2. (a) 2 and 3 tie, then 1, then 4; (b) 3, 2, then 1 and 4 tie (from Eqs. 18–9 and 18–10, assume that the change in area is proportional to initial area)
3. A (see Eq. 18–14)
4. c and e (maximize area enclosed by a clockwise cycle)
5. (a) all tie (ΔE_{int} depends on i and f, not on path); (b) 4, 3, 2, 1 (compare areas under curves); (c) 4, 3, 2, 1 (see Eq. 18–26)
6. (a) zero (closed cycle); (b) negative (W_{net} is negative; see Eq. 18–26)
7. b and d tie, then a, c (P_{cond} identical; see Eq. 18–32)

Chapter 19

1. all but c
2. (a) all tie; (b) 3, 2, 1
3. gas A
4. 5 (greatest change in T), then tie of 1, 2, 3, and 4
5. 1, 2, 3 ($Q_3 = 0$, Q_2 goes into work W_2, but Q_1 goes into greater work W_1 and increases gas temperature)

Chapter 20

1. a, b, c
2. smaller (Q is smaller)
3. c, b, a
4. a, d, c, b
5. b

Chapter 21

1. C and D attract; B and D attract
2. (a) leftward; (b) leftward; (c) leftward
3. (a) a, c, b; (b) less than
4. $-15e$ (net charge of $-30e$ is equally shared)

Chapter 22

1. (a) rightward; (b) leftward; (c) leftward; (d) rightward; (p and e have same charge magnitude and p is farther)
2. all tie
3. (a) toward positive y; (b) toward positive x; (c) toward negative y
4. (a) leftward; (b) leftward; (c) decrease
5. (a) all tie; (b) 1 and 3 tie, then 2 and 4 tie

Chapter 23

1. (a) $+EA$; (b) $-EA$; (c) 0; (d) 0
2. (a) 2; (b) 3; (c) 1
3. (a) equal; (b) equal; (c) equal
4. $+50e$; (b) $-150e$
5. 3 and 4 tie, then 2, 1

Chapter 24

1. (a) negative; (b) increase
2. (a) positive; (b) higher
3. (a) rightward; (b) 1, 2, 3, 5: positive; 4: negative; (c) 3, then 1, 2, and 5 tie, then 4
4. all tie
5. a, c (zero), b
6. (a) 2, then 1 and 3 tie; (b) 3; (c) accelerate leftward

Chapter 25

1. (a) same; (b) same
2. (a) decreases; (b) increases; (c) decreases
3. (a) V, $q/2$; (b) $V/2$, q
4. (a) $q_0 = q_1 + q_{34}$; (b) equal (C_3 and C_4 are in series)
5. (a) same; (b) same (same potential difference across same plate separation)
6. (a) same; (b) decrease; (b) decrease; (c) increase

Chapter 26

1. 8 A, rightward
2. (a)—(c) rightward
3. a and c tie, then b
4. device 2
5. (a) and (b) tie, then (d), then (c)

Chapter 27

1. (a) rightward; (b) all tie; (c) b, then a and c tie; (d) b, then a and c tie
2. (a) all tie; (b) R_1, R_2, R_3
3. (a) less; (b) greater; (c) equal
4. (a) $V/2$, i; (b) V, $i/2$
5. (a) 1, 2, 4, 3; (b) 4, then 1 and 2 tie, then 3

Chapter 28

1. a, $+z$; b, $-x$; c, $\vec{F}_B = 0$
2. (a) 2, then 1 and 3 tie (zero); (b) 4
3. (a) $+z$ and $-z$ tie, then $+y$ and $-y$ tie, then $+x$ and $-x$ tie (zero); (b) $+y$
4. (a) electron; (b) clockwise
5. $-y$
6. (a) all tie; (b) 1 and 4 tie, then 2 and 3 tie

Chapter 29

1. a, c, b
2. b, c, a
3. d, then a and c tie, then b
4. d, a, then b and c tie (zero)

Chapter 30

1. b, then d and e tie, then a and c tie (zero)
2. a and b tie, then c (zero
3. c and d tie, then a and b tie
4. b, out; c, out; d, into; e, into
5. d and e
6. (a) 2, 3, 1 (zero; (b) 2, 3, 1
7. a and b tie, then c

Chapter 31

1. (a) $T/2$; (b) T; (c) $T/2$; (d) $T/4$
2. (a) 5 V; (b) 150 μJ
3. (a) remains the same; (b) remains the same
4. (a) C, B, A; (b) 1, A; 2,B; 3, S; 4, C; (c) A
5. (a) remains the same; (b) increases
6. (a) remains the same; (b) decreases
7. (a) 1, lags; 2, leads; 3, in phase; (b) 3 ($\omega_d = \omega$ when $X_L = X_C$)
8. (a) increase (circuit is mainly capacitive; increase C to decrease X_C to be closer to resonance for maximum P_{avg}); (b) closer
9. (a) greater; (b) step-up

Chapter 32

1. d, b, c, a (zero)
2. (a) 2; (b) 1
3. (a) away; (b) away; (c) less
4. (a) toward; (b) toward; (c) less
5. a, c, b, d (zero)
6. tie of b, c, and d, then a

Chapter 33

1. (a) (Use Fig. 33–5.) On the right side of rectangle, $\vec{E}$ is in the negative y direction; on the left side, $\vec{E}+d\vec{E}$ is greater and in the same direction; (b) $\vec{E}$ is downward. On the right side, $\vec{B}$ is in the negative z direction; on the left side $\vec{B}+d\vec{B}$ is greater and in the same direction.
2. positive direction of x
3. (a) same; (b) decrease
4. a, d, b, c (zero)
5. a
6. (a) no; (b) yes

Chapter 34

1. $0.2d$, $1.8d$, $2.2d$
2. (a) real; (b) inverted; (c) same
3. (a) e; (b) virtual, same
4. virtual, same as object, diverging

Chapter 35

1. b (least n), c, a
2. (a) top; (b) bright intermediate illumination (phase difference is 2.1 wavelengths
3. (a) 3λ, 3; (b) 2.5λ, 2.5
4. a and d tie (amplitude of resultant wave is $4E_0$), then b and c tie (amplitude of resultant wave is $2E_0$)
5. (a) 1 and 4; (b) 1 and 4

Chapter 36

1. (a) expand; (b) expand
2. (a) second side maximum; (b) 2.5
3. (a) red; (b) violet
4. diminish
5. (a) increase; (b) same
6. (a) left; (b) less

Chapter 37

1. (a) same (speed of light postulate); (b) no (the start and end of the flight are spatially separated); (c) no (because his measurement is not a proper time)
2. (a) Sally's; (b) Sally's
3. (a) Eq. 2; (b) $0.90c$; (c) 25 ns; (d) -7.0 m
4. (a) right; (b) more
5. (a) equal; (b) less

Chapter 38

1. b, a, d, c
2. (a) lithium, sodium, potassium, cesium; (b) all tie
3. (a) same; (b) – (d) x rays
4. (a) proton; (b) same; (c) proton
5. same

Chapter 39

1. b, a, c
2. (a) all tie; (b) a, b, c
3. a, b, c, d
4. $E_{1,1}$ (neither n_x nor n_y can be zero
5. (a) 5; (b) 7

Chapter 40

1. 7
2. (a) decrease; (b) – (c) remain the same
3. less
4. A, C, B

Chapter 41

1. (a) larger; (b) same
2. Cleveland: metal; Boca Raton: none; Seattle: semiconductor
3. a, b, and c
4. b

Chapter 42

1. ^{90}As and ^{158}Nd
2. a little more than 75 Bq (elapsed time is a little less than three half-lives)
3. ^{206}Pb

Chapter 43

1. c and d
2. (a) no; (b) yes; (c) no
3. e

Chapter 44

1. (a) the muon family; (b) a particle; (c) $L_\mu = +1$
2. b and e
3. c

SECTION FIVE
ANSWERS TO QUESTIONS

The following are the answers to the end-of-chapter questions.

Chapter 2

1. (a) all tie; (b) 4, tie of 1 and 2, then 3
2. (a) negative; (b) positive; (c) zero; (d) negative; (e) twice
3. (a) positive direction; (b) negative direction; (c) 3 and 5; (d) 2 and 6 tie, then 3 and 5 tie, then 1 and 4 tie (zero)
4. a and c
5. (a) 3, 2, 1; (b) 1, 2, 3; (c) all tie; (d) 1, 2, 3
6. (a) 2, 3; (b) 1, 3; (c) 4
7. 60 km/h, not 0
8. (a) D; (b) E
9. $x = t^2$ and $x = 8(t-2) + 1.5(t-2)^2$, with x in meters and t in seconds

Chapter 3

1. Either the sequence $\vec{d}_2$, $\vec{d}_1$ or the sequence $\vec{d}_2$, $\vec{d}_2$, $\vec{d}_3$
2. yes, when the vectors are in the same direction
3. no
4. (a) $\vec{a}$ and $\vec{b}$ are parallel; (b) $\vec{b} = 0$; (c) $\vec{a}$ and $\vec{b}$ are perpendicular
5. (a) yes; (b) yes; (c) n0
6. (a) −, +; (b) −, −; (c) +, +
7. all but (e)
8. no (the orientations of $\vec{b}$ and $\vec{c}$ can differ
9. (a) positive x direction for (1), positive z direction for (2) and (3); (b) negative x direction for (1); negative z direction for (a) and (3)
10. (a) $\vec{B}$ and $\vec{C}$, $\vec{D}$ and $\vec{E}$; (b) $\vec{D}$ and $\vec{E}$

Chapter 4

1. (a) $(7\,\text{m})\hat{\imath} + (1\,\text{m})\hat{\jmath} - (2\,\text{m})\hat{k}$; (b) $(5\,\text{m})\hat{\imath} - (3\,\text{m})\hat{\jmath} + (1\,\text{m})\hat{k}$; (c) $(-2\,\text{m})\hat{\imath}$
2. a and c tie, then b
3. yes; it is coming down
4. a, b, c
5. (a) all tie; (b) 1 and 2 tie, then 3 and 4 tie
6. (a) A; (b) closer
7. (a) 0; (b) 350 km/h; (c) 350 km/h; (d) the same
8. (a) 3, 2, 1; (b) 1, 2, 3; (c) all tie; (d) 6, 5, 4
9. (a) all tie; (b) all tie; (c) 3, 2, 1; (d) 3, 2, 1
10. (a) c, b, a; (b) a, b, c
11. 2, then 1 and 4 tie, then 3
12. (a) 90° and 270°; (b) 0° and 180°; (c) 90° and 270°
13. (a) yes (just round a curve); (b) no (the direction of the velocity must be changing); (c) yes (if going with constant speed)

Chapter 5

1. (a) 2 and 4; (b) 2 and 4
2. (a) 5; (b) 7; (c) $(2\,\text{N})\hat{\imath}$; (d) $(-6\,\text{N}\hat{\jmath}$; (e) fourth; (f) fourth
3. increase
4. (a) 2 and 3; (b) 2
5. (a) 2, 3, 4; (b) 1, 3, 4; (c) 1: in the positive y direction, 2: in the positive x direction, 3: in the fourth quadrant, 4: in the third quadrant
6. 1, graphs a and e; 2, graphs b and d; 3, graphs b and f; 4, graphs c and f
7. a, then b, c, and d tie
8. (a) 10 kg; (b) 18 kg; (c) 10 kg; (d) all tie; (e) 3, 2, 1
9. (a) increases from mg; (b) decreases from mg to zero
10. d, c, a, b
11. (a) 17 kg; (b) 12 kg; (c) 10 kg; (d) all tie; (e) $\vec{F}$, $\vec{F}_{21}$, $\vec{F}_{32}$
12. A positive and constant; B zero and constant; D zero and constant; E positive and increasing; F positive and decreasing

Chapter 6

1. (a) F_1, F_2, F_3; (b0 all tie
2. (a) same; (b) increases; (c) increases; (d) n0

3. (a) upward; (b) horizontal, away from wall; (c) no change; (d) increases; (e) increases
4. (a) decreases; (b) decreases; (c) increases; (d) increases; (e) increases
5. (a) decreases; (b) decreases; (c) decreases; (d) decreases; (e) decreases
6. At first $\vec{f}_s$ is directed up the ramp and its magnitude decreases from $mg \sin\theta$ to 0 as F increases. Then $\vec{f}_s$ is directed down the ramp and its magnitude increases until it reaches $f_{s\,\text{max}}$. Thereafter the force is one of kinetic friction directed down the ramp and has magnitude f_k, a constant value less than $f_{s\,\text{max}}$.
7. At first $\vec{f}_s$ is directed up the ramp and its magnitude increases from $mg \sin\theta$ until it reaches $f_{s\{sub\ max}$. Thereafter the force is one of kinetic friction directed up the ramp and has magnitude f_k, a constant value less than $f_{s\,\text{max}}$.
8. On opening the parachute produced a large sudden upward force on the diver due to increased air drag and the drag force slowed the diver suddenly. To keep the pumpkin in his grip he had to slow it just as much but this required more force than he could exert. From the diver's viewpoint the apparent weight of the pumpkin suddenly and surprisingly increased and the pumpkin was ripped from his hands. From the pumpkin's viewpoint the sudden upward force on the diver ripped him away from the pumpkin.
9. (a) $5\,\text{m/s}^2$ to $10\,\text{m/s}^2$; (b) 0 to $5\,\text{m/s}^2$
10. 4, 3, then 1, 2, and 5 tie
11. (a) all tie; (b) all tie; (c) 2, 3, 1

Chapter 7

1. all tie
2. (a) positive; (b) negative; (c) negative
3. c, b, a
4. (a) 2; (b) 3; (c) 1
5. all tie
6. b (positive work), a (zero work, c (negative work, d (more negative work
7. (a) A, $\vec{F}_2$; B, $\vec{F}_1$; C, $\vec{F}_3$; D, $\vec{F}_4$; (b) E, A and D; F, B and C; G and H are meaningless
8. (a) A; (b) B
9. e through h
10. (a) 3 m; (b) 3 m; (c) 0 and 6 m; (d) negative x

Chapter 8

1. (a) 12 J; (b) −2 J
2. (a) 3, 2, 1; (b) 1, 2, 3
3. (a) 4; (b) returns to its starting point and repeats the trip; (c) 1; (d) 1
4. 2, 1, 3
5. (a) AB, CD, then BC and DE tie (zero force); (b) 5 J; (c) 5 J; (d) 6 J; (e) FG; (f) DE
6. (a) less; (b) equal
7. +30 J
8. +30 J
9. (a) increasing; (b) decreasing; (c) decreasing; (d) constant in AB and BC, decreasing in CD

Chapter 9

1. (a) ac, cd, bc; (b) bc; (c) bd, ad
2. (a) 2 N, rightward; (b) 2 N, rightward; (c) greater than 2 N, rightward
3. d, c, a, b (zero)
4. (a) x yes, y no; (b) x yes, y no; (c) x no, y yes
5. all tie
6. b, c, a
7. a, c, e, f: the sum of the momenta after the explosion does not equal the momentum before the explosion
8. (a) one was stationary; (b) 2; (c) 5; (d) equal
9. (a) positive; (b) positive; (c) 2 and 3
10. (a) C; (b) B; (c) 3
11. (a) forward; (b) stationary; (c) backward
12. (a) c, kinetic energy cannot be negative; d, total energy cannot increase; (b) a; (c) b

---

Chapter 10

1. (a) positive; (b) zero; (c) negative; (d) negative
2. (a) c, a, then b and d tie; (b) b, then a and c tie, then d
3. (a) 1: counterclockwise (positive); 2: counterclockwise (positive); 3: at $\theta = 0$; (b) 1: before; 2: at $t = 0$; 3: after;

(c) 1: positive; 2: negative; 3: positive
4. c, a, b
5. larger
6. 90°, then 70° and 110° tie
7. $\vec{F}_5$, $\vec{F}_4$, $\vec{F}_2$, $\vec{F}_1$, $\vec{F}_3$ (zero)
8. (a) decrease; (b) clockwise; (c) counterclockwise
9. (a) 1 and 2 tie, then 3; (b) 1 and 3 tie, then 2; (c) 2, 1, 3
10. all tie

Chapter 11

1. (a) 0 or 180°; (b) 90°
2. (a) 1, 2, 3 (zero); (b) 1 and 2 tie, then 3; (c) 1 and 3 tie, then 2
3. (a) 5 and 6; (b) 1 and 4 tie, then the rest tie
4. (a) spins in place; (b) rolls toward you; (c) rolls away from you
5. b, then c and d tie, then a and e tie (zero)
6. a, then b and c tie, then d (zero)
7. (a) 3; (b) 1; (c) 2; (d) 4
8. D, B, then A and C tie
9. (a) 4, 6, 7, 1, then 2, 3, and 5 tie (zero)
10. (a) same; (b) increase; (c) decrease; (d) same, decrease, increase

Chapter 12

1. (a) yes; (b) yes; (c) yes; (d) no
2. a and c (forces and torques balance)
3. (a) 1 and 3 tie, then 2; (b) all tie; (c) 1 and 3 tie, then 2 (zero)
4. (a) 1, 2, 3 (zero), 4, 5, 6; (b) 6, 5, 4, 3, 2, 1
5. (a) 12 kg; (b) 3 kg; (c) 1 kg
6. (a) 15 N (the key is the pulley with the 10-N piñata); (b) 10 N
7. increase
8. (a) same; (b) smaller; (c) smaller; (d) same
9. 4, 1, then 2 and 3 tie
10. A, then tie of B and C

Chapter 13

1. (a) between, closer to less massive particle; (b) no; (c) no
2. b and c tie, then a (zero)
3. Gm^2/r^2, upward
4. $3GM^2/d^2$, leftward
5. (a) c, b, a; (b) a, b, c
6. (a) positive y; (b) yes, rotates counterclockwise until it points toward particle B
7. yes, in the second quadrant, closer to the y axis than to the x axis, at a distance that depends on its mass
8. (a) 1 and 2 tie, then 3 and 4 tie; (b) 1, 2, 3, 4
9. a tie of b, d, and f, then e, c, a
10. 1, tie of 2 and 4, then 3
11. (a) all tie; (b) all tie
12. b, a, c

Chapter 14

1. e, then b and d tie, then a and c tie
2. b, then a and d tie (zero), then c
3. (a) 2; (b) 1, less; 3, equal; 4, greater
4. (a) moves downward; (b) moves downward
5. all tie
6. (a) downward; (b) downward; (c) same
7. a, b, c
8. c, b, a
9. B, C, A
10. (a) 1 and 4; (b) 2; (c) 3

Chapter 15

1. c
2. (a) 2; (b) positive; (c) between 0 and $+x_m$
3. a and b
4. (a) toward $-x_m$; (b) toward $+x_m$; (c) between $-x_m$ and 0; (d) between $-x_m$ and 0; (e) decreasing; (f) increasing
5. (a) $-\pi$, $-180°$; (b) $-\pi/2$, $-90°$; (c) $+\pi/2$, $+90°$
6. (a) between D and E; (b) between $3\pi/2$ rad and 2π rad
7. (a) between B and C; (b) between $\pi/2$ rad and π rad
8. (a) all tie; (b) 3, then 1 and 2 tie; (c) 1, 2, 3 (zero); (d) 1, 2, 3 (zero); (e) 1, 3, 2
9. (a) A, B, C; (b) C, B, A
10. (a) greater; (b) same; (c) same; (d) greater; (e) greater
11. b (infinite period; does not oscillate), c, a
12. one system : $k = 1500\,\text{N/m}$, $m = 500\,\text{kg}$; other system: $k = 1200\,\text{N/m}$, $m = 400\,\text{kg}$

Chapter 16

1. (a) 3, then 1 and 2 tie; (b) all tie; (c) 1 and 2 tie, then 3
2. a, upward; b, upward; c, downward; d, downward; e, downward; f, downward; g, upward; h, upward
3. (a) 1, 4, 2, 3; (b) 1, 4, 2, 3
4. (a) 4; (b) 4; (c) 3
5. intermediate (closer to fully destructive interference)
6. (a) 0, 0.2 wavelength, 0.5 wavelength (zero); (b) $4P_{\text{avg, 1}}$
7. a and d tie, then b and c tie
8. c, a, b
9. (a) 8; (b) antinode; (c) longer; (d) lower
10. d
11. (a) node; (b) antinode

Chapter 17

1. (a) 2.0 wavelengths; (b) 1.5 wavelengths; (c) fully constructive in (a), fully destructive in (b)
2. C, then A and B tie
3. (a) 0, 0.2 wavelength, 0.5 wavelength (zero); (b) $4P_{\text{avg, 1}}$
4. (a) two; (b) antinode
5. 150 Hz and 450 Hz
6. all odd harmonics
7. d, fundamental
8. E, A, D, C, B
9. 1, 4, 3, 2
10. (a) 3, then 1 and 2 tie; (b) 1, then 2 and 3 tie; (c) 3, 2, 1

Chapter 18

1. Z, X, Y
2. c, then the rest tie
3. B, then A and C tie
4. c, b, a
5. (a) both clockwise; (b) both clockwise
6. (a) cycle 2; (b) cycle 2
7. (a) all tie; (b) all tie
8. sphere, hemisphere, cube
9. (a) greater; (b) 1, 2, 3; (c) 1, 3, 2; (d) 1, 2, 3; (e) 2, 3, 1
10. (a) at freezing point; (b) no liquid freezes; (c) ice partly melts
11. (a) f, because the temperature cannot drop once the melting point is reached; (b) b and c end at the freezing point, d ends above the freezing point, e ends below the freezing point; (c) in b the liquid partly freezes and no ice melts; in c no liquid freezes and the ice fully melts; in e the liquid fully freezes and no ice melts

Chapter 19

1. d, tie of a and b, then c
2. (a) 0; (b) 0; (c) negative; (d) positive
3. 20 J
4. (a) 0; (b) 0; (c) negative; (d) positive
5. constant-volume process
6. (a) 0; (b) 0; (c) negative; (d) positive
7. (a) 3; (b) 1; (c) 4; (d) 2; (e) yes
8. (a) same; (b) increases; (c) decreases; (d) increases
9. (a) 1, 2, 3, 4; (b) 1, 2, 3
10. −4 J

Chapter 20

1. and c tie, then b and d tie
2. (a) all tie; (b) all tie; (c) all tie (zero)
3. b, a, c, d
4. 9 and −8, 8 and −5, 5 and −3, 3 and −2
5. unchanged
6. (a) AE; (b) AC; (c) AF; (d) none
7. A: first; B: first and second; C: second; D: neither
8. c, a, b
9. (a) same; (b) increase; (c) decrease
10. (a) same; (b) increase; (c) decrease
11. more than the age of the universe

Chapter 21

1. b and c tie, then a (zero)
2. (a) 3, 1, 2; (b) all tie
3. a and b
4. (a) between; (b) positively charged; (c) unstable
5. 3, 1, 2, 4 (zero)
6. $2kq^2/r^2$, up the page
7. $6kq^2/d^2$, leftward
8. a and d tie, then b and c tie

9. (a) same; (b) less than; (c) cancel; (d) add; (e) adding components; (f) positive y direction; (g) negative y direction; (h) positive x direction; (i) negative x direction
10. (a) neutral; (b) negatively

Chapter 22

1. a, b, c
2. (a) to their left; (b) no
3. 2, 4, 3, 1 (zero)
4. $q/4\pi\epsilon_0 d^2$, leftward
5. (a) yes; (b) toward; (c) no (the field vectors are not along the same line); (d) cancel; (e) add; (f) adding components; (g) toward negative y
6. (a) 3, then 1 and 2 tie (zero); (b) all tie; (c) 1 and 2 tie, then 3
7. e, b, then a and c tie, then d (zero)
8. (a) rightward; (b) $+q_1$ and $-q_3$, increase; $+q_2$, decrease; n, same
9. a, b, c
10. (a) positive; (b) same
11. (a) 4, 3, 1, 2; (b) 3, then 1 and 4 tie, then 2

Chapter 23

1. (a) $8\,\mathrm{N\cdot m^2/C}$; (b) 0
2. (a) all tie; (b) a uniform, b variable, c uniform, d variable
3. all tie
4. a, c, then b and d tie (zero)
5. all tie
6. either 2σ, σ, 3σ or 3σ, σ, 2σ
7. all tie
8. (a) 2, 1, 3; (b) all tie ($+4q$)
9. all tie ($E = 0$); (b) all tie
10. (a) a, b, c, d; (b) a and b tie, then c, d

Chapter 24

1. (a) 1 and 2; (b) none; (c) no; (d) 1 and 2, yes; 3 and 4, no
2. (a) higher; (b) positive; (c) negative; (d) all tie
3. b, then a, c, and d tie
4. $-4q/4\pi\epsilon_0 d$
5. (a) 1, then 2 and 3 tie; (b) 3
6. (a) $Q/4\pi\epsilon_0 R$; (b) $Q/4\pi\epsilon_0 R$; (c) $Q/4\pi\epsilon_0 R$; (d) a, b, c
7. (a) 3 and 4 tie, then 1 and 2 tie; (b) 1 and 2, increase; 3 and 4, decrease
8. (a) 2, 4, and then a tie of 1, 3, and 5 (where $E = 0$); (b) negative x direction; (c) positive x direction
9. (a) 0; (b) 0; (c) 0; (d) all three quantities still 0
10. (a) positive; (b) positive; (c) negative; (d) all tie

Chapter 25

1. a, 2; b, 1; c, 3
2. a, series; b, parallel; c, parallel
3. (a) $V/3$; (b) $CV/3$; (c) $CV/3$ (not CV)
4. (a) $C/3$; (b) $3C$; (c) parallel
5. (a) no; (b) yes; (c) all tie
6. parallel, C_1 alone, C_2 alone, series
7. (a) same; (b) same; (c) more; (d) more
8. (a) less; (b) less; (c) less; (d) less
9. (a) 2; (b) 3; (c) 1
10. (a) increase; (b) same; (c) increase; (d) increase; (e) increase; (f) increase
11. (a) increase; (b) increase; (c) decrease; (d) decrease; (e) same, increase, increase, increase

Chapter 26

1. a, b, and c all tie, then d (zero)
2. a, b, and c all tie, then d
3. b, a, c
4. A, B, and C all tie, then A + B and B + C tie, then A + B + C
5. (a) 1 and 2 tie, then 3; (b) 1 and 2 tie, then 3; (c) 1 and 2 tie, then 3
6. (a) top-bottom, front-back, left-right; (b) top-bottom, front-back, left-right; (c) top-bottom, front-back, left-right; (d) top-bottom, front-back, left-right
7. C, A, B
8. (a) C, B, A; (b) all tie; (c) A, B, C; (d) all tie
9. (a) all tie; (b) B, C, A; (c) B, C, A
10. (a) B, A, C; (b) B, A, C

Chapter 27

1. (a) b and d tie, then a, c, and e tie; (b) b, d, then a, c, and e tie; (c) positive x direction
2. (a) series; (b) parallel; (c) parallel
3. (a) no; (b) yes; (c) all tie
4. (a) equal; (b) more
5. (a) same; (b) same; (c) less; (d) more
6. $60\,\mu C$
7. parallel, R_2, R_1, series
8. 2.0 A
9. (a) less; (b) less; (c) more
10. (a) all tie; (b) 1, 3, 2
11. 1, c; 2, a; 3, d; 4, b

Chapter 28

1. (a) no because $\vec{v}$ and $\vec{F}_B$ must be perpendicular; (b) yes; (c) no because $\vec{B}$ and $\vec{F}_B$ must be perpendicular
2. (a) 3 and 4 tie, then 1 and 2 tie (zero); (b) 4 (assuming that the rightward current is due to leftward motion of electrons in the wire)
3. (a) $\vec{F}_E$; (b) $\vec{F}_B$
4. 2, 5, 6, 9, 10
5. (a) negative; (b) equal; (c) equal; (d) half-circle
6. into page: a, d, e; out of page: b, c, f (the particle is negatively charged)
7. (a) $\vec{B}_1$; (b) $\vec{B}_1$ into page, $\vec{B}_2$ out of page; (c) less
8. 1i, 2e, 3c, 4a, 5g, 6j, 7d, 8b, 9h, 10f, 11k
9. (a) upper plate; (b) lower plate; (c) out of page
10. (a) positive; (b) $1 \to 2$ and $1 \to 3$ tie, then $1 \to 4$ (which is zero)

Chapter 29

1. (a) into; (b) greater
2. c, d, then a and b tie (zero)
3. c, a, b
4. 1, then 3 and 4 tie, then 2 (zero)
5. (a) 1, 3, 2; (b) less
6. b, d, c, a (zero)
7. c and d tie, then b, a
8. d, then a and e tie, then b, c
9. b, a, d, c (zero)
10. (a) c, a, d, b; (b) a, c, b, d; (c) a and c tie, then b and d tie; (d) greater

---

Chapter 30

1. out
2. (a) all tie (zero); (b) 2, then tie of 1 and 3 (zero)
3. (a) into; (b) counterclockwise; (c) larger
4. (a) 2, 1, 3; (b) 2, 1, 3; (c) 1 counterclockwise; 2 clockwise; 3 counterclockwise
5. 1 and 3 tie (clockwise), then 2 and 5 tie (zero), then 4 and 6 tie (counterclockwise
6. d and c tie, then b, a
7. c, b, a
8. (a) all tie (zero); (b) 1 and 2 tie, then 3; (c) all tie (zero)
9. (a) more; (b) same; (c) same; (d) same (zero)
10. a, 2; b, 4; c, 1; d, 3

Chapter 31

1. with n zero or a positive integer, (a) $0 \pm n2\pi$; (c) $\pi/2 \pm n2\pi$; (e) $\pi \pm n2\pi$; (g) $3\pi/2 \pm n2\pi$
2. (a) $T/4$; (b) $T/4$; (c) $T/2$; (d) $T/2$
3. (a) less; (b) greater
4. b, a, c
5. (a) 3, 1, 2; (b) 2, then 1 and 3 tie
6. c, b, a
7. a, inductor; b, resistor; c, capacitor
8. (a) leads; (b) capacitive; (c) less
9. (a) 1 and 4; (b) 2 and 3
10. (a) less; (b) equal; (c) greater
11. (a) rightward, increase (X_L increases, circuit is closer to resonance); (b) rightward, increase (X_C decreases, circuit is closer to resonance); (c) rightward, increase (ω_d/ω increases, circuit is closer to resonance)
12. (a) positive; (b) decreased (to decrease X_L and get closer to resonance); (c) decreased (to increase X_C and get closer to resonance)

Chapter 32

1. a, decreasing; b, decreasing
2. (a) a and b tie, then c, d; (b) none (because plate lacks circular symmetry, $\vec{B}$ not tangent to any circular loop); (c) none
3. (a) rightward; (b) leftward; (c) into

4. 1, a; 2, b; 3, c and d
5. supplied
6. b
7. (a) all down; (b) 1 up, 2 down, 3 zero
8. (a) 1 up, 2 up, 3 down; (b) 1 down, 2 up, 3 zero
9. (a) 1 down, 2 down, 3 up; (b) 1 up, 2 down, 3 zero
10. (a) 1, 3, 2; (b) 2
11. (a) increase; (b) increase

Chapter 33

1. into
2. (a) positive direction of z; (b) x
3. (a) same; (b) increase; (c) decrease
4. (a) and (b) $A = 1$, $n = 4$, $\theta = 30°$
5. 20° and 90°
6. c
7. a, b, c
8. b 30°; c 60°; d 60°; e 30°; f 60°
9. d, b, a, c
10. none
11. n_3, n_2, n_1
12. B

Chapter 34

1. (a) a; (b) c
2. (a) a and c; (b) three times; (c) you
3. (a) from infinity to the focal point; (b) decrease continually
4. (a) I_1 and I_4; (b) I_2 and I_3; (c) I_3; (d) I_3; (e) I_2
5. convex
6. 1 concave, 2 convex, 3 plane
7. d (infinite), tie of a and b, then c
8. (a) I_2 and I_3; (b) I_1 and I_4; (c) I_1; (d) I_1; (e) I_4
9. (a) all but variation 2; (b) for 1, 3, and 4: right, inverted; for 5 and 6: left, same
10. 1 converging, 2 diverging

Chapter 35

1. (a) peak; (b) valley
2. a, c, b
3. (a) $2d$; (b) (odd number)$\lambda/2$; (c) $\lambda/4$
4. (a) 300 nm; (b) exactly out of phase
5. (a) intermediate closer to maximum, $m = 2$; (b) minimum, $m = 3$; (c) intermediate closer to maximum, $m = 2$; (d) maximum, $m = 1$
6. (a) increase; (b) 1λ
7. (a) decrease; (b) decrease; (c) decrease; (d) blue
8. b, 3 and 5; c, 1 and 4; d, 2
9. (a) maximum; (b) minimum; (c) alternates
10. c, d
11. (a) no; (b) $2(0) = 0$; (c) $2L$
12. (a) 0.5 wavelength; (b) 1 wavelength

Chapter 36

1. (a) the $m = 5$ minimum; (b) (approximately) the maximum between the $m = 4$ and $m = 5$ minima
2. (a) contract; (b) contract
3. (a) 1 and 3 tie, then 2 and 4 tie; (b) 1 and 2 tie , then 3 and 4 tie
4. 4
5. (a) A, B, C; (b) A, B, C
6. (a) A, B, C; (b) A, B, C
7. (a) larger; (b) red
8. (a) less; (b) greater; (c) greater
9. (a) decrease; (b) same; (c) in place
10. (a) decrease; (b) decrease; (c) to the right
11. (a) A; (b) left; (c) left; (d) right
12. (a) increase; (b) first order

Chapter 37

1. (a) 4 s; (b) 3 s; (c) 5 s; (d) 4 s; (e) 10 s
2. (a) C_1'; (b) C_1'
3. (a) C_1; (b) C_1
4. c
5. (a) Sam; (b) neither
6. b
7. (a) negative; (b) positive
8. (a) 3, tie of 1 and 2, then 4; (b) 4, tie of 1 and 2, then 3; (c) 1, 4, 2, 3
9. b, a, c, d
10. (a) 3, then 1 and 2 tie; (b) 2, then 1 and 3 tie; (c) 2, 1, 3; (d) 2, 1, 3
11. (a) tie of 3, 4, and 6, then tie of 1, 2, and 5; (b) 1, then tie of 2 and 3, then tie of 5 and 6; (c) 1, 2, 3, 4, 5, 6; (d) 2 and 4; (e) 1, 2, 5

Chapter 38

1. potassium
2. only e
3. (a) microwave; (b) x ray; (c) x ray
4. (a) true; (b) false; (c) false; (d) true; (e) true; (f) false
5. positive charge builds up on the plate, inhibiting further electron emission
6. only b
7. none
8. The fractional wavelength change for visible light is too small
9. (a) B; (b) – (d) A
10. (a) greater; (b) less
11. no essential change
12. (a) decreases by a factor of $1/\sqrt{2}$; (b) decreases by a factor of 1/2
13. electron
14. electron, neutron, alpha particle
15. (a) decreasing; (b) increasing; (c) same; (d) same
16. amplitude of reflected wave is less than that of incident wave
17. a
18. proton
19. all tie
20. (a) zero; (b) yes

Chapter 39

1. a, c, b
2. (a) 1/4; (b) same factor
3. c
4. (a) $(\sqrt{1/L})\sin(\pi/2L)x$; (b) $(\sqrt{4/L})\sin(2\pi/L)x$; (c) $(\sqrt{2/L})\cos(\pi/L)x$
5. (a) 18; (b) 17
6. less
7. equal
8. (a) wider; (b) deeper
9. b, c, and d
10. (a) 3; (b) 4
11. (a) decrease; (b) increase
12. 12 eV ($4 \longrightarrow 2$ in A) matches $1 \longrightarrow 2$ in C; 9 eV ($5 \longrightarrow 4$ in A) matches $1 \longrightarrow 2$ in D; 24 eV ($5 \longrightarrow 1$ in A) matches $1 \longrightarrow 3$ in D; 15 eV ($4 \longrightarrow 1$ in A) matches $1 \longrightarrow 2$ in E
13. $n = 1$, $n = 2$, $n = 3$
14. (a) greater; (b) less; (c) less
15. same
16. (a) first Lyman plus first Balmer; (b) Lyman series limit minus Paschen series limit
17. (a) $n = 3$; (b) $n = 1$; (c) $n = 5$

Chapter 40

1. same number (10)
2. 0, 2, and 3
3. (a) 2; (b) 8; (c) 5; (d) 50
4. 6p
5. −1, 0, 1, and 2
6. (a) bromine; (b) rubidium; (c) hydrogen
7. (a) n; (b) n and ℓ
8. a, c, e, f
9. all true
10. (a) rubidium; (b) krypton
11. (a) 2; (b) 3
12. (a) unchanged; (b) decrease; (c) decrease
13. In addition to the quantized energy, a helium atom has kinetic energy; its total energy can equal 20.66 eV
14. a and b

Chapter 41

1. 4
2. 8
3. much less than
4. (a) anywhere in the lattice; (b) in any silicon-silicon bond; (c) in a silicon ion core, at a lattice site
5. b, c, d (the latter due to thermal expansion)
6. $4s^2$ and $4p^2$
7. b and d
8. a and b
9. none
10. (a) arsenic, antimony; (b) gallium, indium; (c) tin
11. $+4e$
12. zero
13. (a) right to left; (b) back bias
14. silicon with arsenic
15. blue
16. a, b, and c

Chapter 42

1. less
2. more protons than neutrons

3. above
4. ^{240}U
5. (a) ^{196}Pt; (b) no
6. less
7. (a) on the $N = Z$ line; (b) positrons; (c) about 120
8. (a) below; (b) below; (c) radioactive
9. yes
10. yes
11. no
12. (a) A and C tie, then B; (b) B, then A and C tie
13. no effect
14. (a) increases; (b) same
15. ^{209}Po
16. 7h
17. d
18. (a) all except ^{198}Au; (b) ^{132}Sn and ^{208}Pb

Chapter 43

1. a
2. a
3. b, e, a, c, d
4. b
5. c
6. (a) ^{93}Sr; (b) ^{140}I; (c) ^{155}Nd
7. c, a, d, b
8. c
9. d
10. a
11. c
12. c

Chapter 44

1. into
2. d
3. the π^+ pion whose track terminates at point 2
4. b, c, d
5. c, f
6. a, b, c, d
7. baryon number
8. (a) lepton; (b) antiparticle; (c) fermion; (d) yes
9. c
10. 1d, 2e, 3a, 4b, 5c
11. (a) 0; (b) +1; (c) −1; (d) +1; (e) −1
12. b, f, c, d, a, g, e
13. 1b, 2c, 3d, 4e, 5a

SECTION SIX

ANSWERS TO PROBLEMS

The following are the answers to the end-of-chapter problems.

Chapter 1

1. (a) 160 rods; (b) 40 chains
2. 0.18 points
3. (a) 10^9 μm; (b) 10^{-4}; (c) 9.1×10^5 μm
4. (a) 1.9 picas; (b(23 points
5. (a) 4.00×10^4 km; (b) 5.10×10^8 km^2; (c) 1.08×10^{12} km^3
6. (a) 8.33×10^{-2}, 2.08×10^{-2}; 6.94×10^{-3}, 3.47×10^{-3}; (b) 0.250, 8.33×10^{-2}, 4.17×10^{-2}; (c) 0.333, 0.167; (d) 0.500; (e) 14.0 medios; (f) 4.86×10^{-2} cahiz; (g) 3.24×10^4 cm^3
7. 1.9×10^{22} cm^3
8. (a) 60.8 W; (b) 43.3 Z
9. 1.1×10^3 acre-feet
10. (a) 52.6 min; (b) 4.9%
11. 1.21×10^{12} μs
12. 3.1 μm/s
13. C, D, A, B, E; the important criterion is the consistency of the daily variation, not its magnitude
14. 15°
15. (a) 1.43; (b) 0.864
16. (a) 3.88×10^8 rotations; (b) 1557.806 448 872 75 s; (c) $\pm 3 \times 10^{-11}$ s
17. (a) 495 s; (b) 141 s; (c) 198 s; (d) −245 s
18. 2.1 h
19. (a) 1×10^3 kg; (b) 158 kg/s
20. (a) 1.430 m^2; (b) 72.84 km
21. 9.0×10^{49} atoms
22. (a) 2.69×10^5 cm^3; (b) 0.77 y
23. (a) 1.18×10^{-29} m^3; (b) 0.282 nm
24. (a) 2×10^3 m^3, 2×10^4 m^3; (b) 2×10^6 bottles, 2×10^7 bottles; (c) 2×10^6 kg, 2×10^7 kg
25. (a) 22 pecks; (b) 5.5 Imperial bushels; (c) 200 L
26. 1 kilomole
27. $\approx 1 \times 10^{36}$
28. (a) 0.900, 7.50×10^{-2}, 1.56×10^{-3}, 8.32×10^{-6}; (b) 1.00, 8.33×10^{-2}, 1.74×10^{-3}, 9.24×10^{-6}; (c) 12.0, 1.00, 2.08×10^{-2}, 1.11×10^{-4}; (d) 576, 48, 1.00, 5.32×10^{-3}; (e) 1.08×10^5, 9.02×10^3, 188, 1.00; (f) 1.96 m^3
29. (a) 18.8 gallons; (b) 22.5 gallons
30. (a) 2.5 cups, 2 teaspoons; (b) 0.5 quart; (c) 2 teaspoons; (d) 1 teaspoon
31. (a) 14.5 roods; (b) 1.47×10^4 m^2
32. 403 L
33. 0.260 kg
34. (a) 1.0 m^3; (b) 6.0×10^{-4} m^3
35. (a) 11.3 m^3/L; (b) 1.13×10^4 m^{-1}; (c) 2.17×10^{-3} gal/ft^2; (d) number of gallons to cover a square foot if spread uniformly
36. 5.2×10^6 m
37. 0.3 cord
38. (a) 3.88; (b) 7.65; (c) 156 ken^3; (d) 1.19×10^3 m^3
39. (a) 3.9 m, 4.8 m; (b) 3.9×10^3 mm, 4.8×10^3 mm; (c) 2.2 m^3, 4.2 m^3
40. (a) 3.0×10^{-26} kg; (b) 5×10^{46} molecules
41. (a) 293 U.S. bushels; (b) 3.81×10^3 U.S. bushels
42. 1.75×10^3 kg
43. 2×10^4 to 4×10^4 dbugs
44. 5.95 km
45. (a) 3 nebuchadnezzars, 1 methuselah; (b) 0.37 standard bottle; (c) 0.26 L
46. 700 to 1500 oysters
47. 0.12 AU/min
48. 1.3×10^9 kg
49. 6.0×19^{26} atoms
50. 8×10^2 km
51. 3.8 mg/s
52. 9.4×10^{-3}
53. 1.2 m
54. 1.9×10^5 kg
55. 10.7 habaneros
56. 0.020 km^3
57. (a) 4.9×10^{-6} pc; (b) 1.6×10^{-5} ly
58. (a) yes; (b) 8.6 universe seconds
59. (a) 400; (b) 6.4×10^7

Chapter 2

1. 13 m
2. 5.554 s
3. (a) +40 km/h; (b) 40 km/h
4. 48 km/h
5. (a) 0; (b) −2 m; (c) 0; (d) 12 m; (e) +12 m; (f) +7 m/s
6. (a) 1.74 m/s; (b) 2.14 m/s
7. (a) 0; (b) 4.0 m; (c) −0.82 s; (d) 0.82 s; (f) $+20t$; (g) increase
8. 128 km/h
9. 1.4 m
10. 60 km
11. (a) 73 km/h; (b) 68 km/h; (c) 70 km/h; (d) 0
12. (a) −6 m/s; (b) $-x$ direction;(c) 6 m/s; (d) decreasing; (e) 2 s; (f) no
13. (a) 28.5 cm/s; (b) 18.0 cm/s; (c) 40.5 cm/s; (d) 28.1 cm/s; (e) 30.3 cm/s
14. (a) 1.2 s; (b) 0; (c) positive; (d) negative
15. $-20\,\text{m/s}^2$
16. 5.9 m
17. (a) 54 m; (b) 18 m/s; (c) $-12\,\text{m/s}^2$; (d) 64 m; (e) 4.0 s; (f) 24 m/s; (g) 2.0 s; (h) $-24\,\text{m/s}^2$; (i) 18 m/s
18. (a) 1.10 m/s; (b) $6.11\,\text{mm/s}^2$; (c) 1.47 m/s; (d) $6.11\,\text{mm/s}^2$
19. (a) m/s^2; (b) m/s^3; (c) 1.0 s; (d) 82 m; (e) −80 m; (f) 0; (g) −12 m/s; (h) −36 m/s; (i) −72 m/s; (j) $-6\,\text{m/s}^2$; (k) $-18\,\text{m/s}^2$; (l) $-30\,\text{m/s}^2$; (m) $-42\,\text{m/s}^2$
20. (a) 0.100 m
21. (a) +1.6 m/s; (b) +18 m/s
22. (a) 5.00 s; (b) 61.5 m
23. (a) 3.1×10^6 s; (b) 4.6×10^{13} m
24. (a) 30 s; (b) 300 m
25. $1.62 \times 10^{15}\,\text{m/s}^2$
26. $21g$
27. (a) $3.56\,\text{m/s}^2$; (b) 8.43 m/s
28. (a) 2.5 s
29. (a) 10.6 m; (b) 41.5 s
30. (a) 56.6 s; (b) 31.8 m/s
31. (a) $4.0\,\text{m/s}^2$; (b) $+x$
32. (a) 15.0 m; (b) 94 km/h
33. (a) $-2.5\,\text{m/s}^2$; (b) 1; (d) 0; (e) 2
34. (a) 32.9 m/s; (b) 49.1 s; (c) 11.7 m/s
35. 40 m
36. (a) −50 km/h; (b) $-2.0\,\text{m/s}^2$
37. (a) $0.994\,\text{m/s}^2$
38. (a) 1.54 s; (b) 27.1 m/s
39. (a) 29.4 m; (b) 2.45 s
40. 183 m/s; no
41. (a) 31 m/s; (b) 6.4 s
42. (a) 3.70 m/s; (b) 1.74 s; (c) 0.154 m
43. (a) 5.4 s; (b) 41 m/s
44. (a) 0.45 s; (b) 38 m/s; (c) 42 m/s
45. 9.6 m/s
46. 3.0 m/s
47. 4.0m/s
48. (a) 20 m; (b) 59 m
49. $857\,\text{m/s}^2$, upward
50. 26 m
51. $+1.26 \times 10^3\,\text{m/s}^2$; (b) upward
52. (a) 12.3 m/s
53. (a) 89 cm; (b) 22 cm
54. (a) 3.41 s; (b) 57 m
55. 2.34 m
56. (a) 350 ms; (b) 82 ms (each includes both ascent and descent through the 15 cm)
57. 20.4 m
58. (a) $8.0\,\text{m/s}^2$; (b) 20 m/s
59. (a) 2.25 m/s; (b) 3.90 m/s
60. (a) 0.13 m; (b) 0.50 m
61. 100 m
62. 0.56 m/s
63. (a) 82 m; (b) 19 m/s
64. yes, 0, 10 m/s
65. 17 m/s
66. (a) 2.00 s; (b) 12 cm; (c) $-9.00\,\text{cm/s}^2$; (d) right; (e) left; (f) 3.46 s
67. (a) 15.7 m/s; (b) 12.5 m; (c) 82.3 m
68. (a) 5.00 m/s; (b) $1.67\,\text{m/s}^2$; (c) 7.50 m
69. (a) $2.0\,\text{m/s}^2$; (b) 12 m/s; (c) 45 m
70. (a) 3.2 s; (b) 1.3 s
71. (a) either; (b) neither
72. (a) 60.6 s; (b) 36.3 m/s
73. (a) $9.08\,\text{m/s}^2$; (b) $0.926g$; (c) 6.12 s; (d) $15.3T_r$; (e) braking; (f) 5.56 m
74. 8.4 m
75. +47 m/s
76. (a) 3.5; (b) $(5.0\,\text{m})/v_s$
77. 217 m/s
78. (a) 2.5 m/s; (b) 8.0 m/s; (c) $1.0\,\text{m/s}^2$; (d) 0
79. (a) 14 m/s; (b) 18 m/s; (c) 6.0 m/s; (d) $12\,\text{m/s}^2$; (e) 24 m/s; (f) $24\,\text{m/s}^2$
80. (a) 38.1 m; (b) 9.02 m/s; (c) down; (d) 14.5 m/s; (e) up

81. 0.15 m/s
82. (a) 5.0 m/s; (b) $v = 3.0\,\text{m/s} + (0.50\,\text{m/s}^3)t^2$
83. (a) 5.0 m/s^2; (b) 4.0 s; (c) 6.0 s; (d) 90 m
84. (a) 48.5 m/s; (b) 4.95 s; (c) 34.3 m/s; (d) 3.50 s
85. 25g; (b) 400 m
86. 10.2 s; (b) 10.0 m
87. (a) 3.1 m/s^2; (b) 45 m; (c) 13 s
88. (a) 15 m; (b) 2.0 m/s; (c) −2.0 m/s^2; (d) 3.5 m/s; (e) 0
89. (a) 8.85 m/s; (b) 1.00 m
90. (a) 1.23 cm; (b) 4 times; (c) 9 times; (d) 16 times; (e) 25 times
91. 34 m
92. 22.0 m/s
93. 4H
94. 1.5 s
95. (a) 0.74 s; (b) 6.2 m/s^2
96. 414 ms
97. (a) 80 m/s; (b) 110 m/s; (c) 20 m/s^2
98. 3.75 ms
99. 39 m/s
100. 25 km/h
101. 2.3 cm/min
102. 94 m
103. 1.2 h
104. 90 m
105. (a) 17 s; (b) 290 m
106. 1.3 s
107. (a) 0.75 s; (b) 50 m
108. (a) 3.0 s; (b) 9.0 m
109. (a) 18 m/s; (b) 83 m
110. 0.556 s
111. 2.78 m/s^2
112. (a) 0.28 m/s^2; (b) 0.28 m/s^2
113. (a) 32 m; (b) 1.6 m/s (c) 24.5 s; (d) 1.3 m/s
114. (a) $d = \frac{1}{2}at^2$; (c) 7.2 m/s^2
116. (a) 5.44 s; (b) 53.3 m/s; (c) 5.80 m
117. (a) $v = \sqrt{v_0^2 + 2gh}$; (b) $t = [\sqrt{v_0^2 + 2gh} - v_0]/g$; (c) same as (a); (d) $t = [\sqrt{v_0^2 + 2gh} + v_0]/g$, greater

Chapter 3

1. (a) −2.5 m; (b) −6.9 m
2. (a) 0.349 rad; (b) 0.873 rad; (c) 1.75 rad; (d) 18.9°; (e) 120°; (f) 441°
3. (a) 47.2 m; (b) 122°
4. (a) 13 m; (b) 7.5 m
5. (a) 156 km; (b) 39.8° west of north
6. (a) 4.28 m; (b) 11.7 m
7. (a) 6.42 m; (b) no; (c) yes; (d) yes; (e) a possible answer: $(4.30\,\text{m})\hat{\imath} + (3.70\,\text{m})\hat{\jmath} + (3.00\,\text{m})\hat{k}$; (f) 7.96 m
8. (b) 3.2 m; (c) 41° south of west
9. 4.74 km
10. (a) 81 km; (b) 40° north of east
11. (a) $(-9.0\,\text{m})\hat{\imath} + (10\,\text{m})\hat{\jmath}$; (b) 13 m; (c) 132°
12. (a) 12 m; (b) −5.8 m; (c) −2.8 m
13. (a) −70.0 cm; (b) 80.0 cm; (c) 141 cm; (d) −172°
14. (a) $(8.0\,\text{m})\hat{\imath} + (2.0\,\text{m})\hat{\jmath}$; (b) 8.2 m; (c) 14°; (d) $(2.0\,\text{m})\hat{\imath} - (6.0\,\text{m})\hat{\jmath}$; (e) 6.3 m; (f) −72°
15. (a) $(3.0\,\text{m})\hat{\imath} - (2.0\,\text{m})\hat{\jmath} + (5.0\,\text{m})\hat{k}$; (b) $(5.0\,\text{m})\hat{\imath} - (4.0\,\text{m})\hat{\jmath} - (3.0\,\text{m})\hat{k}$; (c) $(-5.0\,\text{m})\hat{\imath} + (4.0\,\text{m})\hat{\jmath} + (3.0\,\text{m})\hat{k}$
16. (a) −80 m; (b) 110 m; (c) 143 m; (d) 168°
17. (a) 38 m; (b) −37.5°; (c) 130 m; (d) 1.2°; (e) 62 m; (f) 130°
18. (a) 26.6 m; (b) −151°
19. (a) 1.59 m; (b) 12.1 m; (c) 12.2 m; (d) 82.5°
20. (a) $(1.28\,\text{m})\hat{\imath} + (6.60\,\text{m})\hat{\jmath}$; (b) 6.72 m; (c) 79.0°; (d) 1.38 rad
21. 5.39 m at 21.8° left of forward
22. 2.6 km
23. (a) 0.84 m; (b) 79° south of west
24. (a) 5.0 km; (b) 4.3° south of west
25. (a) $a\hat{\imath} + a\hat{\jmath} + a\hat{k}$; (b) $-a\hat{\imath} + a\hat{\jmath} + a\hat{k}$; (c) $a\hat{\imath} - a\hat{\jmath} + a\hat{k}$; (c) $a\hat{\imath} - a\hat{\jmath} + a\hat{k}$; (d) $-a\hat{\imath} - a\hat{\jmath} + a\hat{k}$; (e) 54.7°; (f) $\sqrt{3}a$
26. (a) 9.51 m; (b) 14.1 m; (c) 13.4 m; (d) 10.5 m
27. (a) −18.8; (b) 26.9, in the positive z direction
28. (a) 2.0$\hat{k}$; (b) 26; (c) 46; (d) 5.8
29. (a) −21; (b) −9; (c) $5\hat{\imath} - 11\hat{\jmath} - 9\hat{k}$
30. 0
31. 22°
32. $-3.0\hat{\imath} - 3.0\hat{\jmath} - 4.0\hat{k}$
33. 70.5°
34. 540
35. (a) 3.00 m; (b) 0; (c) 3.46 m; (d) 2.00 m; (e) −5.00 m; (f) 8.66 m; (g) −6.67; (h) 4.33
36. (a) 31$\hat{k}$; (b) 8.0; (c) 33; (d) 1.6
37. (a) 27.8 m; (b) 13.4 m

38. (a) 57°; (b) 2.2 m; (c) −4.5 m; (d) −2.2 m; (e) 4.5 m
39. (a) 168 cm; (b) 32.5°
40. (a) $-(40\,\text{m})\hat{\text{i}} - (20\,\text{m})\hat{\text{j}} + (25\,\text{m})\hat{\text{k}}$; (b) 45 m
41. 4.1
42. 3.2
43. (a) 0; (b) −16; (c) −9
44. (a) 12; (b) $+z$; (c) 12; (d) $-z$; (e) 12; (f) $+z$
45. (a) 30; (b) 52
46. (a) $(9.19\,\text{m})\hat{\text{i}}' + (7.71\,\text{m})\hat{\text{j}}'$; (b) $(14.0\,\text{m})\hat{\text{i}}' + (3.41\,\text{m})\hat{\text{j}}'$
47. (a) 103 km; (b) 60.9° north of west
48. (a) $(10.0\,\text{m})\hat{\text{i}} + (1.63\,\text{m})\hat{\text{j}}$; (b) 10.2 m; (c) 9.24°
49. (a) 5.0 m; (b) −37°; (c) 10 m; (d) 53°; (e) 11 m; (f) 27°; (g) 11 m; (h) 80°; (i) 11 m; (j) 260°; (k) 180°
50. (a) −28 cm; (b) −28 cm; (c) 50 cm; (d) 0; (e) 30 cm; (f) 52 cm; (g) 52 cm; (h) 24 cm; (i) 57 cm; (j) 25° north of east; (k) 57 cm; (l) 25° south of west
51. (a) 370 m; (b) 36° north of east; (c) 425 m; (d) the distance
52. (a) $2.81\,\text{m}^2$; (b) $(1.43\,\text{m}^2)\hat{\text{i}} + (4.86\,\text{m}^2)\hat{\text{j}} - (2.48\,\text{m}^2)\hat{\text{k}}$; (c) 63.5°
53. (a) $(9.0\,\text{m})\hat{\text{i}} + (6.0\,\text{m})\hat{\text{j}} - (7.0\,\text{m})\hat{\text{k}}$; (b) 123°; (c) −3.2 m; (d) 8.2 m
54. (a) 140°; (b) 90.0°; (c) 99.1°
55. (a) −83.4; (b) $(1.14\times10^3)\hat{\text{k}}$; (c) 1.14×10^3, θ not defined, $\phi = 0°$; (d) 90.0°; (e) $-5.14\hat{\text{i}} + 6.13\hat{\text{j}} + 3.00\hat{\text{k}}$; (f) 8.54, $\theta = 130°$, $\phi = 69.4°$
56. (a) $+x$ direction; (b) $+y$ direction; (c) 0; (d) 0; (e) $+z$ direction; (f) $-z$ direction; (g) d_1d_2; (h) d_1d_2; (i) $d_1d_2/4$; (j) $+z$ direction
57. (a) 15 m; (b) south; (c) 6.0 m; (d) north
58. (a) $8\hat{\text{i}} + 16\hat{\text{j}}$; (b) $2\hat{\text{i}} + 4\hat{\text{j}}$
59. (a) $(-3.18\,\text{m})\hat{\text{i}} + (4.72\,\text{m})\hat{\text{j}}$; (b) 5.69 m; (c) +124°
60. 2.2 m
61. (a) $3.0\,\text{m}^2$; (b) $52\,\text{m}^3$; (c) $(11\,\text{m}^2)\hat{\text{i}} + (9.0\,\text{m}^2)\hat{\text{j}} + (3.0\,\text{m}^2)\hat{\text{k}}$
62. (a) 4.2 m; (b) 50° north of east; (c) 8.0 m; (d) 24° north of west
63. (a) 1.8 m; (b) 69° north of east
64. (a) −2.83; (b) −2.83 m; (c) 5.00 m; (d) 0; (e) 3.00 m; (f) 5.20 m; (g) 5.17 m; (h) 2.37 m; (i) 5.69 m; (j) 25° north of east; (k) 5.69 m; (l) 25° south of west
65. (a) 2.97; (b) $1.51\hat{\text{i}} + 2.67\hat{\text{j}} - 1.36\hat{\text{k}}$; (c) 48°
66. (a) $11\hat{\text{i}} + 5.0\hat{\text{j}} - 7.0\hat{\text{k}}$; (b) 120°; (c) −4.9; (d) 7.3
67. 3.6 m
68. (a) $+y$; (b) $-y$; (c) 0; (d) 0; (e) $+z$; (f) $-z$; (g) ab for both; (h) ab/d; (i) $+z$
69. (a) 10 m; (b) north; (c) 7.5 m; (d) south
70. (a) $(1000\,\text{m})\hat{\text{i}} + (2000\,\text{m})\hat{\text{j}} - (500\,\text{m})\hat{\text{k}}$; (b) 0
71. 70.5°
72. (a) $9\hat{\text{i}} + 12\hat{\text{j}}$; (b) $3\hat{\text{i}} + 4\hat{\text{j}}$
73. (a) 0; (b) 0; (c) −1; (d) west; (e) up; (f) west
74. (a) parallel; (b) antiparallel; (c) perpendicular
75. Walpole (where the state prison is located)
76. (b) $a^2 b \sin\phi$

Chapter 4

1. (a) 6.2 m
2. (a) $(-5.0\,\text{m})\hat{\text{i}} + (8.0\,\text{m})\hat{\text{j}}$; (b) 9.4 m; (c) 122°; (e) $(8.0\,\text{m})\hat{\text{i}} - (8.0\,\text{m})\hat{\text{j}}$; (f) 11 m; (g) −45°
3. $(-2.0\,\text{m})\hat{\text{i}} + (6.0\,\text{m})\hat{\text{j}} - (10\,\text{m})\hat{\text{k}}$
4. (a) 14 cm; (b) −135°; (c) 20 cm; (d) 90°; (e) 0; (f) 0
5. (a) 7.59 km/h; (b) 22.5° east of north
6. (a) $(3.00\,\text{m/s})\hat{\text{i}} - (8.00\,\text{m/s}^2)t\hat{\text{j}}$; b) $(3.00\,\text{m/s})\hat{\text{i}} - (16.0\,\text{m/s})\hat{\text{j}}$; (c) 16.3 m/s; (d) −79.4°
7. $(-0.70\,\text{m/s})\hat{\text{i}} + (1.4\,\text{m/s})\hat{\text{j}} - (0.40\,\text{m/s})\hat{\text{k}}$
8. (a) 1.08×10^3 km; (b) 26.6° east of south; (c) 480 km/h; (d) 26.6° east of south; (e) 644 km/h
9. (a) $(8\,\text{m/s}^2)t\hat{\text{j}} + (1\,\text{m/s})\hat{\text{k}}$; (b) $(8\,\text{m/s}^2)\hat{\text{j}}$
10. (a) 56.6 m; (b) 45° north of west; (c) 1.89 m/s; (d) 45° north of west; (e) $0.471\,\text{m/s}^2$; (f) 45° north of east
11. (a) $(6.00\,\text{m})\hat{\text{i}} - (106\,\text{m})\hat{\text{j}}$; (b) $(19.0\,\text{m/s})\hat{\text{i}} - (224\,\text{m/s})\hat{\text{j}}$; (c) $(24.0\,\text{m/s}^2)\hat{\text{i}} - (336\,\text{m/s}^2)\hat{\text{j}}$; (d) −85.2°
12. (a) 15.8 m/s; (b) 42.6°
13. $(32\,\text{m/s})\hat{\text{i}}$
14. (a) $(-18\,\text{m/s}^2)\hat{\text{i}}$; (b) 0.75 s; (c) never; (d) 2.2 s
15. (a) $(-1.50\,\text{m/s})\hat{\text{j}}$; (b) $(4.50\,\text{m})\hat{\text{i}} - (2.25\,\text{m})\hat{\text{j}}$
16. 60°
17. (a) 3.03 s; (b) 758 m; (c) 29.7 m/s

18. 25.9 cm
19. (a) 11 m; (b) 23 m; (c) 17 m/s; (d) 63°
20. (a) 0.495 s; (b) 3.07 m/s
21. (a) 18 cm; (b) 1.9 m
22. (a) 51.8 m; (b) 27.4 m/s; (c) 67.5 m
23. (a) 10.0 s; (b) 897 m
24. (a) 16.9 m; (b) 8.21 m; (c) 27.6 m; (d) 7.26 m; (e) 40.2 m; (f) 0
25. (a) 1.60 m; (b) 6.86 m; (c) 2.86 m
26. (a) yes; (b) 20 cm; (c) no; (d) 86 cm
27. (a) 202 m/s; (b) 806 m; (c) 161 m/s; (d) −171 m/s
28. 5.8 m/s
29. 78.5°
30. (a) 12.0 m; (b) 19.2 m/s; (c) 4.80 m/s; (d) no
31. 4.84 cm
32. (a) 0.205 s; (b) 0.205 s; (c) 20.5 cm; (d) 61.5 cm
33. (a) 32.2 m; (b) 21.9 m/s; (c) 40.4°; (d) below the horizontal.
34. (a) 95 m; (b) 31 m
35. (a) it lands on the ramp; (b) 5.82 m; (c) 31.0°
36. (a) 33.7 m; (b) 26.0 m/s; (c) 71.1°
37. (a) yes; (b) 2.56 m
38. 14°
39. (a) 31°; (b) 63°
40. 42 m/s
41. the third
42. (a) 20 m/s; (b) 36 m/s; (c) 74 m
43. (a) 75.0 m; (b) 31.9 m/s; (c) 66.9°; (d) 25.5 m
44. $4.0\,\mathrm{m/s^2}$
45. (a) 12 s; (b) $4.1\,\mathrm{m/s^2}$; (c) down; (d) $4.1\,\mathrm{m/s^2}$; (e) up
46. (a) 0; (b) 0
47. (a) 7.32 m; (b) west; (c) north
48. (a) 0.94 m; (b) 19 m/s; (c) $2.4\,\mathrm{km/s^2}$; (d) 50 ms
49. $(3.00\,\mathrm{m/s^2})\hat{\mathrm{i}} + (6.00\,\mathrm{m/s^2})\hat{\mathrm{j}}$
50. (a) 8.82 m; (b) 6.00 m
51. 2.92 m
52. (a) 4.00 m; (b) 6.00 m
53. $160\,\mathrm{m/s^2}$
54. (a) $5.24\,\mathrm{m/s^2}$; (b) $3.33\,\mathrm{m/s^2}$
55. (a) $13\,\mathrm{m/s^2}$; (b) eastward; (c) $13\,\mathrm{m/s^2}$; (d) eastward
56. (a) 5 km/h; (b) positive x; (c) 1 km/h; (d) negative x
57. 5/3
58. 130°
59. 60°
60. (a) 185 km/h; (b) 22° south of west
61. (a) 38 knots; (b) 1.5° east of north; (c) 4.2 h; (d) 1.5° west of south
62. 240 km/h
63. 32 m/s
64. (a) 24.8 m/s; (b) 83.8° north of east; (c) $0.40\,\mathrm{m/s^2}$; (d) 60.0° north of east
65. (a) $(-32\,\mathrm{km/h})\hat{\mathrm{i}} - (46\,\mathrm{km/h})\hat{\mathrm{j}}$; (b) $[(2.5\,\mathrm{km}) - (32\,\mathrm{km/h})t]\,\hat{\mathrm{i}} + [(4.0\,\mathrm{km}) - (46\,\mathrm{km/h})t]\,\hat{\mathrm{j}}$; (c) 0.084 h; (d) 0.20 km
66. (a) 37° west of north; (b) 62.6 s
67. (a) 2.7 km; (b) 76° clockwise
68. (a) A: 10.1 km, 0.556 km;
    B: 12.1 km, 1.51 km;
    C: 14.3 km, 2.68 km;
    D: 16.4 km, 3.99 km;
    E: 18.5 km, 5.53 km;
    (b) the rocks form a curtain that curves upward and away from you
69. (a) 55.6°; (b) 6.85 m; (c) 6.78 m/s
70. 2.64 m
71. (a) 0.83 cm/s; (b) 0; (c) 0.11 m/s; (d) −63°
72. (a) 8.43 m; (b) −129°
73. $(-2.69\,\mathrm{m/s})\hat{\mathrm{i}} + (-1.80\,\mathrm{m/s})\hat{\mathrm{j}}$
74. (a) 2.5 m; (b) 0.82 m; (c) $9.8\,\mathrm{m/s^2}$; (d) $9.8\,\mathrm{m/s^2}$
75. (a) 10 m/s; (b) 19.6 m/s; (c) 40 m; (d) 40 m
76. (a) $(72.0\,\mathrm{m})\hat{\mathrm{i}} + (90.7\,\mathrm{m})\hat{\mathrm{j}}$; (b) 49.5°
77. (a) 6.29°; (b) 83.7°
78. (a) 3.50 m/s; (b) $-0.125\,\mathrm{m/s^2}$
79. (a) −30°; (b) 69 min; (c) 80 min; (d) 80 min; (e) 0; (f) 60 min
80. (a) 2.6×10^2 m/s; (b) 45 s; (c) increase
81. (a) 63 km; (b) 18° south of east; (c) 0.70 km/h; (d) 18° south of east; (e) 1.6 km/h; (f) 1.2 km/h; (g) 33° north of east
82. (a) 1.3×10^5 m/s; (b) $7.9 \times 10^5\,\mathrm{m/s^2}$; (c) increase
83. (c) 2.10 s; (d) 25.7 m; (e) 25.7 m; (f) 0; (g) 1.71 s; (h) 13.5 m; (i) 4.76 m; (j) 12.6 m
84. 143 km/h
85. (a) 1030 m; (b) west

86. (a) $(80\,\text{km/h})\hat{\imath} - (60\,\text{km/h})\hat{\jmath}$
87. (a) 62 ms; (b) 4.8×10^2 m/s
88. (a) 7.3 km; (b) 80 km/h
89. (a) 6.7×10^6 m/s; (b) 1.4×10^{-7} s
90. (a) 7.2 m/s; (b) 16° west of north; (c) 29 s (not 28 s)
91. 3×10^1 m
92. (a) $(-7.0\,\text{m})\hat{\imath} + (12\,\text{m})\hat{\jmath}$
93. (a) 5.4×10^{-13} m; (b) decrease
94. (a) 14 m/s; (b) 14 m/s; (c) −10 m; (d) −4.9 m; (e) +10 m; (f) −4.9 m
95. (a) $(-1.5\,\text{m/s}^2)\hat{\imath} + (0.50\,\text{m/s}^2)\hat{k}$; (b) 1.6 m/s^2; (c) 162°
96. (a) 4.2 m, 45°; (b) 5.5 m, 68°; (c) 6.0 m, 90°; (d) 4.2 m, 135°; (e) 0.85 m/s, 135°; (f) 0.94 m/s, 90°; (g) 0.94 m/s, 180°; (h) 0.30 m/s^2, 180°; (i) 0.30 m/s^2, 270°
97. $(-2.1\,\text{m/s}^2)\hat{\imath} + (2.8\,\text{m/s}^2)\hat{\jmath}$
98. (a) $(6.0\,\text{m/s})\hat{\imath} + (4.2\,\text{m/s})\hat{\jmath}$; (b) $(18\,\text{m})\hat{\imath} + 6.3\,\text{m})\hat{\jmath}$
99. (a) 45 m; (b) 22 m/s
100. (a) 6.79 km/h; (b) 6.96°
101. 67 km/h
102. (a) 22 m; (b) 15 s
103. 7.0 m/s
104. (a) 2.00 ns; (b) 2.00 mm; (c) 1.00×10^7 m/s; (d) 2.00×10^6 m/s
105. (a) 16 m/s; (b) 23°; (c) above; (d) 27 m/s; (e) 57°; (f) below
106. (a) 38 ft/s; (b) 32 ft/s; (c) 9.3 ft
107. 48 s
108. (a) 24 m/s; (b) 65°
109. (a) from 75° east of south; (b) 30° east of north; for a second set of solutions substitute west for east in both answers
110. (a) 20.3 m/s; (b) 21.7 m/s
111. (a) 1.5; (b) (36 m, 54 m)
112. (a) 7.49 km/s; (b) 8.00 m/s^2
113. (a) 0, 0; 2.0 m, 1.4 m; 4.0 m, 2.0 m; 6.0 m, 1.4 m; 8.0 m, 0; (b) 2.0 m/s, 1.1 m/s; 2.0 m/s, 0; 2.0 m/s, −1.1 m/s; (c) 0, −0.87 m/s^2; 0, −1.2 m/s^2; 0, −.87 m/s^2
114. (a) 11 m; (b) 45 m/s
115. (a) 19 m/s; (b) 35 rev/min; (c) 1.7 s
116. (a) 0.035 m/s^2; (b) 84 min
117. (a) 76 m; (b) 4.2 s
118. 36 s, no
119. (a) $(10\,\text{m/s})\hat{\imath} + (10\,\text{m/s})\hat{\jmath}$; (b) 8.0 m/s^2; (c) 2.7 s; (d) 2.2 s
120. (a) $y = 7.5 - 4.0t + 0.5t^2$, with y in meters and t in seconds; (b) 3.0 s, 5.0 s; (c) 3.0 s; (d) 21 m; (e) $(-1.9\,\text{m/s}^2)\hat{\imath} + (1.1\,\text{m/s}^2)\hat{\jmath}$
121. (a) 2.1 m/s; (b) not accidental because horizontal launch speed is about 20% of world-class sprint speed
122. (a) 73 ft; (b) 7.6°; (c) 1.0 s
123. (a) yes; (b) 0.16 s
124. 0.421 m/s at 3.1° west of north
125. (a) $(1.00\,\text{m})\hat{\imath} - (2.00\,\text{m})\hat{\jmath} + (2.50\,\text{m})\hat{k}$; (b) 2.45 m; (c) $(2.50\,\text{cm/s})\hat{\imath} - (5.00\,\text{cm/s})\hat{\jmath} + (2.50\,\text{cm/s})\hat{k}$; (d) insufficient information
126. (a) 32.4 m; (b) 37.7 m
127. (a) 44 m; (b) 13 m; (c) 8.9 m
128. (a) 4.6×10^{12} m; (b) 2.4×10^5 s
129. (a) 48 m, west of center; (b) 48 m, west of center
130. (a) 1.63 s; (b) no (18 cm); (c) 14.3 m/s; (d) yes
131. longer by about 1 cm
132. 23 ft/s
133. (a) 5.8 m/s; (b) 17 m; (c) 67°
134. (a) 2.7g; (b) 3.8g
135. (a) 96.2 m; (b) 4.31 m; (c) 86.5 m forward; (d) 25.1 m up
136. 93° from the car's direction of motion

Chapter 5

1. 2.9 m/s^2
2. (a) 1.88 N; (b) 0.684 N; (c) $(1.88\,\text{N})\hat{\imath} + (0.684\,\text{N})\hat{\jmath}$
3. (a) 0; (b) $(4.0\,\text{m/s}^2)\hat{\jmath}$; (c) $(3.0\,\text{m/s}^2)\hat{\imath}$
4. (a) $(0.86\,\text{m/s}^2)\hat{\imath} - (0.16\,\text{m/s}^2)\hat{\jmath}$; (b) 0.88 m/s^2; (c) −11°
5. (a) $(-32.0\,\text{N})\hat{\imath} - (20.8\,\text{N})\hat{\jmath}$; (b) 38.2 N; (c) −147°
6. $(-34\,\text{N})\hat{\imath} - (12\,\text{N})\hat{\jmath}$
7. $(-2\,\text{N})\hat{\imath} + (6\,\text{N})\hat{\jmath}$
8. 56°
9. (a) 108 N; (b) 108 N; (c) 108 N
10. (a) 2.0 N; (b) down
11. (a) 4.0 kg; (b) 1.0 kg; (c) 4.0 kg; (d) 1.0 kg
12. (a) 0.26; (b) decrease
13. (a) 42 N; (b) 72 N; (c) 4.9 m/s^2
14. (a) 180 N; (b) east; (c) 4.0 m/s^2; (d) west; (e) 2.0 m/s^2; (f) east

15. (a) 11.7 N; (b) −59.0°
16. (a) 0; (b) (20 N) $\hat{\imath}$; (c) (−20 N) $\hat{\imath}$; (d) (−40 N) $\hat{\imath}$; (e) (−60 N) $\hat{\imath}$
17. 1.2×10^5 N
18. 3.1×10^2 N
19. (a) 0.022 m/s^2; (b) 8.3×10^4 km; (c) 1.9×10^3 m/s
20. 6.8×10^3 N
21. (a) 494 N; (b) up; (c) 494 N; (d) down
22. (a) 5.5 kN; (b) 2.7 s; (c) 4.0; (d) 2.0
23. 1.5 mm
24. (a) 566 N; (b) 1.13 kN
25. (a) (285 N) $\hat{\imath}$ + (705 N) $\hat{\jmath}$; (b) (285 N) $\hat{\imath}$ −(115 N) $\hat{\jmath}$; (c) 307 N; (d) −22.0°; (e) 3.67 m/s^2; (f) −22.0°
26. (a) (1.70 N) $\hat{\imath}$ + (3.06 N) $\hat{\jmath}$; (b) (1.70 N) $\hat{\imath}$ + (3.06 N) $\hat{\jmath}$; (c) (2.02 N) $\hat{\imath}$ + (2.71 N) $\hat{\jmath}$
27. (a) 0.62 m/s^2; (b) 0.13 m/s^2; (c) 2.6 m
28. (a) +68 N; (b) +28 N; (c) −12 N
29. (a) 1.18 m; (b) 0.674 s; (c) 3.50 m/s
30. 47.4 N
31. (a) 2.2×10^{-3} N; (b) 3.7×10^{-3} N
32. (a) 68 N; (b) 73 N
33. 1.8×10^4 N
34. (a) 7.3 kg; (b) 89 N
35. (a) 31.3 kN; (b) 24.3 kN
36. 16.0 kN
37. (a) 1.4 m/s^2; (b) 4.1 m/s
38. 176 N
39. (a) 1.23 N; (b) 2.46 N; (c) 3.69 N; (d) 4.92 N; (e) 6.15 N; (f) 0.250 N
40. (a) 6.8 kN; (b) −21°
41. (a) 2.18 m/s^2; (b) 116 N; (c) 21.0 m/s^2
42. 23 kg
43. (a) 1.1 N
44. (a) 2.50 m/s^2; (b) 30.0 N
45. (a) 0.970 m/s^2; (b) 11.6 N; (c) 34.9 N
46. (a) 36.8 N; (b) 19.1 cm
47. (a) 3.6 m/s^2; (b) 17 N
48. 5.1 m/s
49. (a) 4.9 m/s^2; (b) 2.0 m/s^2; (c) up; (d) 120 N
50. (a) 466 N; (b) 527 N; (c) 931 N; (d) 1.05 kN; (e) 931 N; (f) 1.05 kN; (g) 1.86 kN; (h) 2.11 kN
51. (a) 0.735 m/s^2; (b) down; (c) 20.8 N
52. 81.7 N
53. $2Ma/(a+g)$
54. (a) 3.1 N; (b) 15 N
55. (a) 8.0 m/s; (b) positive x direction
56. 18 kN
57. (a) 13 597 kg; (b) 4917 L; (c) 20 075 L; (d) 45%
58. 2.2 kg
59. 9.0 m/s^2
60. (b) $F/(m+M)$; (c) $FM/(m+M)$; (d) $F(m+2M)/2(m+M)$
61. (a) 0; (b) 0.83 m/s^2; (c) 0
62. (3 N) $\hat{\imath}$ − (11 N) $\hat{\jmath}$ + (4 N) $\hat{k}$
63. (a) 0.74 m/s^2; (b) 7.3 m/s^2
64. 16 N
65. (a) 3.5 N; (b) west; (c) 2.7 N; (d) 22° west of south
66. 2.4 N
67. 16 N
68. (a) 2.2×10^5 N; (b) 5.0×10^4 N
69. (a) rope breaks; (b) 1.6 m/s^2
70. (a) (1.0 m/s^2) $\hat{\imath}$ − (1.3 m/s^2) $\hat{\jmath}$; (b) 1.6 m/s^2; (c) −50°
71. 12 N
72. (a) 3260 N; (b) 2.7×10^3 kg; (c) 1.2 m/s^2
73. (a) 4.6 m/s^2; (b) 2.6 m/s^2
74. (a) 3.0 N; (b) 0.34 kg
75. 4.6 N
76. (a) 1.1 kN; (b) up; (c) 9.8×10^2 N; (d) up; (e) 8.1×10^2 N; (f) up
77. (a) 2.6 N; (b) 17°
78. (a) 65 N; (b) 49 N
79. (b) 313 N; (c) 0; (d) no; (e) yes
80. (a) 7.4×10^2 N; (b) 2.9×10^2 N; (c) 0; (d) 75 kg
81. (a) 11 N; (b) 2.2 kg; (c) 0; (d) 2.2 kg
82. (a) 1.8×10^2 m/s^2; (b) $12g$; (c) 1.4×10^8 N; (d) 4.2 y
83. (a) 1.8×10^2 N; (b) 6.4×10^2 N
84. 6.35×10^2 N
85. (a) 620 N; (b) 580 N
86. (a) 4.6×10^3 N; (b) 5.8×10^3 N
87. (a) (5.0 m/s) $\hat{\imath}$ + (4.3 m/s) $\hat{\jmath}$; (b) (15 m) $\hat{\imath}$ + (6.4 m) $\hat{\jmath}$
88. (a) 590 N, up; (b) 340 N, up; (c) 590 N, down
89. (a) $\cos\theta$; (b) $\sqrt{\cos\theta}$
90. (a) (−6.26 N) $\hat{\imath}$ − (3.23 N) $\hat{\jmath}$; (b) 7.04 N; (c) 207°
91. (a) 4.9×10^5 N; (b) 1.5×10^6 N
92. (a) 245 m/s^2; (b) 20.4 kN

93. (a) 4.1 m/s^2; (b) 836 N
94. (a) 4 kg; (b) 6.5 m/s^2; (c) 13 N
95. (a) $(1.0 \text{ N})\hat{\imath} - (2.0 \text{ N})\hat{\jmath}$; (b) 2.2 N; (c) $-63°$; (d) 2.2 m/s^2; (e) $-63°$
96. 195 N
97. (a) 1.1×10^{-15} N; (b) 8.9×10^{-30} N
98. 10 m/s^2
99. (a) 44 N; (b) 78 N; (c) 54 N; (d) 152 N
100. (a) 27 N; (b) 27° north of west
101. (a) 2.8 N, west; (b) 2.2 N, 22° west of south

Chapter 6

1. 2°
2. 0.61
3. (a) 2.0×10^2 N; (b) 1.2×10^2 N
4. 1.6×10^2 N
5. (a) 1.9×10^2 N; (b) 0.56 m/s^2
6. 36 m
7. (a) 11 N; (b) 0.14 m/s^2
8. 0.53
9. 0.58
10. (a) 6.0 N; (b) 3.6 N; (c) 3.1 N
11. (a) 1.1×10^2 N; (b) 1.3×10^2 N; (c) no; (d) 46 N; (e) 17 N
12. 2.8×10^2 N
13. (a) 3.0×10^2 N; (b) 1.3 m/s^2
14. (b) 3.0×10^7 N
15. (a) no; (b) $(-12 \text{ N})\hat{\imath} + (5.0 \text{ N})\hat{\jmath}$
16. 0.54
17. (a) 19°; (b) 3.3 kN
18. (a) 12.1 m/s; (b) 19.4 m/s
19. (a) $(17 \text{ N})\hat{\imath}$; (b) $(20 \text{ N})\hat{\imath}$; (c) $(15 \text{ N})\hat{\imath}$
20. (a) 8.6 N; (b) 46 N; (c) 39 N
21. 1.0×10^2 N
22. 8.5 N
23. 0.37
24. (a) 147 N; (b) same
25. (a) 3.5 m/s^2; (b) 0.21 N
26. 3.3 kg
27. (a) 0; (b) $(-3.9 \text{ m/s}^2)\hat{\imath}$; (c) $(-1.0 \text{ m/s}^2)\hat{\imath}$
28. (a) 66 N; (b) 2.3 m/s^2
29. 4.9×10^2 N
30. (a) $(-6.1 \text{ m/s}^2)\hat{\imath}$; (b) $(-0.98 \text{ m/s}^2)\hat{\imath}$
31. 9.9 s
32. 3.75
33. 2.3
34. (a) 2×10^4 N; (b) $18g$
35. (a) 3.2×10^2 km/h; (b) 6.5×10^2 km/h; (c) no
36. 48 km/h
37. 21 m
38. $9.7g$
39. 0.60
40. (a) 3.7 kN; (b) up; (c) 1.3 kN; (d) down
41. 1.37×10^3 N
42. (a) 547 N; (b) 9.53°
43. (a) 10 s; (b) 4.9×10^2 N; (c) 1.1×10^3 N
44. (a) 3.7 kN; (b) up; (c) 2.3 kN; down
45. (a) light; (b) 778 N; (c) 223 N; (d) 1.11 kN
46. 12°
47. 2.2 km
48. 2.6×10^3 N
49. 1.81 m/s
50. 0.078
51. (a) 8.74 N; (b) 37.9 N; (c)6.45 m/s; (d) radially inward
52. (a) 0.40 N; (b) 1.9 s
53. (a) 69 km/h; (b) 139 km/h; (c) yes
54. (a) $\sqrt{Rg(\tan\theta + \mu_s)/(1 - \mu_s \tan\theta)}$; (b) 149 km/h; (c) 76.2 km/h
55. (a) 222 N; (b) 334 N; (c) 311 N; (d) 311 N; (e) c, d
56. 3.4%
57. (a) 7.5 m/s^2; (b) down; (c) 9.5 m/s^2; (d) down
58. (b) 55°; (c) increase; (d) 59°
59. (a) $\mu_s mg/(\sin\theta - \mu_s \cos\theta)$; (b) $\theta_0 = \tan^{-1}\mu_s$
60. (a) lowest point; (b) 8.73 m/s
61. (a) 27 N; (b) 3.0 m/s^2
62. 8.8 N
63. (a) 35.3 N; (b) 39.7 N; (c) 320 N
64. 0.74
65. (a) 3.0 N; (b) 3.0 N; (c) 1.6 N; (d) 4.4 N; (e) 1.0 N; (f) e
66. 9.4 N
67. $g(\sin\theta - \sqrt{2}\mu_s \cos\theta)$
68. (a) 1.05 N; (b) 3.62 m/s^2; (c) answers are the same except that the rod is under compression
69. (a) 13 N; (b) 1.6 m/s^2
70. (a) 0.58; (b) 0.54
71. 118 N
72. (a) 11°; (b) 0.19
73. (a) $v_0^2/(4g\sin\theta)$; (b) no
74. (a) 12 N; (b) 10 N; (c) 26 N; (d) 23 N; (e) 32 N; (f) 23 N; (g) d; (h) f; (i) a, c, d

75. 0.76
76. (a) 3.0×10^5 N; (b) 1.2°
77. (a) 30 cm/s; (b) 180 cm/s^2; (c) inward; (d) 3.6×10^{-3} N; (e) inward; (f) 0.37
78. 0.12 m
79. 4.6 N
80. (a) 6.80 s; (b) 6.76 s
81. 20°
82. (a) 2.2 m/s^2; (b) 53 N
83. (a) 0.11 m/s^2; (b) 0.23 m/s^2; (c) 0.041; (d) 0.029%
84. 147 m/s
85. (a) 0.34; (b) 0.24
86. (a) 190 N; (b) 320 N
87. (a) 3.21×10^3 N; (b) 3.75×10^3 N
88. (a) 0.96 m/s; (b) 0.021
89. 178 km/h
90. 3.4 m/s^2
91. 0.18
92. (a) 0.37; (b) $0.37 < \mu_s < 0.47$
93. (a) 100 N; (b) 245 N; (c) 86.6 N; (d) 195 N; (e) 50.0 N; (f) 158 N; (g) at rest; (h) slides; (i) at rest
94. (a) 90 N; (b) 70 N; (c) 0.88 m/s^2
95. 0.56
96. (a) 56 N; (b) 59 N; (c) 1.1×10^3 N
97. (a) 2.1 m/s^2; (b) down the plane; (c) 3.9 m; (d) it stays there
98. (a) 210 N; (b) 44.0 m/s
99. (a) 275 N; (b) 877 N
100. 6.2 kN
101. 874 N
102. (a) 240 N; (b) 0.60
103. (a) 84.2 N; (b) 52.8 N; (c) 1.87 m/s^2
104. (a) 0.13 N; (b) 0.12
105. (a) 74 N; (b) $(76\,\text{N})/(\cos\theta + 0.42\sin\theta)$; (c) 23°; (d) 70 N
106. (a) 0.0338 N; (b) 9.77 N
107. (a) bottom of circle; (b) 9.5 m/s
108. (a) 5.1 m/s^2; (b) 4.8 N; (c) 10 N

Chapter 7

1. 1.8×10^{13} J
2. (a) 5×10^{14} J; (b) 0.1 megaton TNT; (c) 8 bombs
3. (a) 2.9×10^7 m/s; (b) 2.1×10^{-13} J
4. 7.1 J
5. (a) 2.4 m/s; (b) 4.8 m/s
6. 20 J
7. 6.8 J
8. 5.0 kJ
9. 0.96 J
10. 3.5 m/s
11. (a) 1.7×10^2 N; (b) 3.4×10^2 m; (c) -5.8×10^4 J; (d) 3.4×10^2 N; (e) 1.7×10^2 N; (f) -5.8×10^4 J
12. (a) 3.00 N; (b) 9.00 J
13. (a) 1.50 J; (b) increases
14. 15.3 J
15. (a) 62.3°; (b) 118°
16. (a) 36 kJ; (b) 2.0×10^2 J
17. (a) 12 kJ; (b) −11 kJ; (c) 11 kJ; (d) 5.4 m/s
18. (a) 1.31 J; (b) 0.935 m/s
19. (a) $-3Mgd/4$; (b) Mgd; (c) $Mgd/4$; (d) $\sqrt{gd/2}$
20. 4.41 J
21. 25 J
22. (a) 8.84 kJ; (b) 7.84 kJ; (c) 6.84 kJ
23. (a) 25.9 kJ; (b) 2.45 N
24. 1.25 kJ
25. $x = -4.9$ cm and $x = +4.9$ cm
26. (a) 7.2 J; (b) 7.2 J; (c) 0; (d) −25 J
27. (a) 16 J; (b) 16 J; (c) 0; (d) −14 J
28. (a) 0.905 J; (b) 2.15 J; (c) 0
29. (a) 6.6 m/s; (b) 4.7 m
30. (a) 0.12 m; (b) 0.36 J; (c) −0.36 J; (d) 0.060 m; (e) 0.090 J
31. 8.0×10^2 J
32. 25 J
33. (a) 0; (b) 0
34. 0.21 J
35. 5.3×10^2 J
36. (a) 2.3 J; (b) 2.6 J
37. (a) 42 J; (b) 30 J; (c) 12 J; (d) 6.5 m/s, positive x direction; (e) 5.5 m/s, positive x direction; (f) 3.5 m/s, positive x direction
38. 4.00 N/m
39. +41.7 J
40. 2.7×10^5 W
41. 4.9×10^2 W
42. (a) 9.0×10^2 J; (b) 1.1×10^2 W; (c) 2.3×10^2 W
43. (a) 0.83 J; (b) 2.5 J; (c) 4.2 J; (d) 5.0 W
44. (a) 28 W; (b) $(6\,\text{m/s})\hat{\text{j}}$
45. 7.4×10^2 W
46. (a) 0; (b) -3.5×10^2 W

47. (a) 1.0×10^2 J; (b) 8.4 W
48. (a) 32.0 J; (b) 8.00 W; (c) 78.2°
49. (a) 12 J; (b) 4.0 m; (c) 18 J
50. (a) 0.29 J; (b) −1.8 J; (c) 3.5 m/s; (d) 23 cm
51. (a) 2.7×10^2 N; (b) -4.0×10^2 J; (c) 4.0×10^2 J; (d) 0; (e) 0
52. (a) 590 J; (b) 0; (c) 0; (d) 590 J
53. (a) 11 J; (b) −21 J
54. (a) $v_f = \sqrt{\cos\theta}$, with v_f in meters per second; (b) $v_f = \sqrt{1 + \cos\theta}$; (c) $v_f = \sqrt{1 - \cos\theta}$
55. (a) 0.6 J; (b) 0; (c) −0.6 J
56. (a) 6.0 N; (b) −2.5 N; (c) 15 N
57. (a) 1.20 J; (b) 1.10 m/s
58. (a) 13 J; (b) 13 J;
59. (a) 314 J; (b) −155 J; (c) 0; (d) 158 J;
60. (a) 32 J; 8.0 W; (c) 789°
61. (a) 8.0 N; (b) 8.0 N/m
62. −37 J;
63. (a) 98 N; (b) 4.0 cm; (c) 3.9 J; (d) −3.9 J
64. (a) 1.0×10^2 J; (b) 67 W; (c) 33 W
65. −6 J
66. 165 kW
67. (a) 1.7 W; (b) 0; (c) −1.7 W
68. 1.5 kJ
69. (a) 2.1×10^2 J; (b) 2.1×10^2 J
70. (a) 797 W; (b) 0; (c) −1.55 kJ; (d) 0; (e) 1.55 kJ; (f) because F varies during displacement
71. (a) 23 mm; (b) 45 N
72. (a) $c = 4$ m; (b) $c < 4$ m; (c) $c > 4$ m
73. 235 kW
74. (a) 6 J; (b) 6.0 J
75. (b) $x = 3.00$ m; (c) 13.5 J; (d) $x = 4.50$ m; (e) $x = 4.50$ m
76. 0.47 J
77. (a) 1.8×10^5 ft · lb; (b) 0.55 hp
78. (a) 2.5 kJ; (b) −2.1 kJ
79. (a) 1×10^5 megatons TNT; (b) 1×10^7 bombs
80. 6.67×10^5 J

Chapter 8

1. 89 N/cm
2. (a) 167 J; (b) −167 J; (c) 196 J; (d) 29 J; (e) 167 J; (f) −167 J; (g) 296 J; (h) 129 J
3. (a) 4.31 mJ; (b) −4.31 mJ; (c) 4.31 mJ; (d) −4.41 mJ; (e) all increase
4. (a) 1.51 J; (b) −1.51 J; (c) 0; (d) −1.51 J; (e) 1.51 J; (f) 0; (h) same
5. (a) 0; (b) 170 kJ; (c) 340 kJ; (d) 170 kJ; (e) 340 kJ; (f) increase
6. (a) 184 J; (b) −184 J; (c) −184 J
7. (a) 0.15 J; (b) 0.11 J; (c) 0.19 J; (d) 38 mJ; (e) 75 mJ; (f) all the same
8. (a) 13.1 J; (b) −13.1 J; (c) 13.1, J; (d) all increase
9. (a) 2.08 m/s; (b) 2.08 m/s; (c) increase
10. (a) 12.9 m/s; (b) 12.9 m/s; (c) increase
11. (a) 17.0 m/s; (b) 26.5 m/s; (c) 33.4 m/s; (d) 56.7 m; (e) all the same
12. (a) 2.98 m/s; (b) 4.21 m/s; (c) 2.98 m/s; (d) all the same
13. (a) 2.6×10^2 m; (b) same; (c) decrease
14. (a) 21.0 m/s; (b) 21.0 m/s; (c) 21.0 m/s
15. (a) 3.0 m; (b) 0.81 m; (c) 11 m/s; (d) 6.3 m/s; (f) 0.51 m
16. (a) 7.2 J; (b) −7.2 J; (c) 86 cm; (d) 26 cm
17. (a) 0.98 J; (b) −0.98 J; (c) 3.1 N/cm
18. 10 cm
19. (a) $U = 27 + 12x - 3x^2$; (b) 39 J; (c) −1.6 m; (d) 5.6 m
20. (a) 2.29 m/s; (b) same
21. (a) 2.5 N; (b) 0.31 N; (c) 30 cm
22. (a) no; (b) 9.3×10^2 N
23. (a) 4.85 m/s; (b) 2.42 m/s
24. -3.2×10^2 J
25. (a) 4.4 m; (b) same
26. (a) 8.35 m/s; (b) 4.33 m/s; (c) 7.45 m/s; (d) both decrease
27. (a) 5.0 m/s; (b) 79°; (c) 64 J
28. (a) 784 N/m; (b) 62.7 J; (c) 62.7 J; (d) 80.0 cm
29. (a) 35 cm; (b) 1.7 m/s
30. (a) 2.40 m/s; (b) 4.19 m/s
31. (a) 39.2 J; (b) 39.2 J; (c) 4.00 m
32. (a) 0.81 m/s; (b) 0.21 m; (c) 6.3 m/s^2; (d) up
33. (a) 2.8 m/s; (b) 2.7 m/s
34. 1.25 cm
35. −18 mJ
36. 9.20 m
37. (a) 2.1 m/s; (b) 10 N; (c) positive x direction; (d) 5.7 m; (e) 30 N; (f) negative x direction
38. (a) $1.12(A/B)^{1/6}$; (b) repulsive; (c) attractive

39. (a) -3.7 J; (c) 1.3 M; (d) 9.1 m; (e) 2.2 J; (f) 4.0 m; (g) $(4 - x)e^{-x/4}$; (h) 4.0 m
40. (a) 5.6 J; (b) 3.5 J
41. (a) 5.6×10^2 J; (b) 5.6×10^2 J
42. (a) 105 J; (b) 30.6 J; (c) 34.4 J
43. (a) 30.1 J; (b) 30.1 J; (c) 0.225
44. 11 kJ
45. (a) -2.9 kJ; (b) 3.9×10^2 J; (c) 2.1×10^2 N
46. 0.53 J
47. 20 ft · lb
48. (a) 1.5 MJ; (b) 0.51 MJ; (c) 1.0 MJ; (d) 63 m/s
49. 75 J
50. 1.2 m
51. (a) 67 J; (b) 67 J; (c) 46 cm
52. 0.15
53. (a) 0.292 m; (b) 14.2 J
54. 4.3 m
55. (a) 1.5×10^2 J; (b) 5.5 m/s
56. (a) 13 cm; (b) 2.7 m/s; (c) both increase
57. (a) -0.90 J; (b) 0.46 J; (c) 1.0 m/s
58. (a) 19.4 m; (b) 19.0 m/s
59. 20 cm
60. $H = 30$ cm
61. 3.5 m/s
62. (a) 7.4 m/s; (b) 90 cm; (c) 2.8 m; (d) 15 m
63. (a) 39.6 cm; (b) 3.64 cm
64. 0.72 m
65. (a) 10 m; (b) 49 N; (c) 4.1 m; (d) 1.2×10^2 N
66. (a) 216 J; (b) 1.18 kN; (c) 432 J; (d) motor also supplies thermal energy to crate and belt
67. 4.33 m/s
68. (a) -3.8 kJ; (b) 31 kN
69. (a) 4.9 m/s; (b) 4.5 N; (c) 71°; (d) same
70. 1.0 mJ
71. (a) 4.8 N; (b) positive x direction; (c) 1.5 m; (d) 13.5 m; (e) 3.5 m/s
72. (a) 31.0 J; (b) 5.35 m/s; (c) conservation
73. (a) 5.5 m/s; (b) 5.4 m/s; (c) same
74. (a) 1.5×10^{-2} N; (b) $(3.8 \times 10^2)g$
75. 69 hp
76. (a) 1.4 m/s; (b) 1.9 m/s; (c) 28°
77. (a) 13 m/s; (b) 11 m/s; (c) no, 9.3 m
78. (a) 18 J; (B) 0; (c) 30 J; (d) 0; (e) b and d
79. (a) 109 J; (b) 60.3 J; (c) 68.2 J; (d) 41.0 J
80. (a) 5.00 J; (b) 9.00 J; (c) 11.0 J; (d) 3.00 J; (e) 12.0 J; (f) 2.00 J; (g) 13.0 J; (h) 1.00 J; (i) 13.0 J; (j) 1.00 J; (l) 11.0 J; (m) 10.8 m; (n) it returns to $x = 0$ and stops
81. (a) 0.950 m/s; (b) 11.0 m
82. (a) 7 J; (b) 16 J
83. (a) 24 kJ; (b) 4.7×10^2 N
84. (a) 3.0×10^5 J; (b) 10 kW; (c) 20 kW
85. (a) 2.1×10^6 kg; (b) $\sqrt{100 + 1.5t}$ m/s; (c) $(1.5 \times 10^6)/\sqrt{100 + 1.5t}$ N; (d) 6.7 km
86. (a) 2.6 m; (b) 1.5 m; (c) 26 J; (d) 2.1 m/s
87. (a) 6.75 J; (b) -6.75 J; (c) 6.75 J; (d) 6.75 J; (e) -6.75 J; (f) 0.4459 m
88. (a) 54 m/s; (b) 52 m/s; (c) -76 m
89. 3.7 J
90. (a) 300 J; (b) 93.8 J; (c) 6.38 m
91. 5.4 kJ
92. 15 J
93. (a) 2.2 kJ; (b) 7.7×10^2 J
94. (a) 5.6 J; (b) 12 J; (c) 12 J
95. (a) 2.7 J; (b) 1.8 J; (c) 0.39 m
96. 56 m/s
97. 80 mJ
98. (a) 3.5 kJ; (b) 3.5 kJ
99. (a) 7.0 J; (b) 22 J
100. (a) 7.4×10^2 J; (b) 2.4×10^2 J
101. (a) 94 J; (b) 94 J; (c) 7.7 m/s
102. (a) -0.80 J; (b) -0.80 J; (c) $+1.1$ J
103. 5.5×10^6 N
104. (a) 12 m/s; (b) 11 cm
105. 25 J
106. (a) 44 m/s; (b) 0.036
107. 24 W
108. 100 m
109. (a) 2.35×10^3 J; (b) 352 J
110. (a) 0.2 to 0.3 MJ; (b) same amount
111. -12 J
112. (a) 7.8 MJ; (b) 6.2 bars
113. (a) 7.8 MJ; (b) 2.6 kJ; (c) 1.6 kW
114. 17 kW
115. (a) 3.7 J; (b) 4.3 J; (c) 4.3 J
116. 8580 J
117. (a) 3.0 mm; (b) 1.1 J; (d) yes; (e) ≈ 40 J; (f) no
118. (a) 9.2 m/s; (b) 4.8 m/s
119. (a) 6.0 kJ; (b) 8.6×10^2 W; (c) 3.0×10^2 W; (d) 9.0×10^2 W
120. (a) 8.6 kJ; (b) 8.6×10^2 W; (c) 4.3×10^2 W; (d) 1.3 kW
121. 3.1×10^{11} W
122. (a) 19 J; (b) 6.4 m/s; (c) 11 J, 6.4 m/s

123. (a) 0.75 J; (b) −1.0 J; (c) 0.25 J; (d) 1.0 J; (e) −2.0 J; (f) 1.0 J; (g) 0.75 J; (h) −3.0 J; (i) 2.3 J; (j) 0; (k) −4.0 J; (l) 4.0 J
124. (a) 6.4 m/s; (b) 4.9 m/s; (c) same
125. 880 MW
126. (a) 39 kW; (b) 39 kW
127. (a) 1.2 J; (b) 11 m/s; (c) no; (d) no
128. 738 m
129. (a) $v_0 = \sqrt{2gL}$; (b) $5mg$; (c) $-mgL$; (d) $-2mgL$
130. 181 W
131. (a) 2.7×10^9 J; (b) 2.7×10^9 W; (c) \$$2.4\times10^8$
132. 54%
133. (a) turning point on left, none on right, molecule breaks apart; (b) turning points on both left and right, molecule does not break apart; (c) -1.1×10^{-19} J; (d) 2.1×10^{-19} J; (e) $\approx 1 \times 10^{-9}$ N on each, directed toward the other; (f) $r < 0.2$ nm; (g) $r > 0.2$ nm; (h) $r = 0.2$ nm
135. (a) $U(x) = -Gm_1m_2/x$; (b) $Gm_1m_2d/x_1(x_1 + d)$
136. because your force on the cabbage (as you lower it) does work

Chapter 9

1. (a) 1.1 m; (b) 1.3 m; (c) toward
2. (a) −1.50 m; (b) −1.43 m
3. (a) 11 cm; (b) −4.4 cm
4. (a) −0.45 cm; (b) −2.0 cm
5. (a) 0; (b) 3.13×10^{-11} m
6. (a) −6.5 cm; (b) 8.3 cm; (c) 1.4 cm
7. (a) 20 cm; (b) 20 cm; (c) 16 cm
8. (a) 6.0 cm; (b) 6.0 cm; (c) descends to lowest point and then ascends to 6.0 cm; (d) 2.8 cm
9. (a) 28 cm; (b) 2.3 m/s
10. 6.2 m
11. (a) 22 m; (b) 9.3 m/s
12. $(-4.0\,\text{m})\hat{\mathrm{i}} + (4.0\,\text{m})\hat{\mathrm{j}}$
13. (a) $(2.35\,\text{m/s}^2)\hat{\mathrm{i}} - (1.57\,\text{m/s}^2)\hat{\mathrm{j}}$; (b) $\left[(2.35\,\text{m/s}^2)\hat{\mathrm{i}} - (1.57\,\text{m/s}^2)\hat{\mathrm{j}}\right] t$; (d) straight, at a downward angle of 34°
14. (a) 5.74 m; (b) $(10.0\,\text{m/s})\hat{\mathrm{i}}$; (c) $(-3.68\,\text{m/s}^2)\hat{\mathrm{j}}$
15. 53 m
16. 58 kg
17. 4.2 m
18. 4.2 kg · m/s
19. (a) 7.5×10^4 J; (b) 3.8×10^4 kg·m/s; (c) 39° south of east
20. (a) 30.0°; (b) $(-0.572\,\text{kg}\cdot\text{m/s})\hat{\mathrm{j}}$
21. 48°
22. (a) 5.0 kg · m/s; (b) 10 kg · m/s
23. (a) 67 m/s; (b) $-x$; (c) 1.2 kN; (d) $-x$
24. 1.0×10^3 to 1.2×10^3 kg · m/s
25. (a) 1.1 m; (b) 4.8×10^3 kg · m/s
26. (a) 42 N · s; (b) 2.1 kN
27. 5 N
28. (a) $(30\,\text{kg}\cdot\text{m/s})\hat{\mathrm{i}}$; (b) $(38\,\text{kg}\cdot\text{m/s})\hat{\mathrm{i}}$; (c) $(6.0\,\text{m/s})\hat{\mathrm{i}}$
29. (a) 5.86 kg · m/s: (b) 59.8°; (c) 2.93 kN; (d) 59.8°
30. (a) 4.50×10^{-3} N·s; (b) 0.529 N·s; (c) push
31. (a) 1.00 N · s; (b) 100 N; (c) 20 N
32. 9.9×10^2 N
33. (a) $(1.8\,\text{N}\cdot\text{s})\hat{\mathrm{j}}$; (b) $(-180\,\text{N})\hat{\mathrm{j}}$
34. (a) 7.17 N · s; (b) 16.0 kg · m/s
35. 3.0 mm/s
36. $(-1.4\,\text{m/s})\hat{\mathrm{i}}$
37. 4.4×10^3 km/h
38. (a) $(-0.15\,\text{m/s})\hat{\mathrm{i}}$; (b) 0.18 m
39. 3.5 m/s
40. $mv^2/6$
41. (a) 14 m/s; (b) −45°
42. 3.4 kg
43. (a) $(1.00\,\text{km/s})\hat{\mathrm{i}} - (0.167\,\text{km/s})\hat{\mathrm{j}}$; (b) 3.23 MJ
44. (a) 20 J; (b) 40 J
45. (a) 1.81 m/s; (b) 4.96 m/s
46. 3.1×10^2 m/s
47. (a) $(2.67\,\text{m/s})\hat{\mathrm{i}} - (3.00\,\text{m/s})\hat{\mathrm{j}}$; (b) 4.01 m/s; (c) 48.4°
48. (a) 4.6 m/s; (b) 3.9 m/s; (c) 7.5 m/s
49. (a) 721 m/s; (b) 937 m/s
50. 7.3 cm
51. (a) +2.0 m/s; (b) −1.3 J; (c) +40 J; (d) energy entered system from some source such as a small explosion
52. 2.6 m
53. 25 cm
54. 33 cm
55. (a) 99 g; (b) 1.9 m/s; (c) 0.93 m/s
56. (a) 1.9 m/s; (b) right; (c) yes
57. (a) 100 g; (b) 1.0 m/s
58. −28 cm
59. (a) 1.2 kg; (b) 2.5 m/s

60. (a) 2.47 m/s; (b) 1.23 m/s
61. (a) 3.00 m/s; (b) 6.00 m/s
62. (a) 2.22 m; (b) 0.556 m
63. (a) 0.21 kg; (b) 7.2 m
64. 1.0 kg
65. (a) 4.15×10^5 m/s; (b) 4.84×10^5 m/s
66. (a) $(10\,\text{m/s})\hat{\imath} + (15\,\text{m/s})\hat{\jmath}$; (b) −500 J
67. (a) 433 m/s; (b) 250 m/s
68. (a) 27°
69. 120°
70. (a) 2.7; (b) 7.4
71. (a) 1.57×10^6 N; (b) 1.35×10^5 kg; (c) 2.08 km/s
72. 108 m/s
73. (a) 46 N; (b) none
74. (a) stuck-together particles travel along the x axis; (b) one particle along line 2, the other along line 3; (c) one particle through region B, the other through region C, with paths symmetric about the x axis; (c) 3.06 m/s; (e) 4.00 m/s, each particle
75. (a) 7.11 m/s; (b) greater; (c) less; (d) less
76. (a) 1.78 m/s; (b) less; (c) less; (d) greater
77. (a) 1.92 m; (b) 0.640 m
78. (a) 40 m/s; (b) 0; (c) 60 m/s; (d) M
79. 28.8 N
80. 1.10 m/s
81. (a) 25 mm; (b) 26 mm; (c) down; (d) 1.6×10^{-2} m/s^2
82. (a) −0.25 m; (b) 0
83. (a) 11.4 m/s; (b) 95.1°
84. (a) $(-3.8\,\text{m/s})\hat{\imath}$; (b) $(7.2\,\text{m/s})\hat{\imath}$
85. (a) 7290 m/s; (b) 8200 m/s; (c) 1.271×10^{10} J; (d) 1.275×10^{10} J
86. (a) 0.800 kg · m/s; (b) 0.400 kg · m/s
87. (a) $(-4.0 \times 10^4$ kg · m/s; (b) west; (c) 0
88. (a) 0.60 cm; (b) 4.9 cm; (c) 9.0 cm; (d) 0
89. (a) down; (b) 0.50 m/s; (c) 0
90. (a) $(-0.450\,\text{kg}\cdot\text{m/s})\hat{\imath} - (0.450\,\text{kg}\cdot\text{m/s})\hat{\jmath} - (1.08\,\text{kg}\cdot\text{m/s})\hat{k}$; (b) $(-0.450\,\text{N}\cdot\text{s})\hat{\imath} - (0.450\,\text{N}\cdot\text{s})\hat{\jmath} - (1.08\,\text{N}\cdot\text{s})\hat{k}$; (c) $(0.450\,\text{N}\cdot\text{s})\hat{\imath} + (0.450\,\text{N}\cdot\text{s})\hat{\jmath} + (1.08\,\text{N}\cdot\text{s})\hat{k}$
91. (a) 0; (b) 0; (c) 0
92. (a) $(8.25\,\text{kg}\cdot\text{m/s})\hat{\jmath}$; (b) $(8.25\,\text{N}\cdot\text{s})\hat{\jmath}$; (c) $(-8.25\,\text{N}\cdot\text{s})\hat{\jmath}$
93. (a) 0; (b) 4.0 m/s
94. (a) 30 cm; (b) 3.3 m
95. (a) 0.745 mm; (b) 153°; (c) 1.67 mJ
96. (a) $(-1.00 \times 10^{-19}\,\text{kg}\cdot\text{m/s})\hat{\imath} + (0.67 \times 10^{-19}\,\text{kg}\cdot\text{m/s})\hat{\jmath}$; (b) 1.19×10^{-12} J
97. (a) 0.841 m/s; (b) 0.975 m/s
98. (a) 1.14×10^{-3}; (b) same
99. (a) 1.0 kg · m/s; (b) 2.5×10^2 J; (c) 10 N; (d) 1.7 kN; (e) answer for (c) includes time between pellet collisions
100. 41.7 cm/s
101. (a) $(7.4 \times 10^3\,\text{N}\cdot\text{s})\hat{\imath} - (7.4 \times 10^3\,\text{N}\cdot\text{s})\hat{\jmath}$; (b) $(-7.4 \times 10^3\,\text{N}\cdot\text{s})\hat{\imath}$; (c) 2.3×10^3 N; (d) 2.1×10^4 N; (e) −45°
102. 6.46×10^{-11} m
103. (a) 3.7 m/s; (b) 1.3 N · s; (c) 1.8×10^2 N
104. 72 km/h
105. (a) 9.0 kg · m/s; (b) 3.0 kN; (c) 4.5 kN; (d) 20 m/s
106. 0.57 m/s
107. 1.18×10^4 kg
108. +4.4 m/s
109. (a) 4.4 m/s; (b) 0.80
110. (a) 1.4×10^{-22} kg · m/s; (b) 28°; (c) 1.6×10^{-19} J
111. 0.22%
112. (a) 8.0×10^4 N; (b) 27 kg/s
113. 2.2 kg
114. 2.2×10^{-3}
115. 61.2 kJ
116. 3.0 m
117. (a) $(1.3\,\text{m/s})\hat{\imath} + (1.3\,\text{m/s})\hat{\jmath}$; (b) 1.9 m/s; (c) 45°
118. (a) 1; (b) 1.83×10^3; (c) 1.83×10^3; (d) all the same
119. (a) 2.18 kg · m/s; (b) 575 N
120. (a) $(-4.9\,\text{m/s}^2)\hat{\jmath}$; (b) $(-9.8\,\text{m/s}^2)\hat{\jmath}$; (c) $(-4.9\,\text{m/s}^2)\hat{\jmath}$; (d) 1.23 m/s; (e) 4.90 m/s; (f) 6.13 m/s
121. 5.0 kg
122. 2.5×10^{-3}
123. (a) $(24.0\,\text{kg}\cdot\text{m/s})\hat{\imath} - (180\,\text{kg}\cdot\text{m/s})\hat{\jmath} + (30.0\,\text{kg}\cdot\text{m/s})\hat{k}$; (b) 4.23 kJ; (c) 4.30 kJ
124. (a) 4.4 m/s; (b) 38 J
125. 190 m/s
126. 29 J
127. (a) 0.54 m/s; (b) 0; (c) 1.1 m/s
128. (a) 4.0 kg · m/s^2; (b) 8.0 kg · m/s
129. (a) $5mg$; (b) $7mg$; (c) 5 m
130. (a) 6.9 m/s; (b) 30°; (c) 6.9 m/s; (d) −30°; (e) 2.0 m/s; (f) −180°
131. (a) 1.9 m/s; (b) −30°; (c) elastic

132. (a) 41.0°; (b) 4.75 m/s; (c) no
133. (a) 4.6×10^4 km; (b) 73%
134. (a) −0.50 m; (b) −1.8 cm; (c) 0.50 m
135. (a) 50 kg/s; (b) 1.6×10^2 kg/s
136. 5.0×10^6 N
137. (a) 0; (b) 2.25 kJ; (b) 2.25 kJ; (c) 1.61 m/s; (d) 1.00 m/s
138. (a) 0; (b) 0.75 m
139. (a) 8.1 m/s; (b) 38° south of east
140. (a) 2.0 kg · m/s, east; (b) 1.0 kg · m/s, west; (c) 4.0 kg · m/s, west

---

Chapter 10

1. (a) 0.105 rad/s; (b) 1.75×10^{-3} rad/s; (c) 1.45×10^{-4} rad/s
2. 14 rev
3. (a) 12 : 00; (b) 12 : 00; (c) 3 : 00; (d) 6 : 00; (e) 9 : 00; (f) 12 : 00; (g) 2 : 24; (h) 4 : 48; (i) 7 : 12; (j) 9 : 36; (k) 12 : 00
4. 4.0 rad/s; (b) 28 rad/s; (c) 12 rad/s^2; (d) 6.0 rad/s^2; (d) 18 rad/s^2
5. 11 rad/s
6. (a) 2.0 rad; (b) 0; (c) 1.3×10^2 rad/s; (d) 32 rad/s^2; (e) no
7. (a) 4.0 m/s; (b) no
8. (a) 3.00 s; (b) 18.9 rad
9. (a) 9.0×10^3 rev/min^2; (b) 4.2×10^2 rev
10. (a) 30 s; (b)1.8×10^3 rad
11. (a) 2.0 rad/s^2; (b) 5.0 rad/s; (c) 10 rad/s; (d) 75 rad
12. (a) 4.09 s; (b) 1.70 s
13. 8.0 s
14. (a) 40 s; (b) 2.0 rad/s^2
15. (a) 3.4×10^2 s; (b) -4.5×10^{-3} rad/s^2; (c) 98 s
16. (a) 1.0 rev/s^2; (b) 4.8 s; (c) 9.6 s; (d) 48 rev
17. (a) 44 rad; (b) 5.5 s; (c) 32 s; (d) −2.1 s; (e) 40 s
18. (a) 13.5 s; (b) 27.0 rad/s
19. 6.9×10^{-13} rad/s
20. 199 hits/s
21. (a) 20.9 rad/s; (b) 12.5 m/s; (c) 800 rev/min^2; (d) 600 rev
22. (a) 3.0 rad/s; (b) 30 m/s; (c) 6.0 m/s^2; (d) 90 m/s^2
23. (a) 2.50×10^{-3} rad/s; (b) 20.2 m/s^2; (c) 0
24. (a) 3.1×10^2 m/s; (b) 3.4×10^2 m/s
25. (a) 6.4 cm/s^2; (b) 2.6 cm/s^2
26. (a) 40.2 cm/s^2; (b) 2.36×10^3 m/s^2; (c) 83.2 m
27. (a) 7.3×10^{-5} rad/s; (b) 3.5×10^2 m/s; (c) 7.3×10^{-5} rad/s; (d) 4.6×10^2 m/s
28. 16 s
29. (a) 3.8×10^3 rad/s; (b) 1.9×10^2 m/s
30. (a) −1.1 rev/min^2; (b) 9.9×10^3 rev; (c) −0.99 mm/s^2; (d) 31 m/s^2
31. (a) 73 cm/s^2; (b) 0.075; (c) 0.11
32. (a) -2.3×10^{-9} rad/s^2; (b) 2.6×10^3 y; (c) 24 ms
33. 12.3 kg · m^2
34. (a) 1.5 rad/s^2; (b) 0.40 J
35. (a) 1.1 kJ; (b) 9.7 kJ
36. (a) 7.1%; (b) 64%
37. 0.097 kg · m^2
38. 2.5 kg
39. (a) 0.023 kg · m^2; (b) 11 mJ
40. (a) 8.352×10^{-3} kg · m^2; (b) −0.22%
41. 4.7×10^{-4} m^2
42. (a) 2.0 kg · m^2; (b) 6.0 kg · m^2; (c) 2.0 kg · m^2
43. (a) 1.3×10^3 g · cm^2; (b) 5.5×10^2 g · cm^2; (c) 1.9×10^3 g · cm^2; (d) $A + B$
44. (a) 49 MJ; (b) 1.0×10^2 min
45. 4.6 N · m
46. (a) 8.4 N · m; (b) 17 N · m; (c) 0
47. −3.85 N · m
48. 12 N · m
49. (a) 28.2 rad/s^2; (b) 338 N · m
50. 1.28 kg · m^2
51. 0.140 N
52. (a) 3.0 rad/s^2; (b) 9.4 rad/s^2
53. (a) 9.7 rad/s^2; (b) counterclockwise
54. (a) 1.7 m/s^2; (b) 6.9 m/s^2
55. (a) 6.00 cm/s^2; (b) 4.87 N; (c) 4.54 N; (d) 1.20 rad/s^2; (e) 0.0138 kg · m^2
56. 2.51×10^{-4} kg · m^2
57. (a) 4.2×10^2 rad/s^2; (b) 5.0×10^2 rad/s
58. 396 N · m
59. (a) 1.4 m/s; (b) 1.4 m/s
60. (a) 19.8 kJ; (b) 1.32 kW
61. (a) 0.63 J; (b) 0.15 m
62. (a) 11.2 mJ; (b) 33.6 mJ; (c) 56.0 mJ; (d) 2.80×10^{-5} J · s^2/rad^2
63. 5.42 m/s
64. (a) 0.15 kg · m^2; (b) 11 rad/s
65. 9.82 rad/s

66. (a) 0.227 rad/s; (b) 5.32 m/s^2; (c) 8.43 rad/s^2; (d) 41.8°
67. 1.4 m/s
68. (a) $1.2t^5 - 1.3t^3 + 2.0$; (b) $0.20t^6 - 0.33t^4 + 2.0t + 1.0$
69. (a) 314 rad/s^2; (b) 7.54 m/s^2; (c) 14.0 N; (d) 4.36 N
70. 3×10^5 J
71. 6.16×10^{-5} $kg \cdot m^2$
72. 146 rad/s
73. (a) 5.1 h; (b) 8.1 h
74. 25 N
75. (a) 0.32 rad/s; (b) 1.0×10^2 km/h
76. (a) 8.6 s; (b) no
77. (a) 3.3 J; (b) 2.9 J
78. (a) 1.57 m/s^2; (b) 4.55 N; (c) 4.94 N
79. (a) −7.66 rad/s^2; (b) −11.7 N·m; (c) 4.59×10^4 J; (d) 624 rev; (e) 4.59×10^4 J
80. (a) 4.81×10^5 N; (b) 1.12×10^4 N · m; (c) 1.25×10^6 J
81. (a) 1.5×10^2 cm/s; (b) 15 rad/s; (c) 15 rad/s; (d) 75 cm/s; (e) 3.0 rad/s
82. 30 rev
83. 4.6 rad/s^2
84. 6.06 rad/s
86. (a) yes; (b) 1.1×10^2 kg
87. (a) 0.689 N · m; (b) 3.05 N; (c) 9.84 N · m; (d) 11.5 N
88. 0.054 $kg \cdot m^2$
89. 3.1 rad/s
90. (a) 0.20 $kg \cdot m^2$; (b) 6.3 rad/s
91. (a) −1.25 rad/s^2; (b) 250 rad; (c) 39.8 rev
92. (a) 5.92×10^4 m/s^2; (b) 4.39×10^4 s^{-2}
93. (a) 0.791 $kg \cdot m^2$; (b) 1.79×10^{-2} N · m
94. 1.6 $kg \cdot m^2$
95. 1.5×10^3 rad
96. 18 rad
97. (a) 2.8 rad; (b) 0.42 m/s^2
98. (a) 0.019 $kg \cdot m^2$; (b) 0.019 $kg \cdot m^2$
99. (a) 0.17 $kg \cdot m^2$; (b) 0.22 $kg \cdot m^2$; (c) 0.10 $kg \cdot m^2$
100. (a) 3.4×10^5 $g \cdot cm^2$; (b) 2.9×10^5 $g \cdot cm^2$; (c) 6.3×10^5 $g \cdot cm^2$; (d) $(1.2\,cm)\hat{i} + (5.9\,cm)\hat{j}$
101. (a) 10 J; (b) 0.27 m
102. (a) 3.1 rad/s; (b) same
103. (a) 11 rad/s
104. 2.6 J
105. (a) 5.00 rad/s; (b) 1.67 rad/s^2; (c) 2.50 rad
106. (a) 5.5×10^{15} s; (b) 26
107. (a) −67 rev/min^2; (b) 8.3 rev
108. (a) 155 $kg \cdot m^2$; (b) 64.4 kg
109. (a) $\omega_0 + at^4 - bt^3$; (b) $\theta_0 + \omega_0 t + at^5/5 - bt^4/4$
110. (a) $a + 3bt^2 - 4ct^3$; (b) $6bt - 12ct^2$
111. 17
112. 2.1×10^{-22} J
113. 1.4×10^2 N · m
114. (a) 2.0 rev/s; (b) 3.8 s
115. 5.6 rad/s^2
116. (a) 7.0 $kg \cdot m^2$; (b) 7.2 m/s; (c) 71°
117. (a) 1.94 m/s^2; (b) 75.1°
118. (a) 1.4×10^2 rad; (b) 14 s
119. 200 rev/min
120. (a) 221 $kg \cdot m^2$; (b) 1.10×10^4 J
121. (a) 3.5 rad/s; (b) 52 cm/s; (c) 26 cm/s
122. 0.13 rad/s
123. 6.75×10^{12} rad/s
124. (a) 8.2×10^{28} N · m; (b) 2.6×10^{29} J; (c) 3.0×10^{21} kW
125. (a) 9.71×10^{37} $kg \cdot m^2$; (c) 1×10^9 y

Chapter 11

1. (a) 59.3 rad/s; (b) 9.31 rad/s^2; (c) 70.7 m
2. (a) 0; (b) $(22\,m/s)\hat{i}$; (c) $(-22\,m/s)\hat{i}$; (d) 0; (e) 1.5×10^3 m/s^2; (f) 1.5×10^3 m/s^2; (g) $(22\,m/s)\hat{i}$; (h) $(44\,m/s)\hat{i}$; (i) 0; (j) 0; (k) 1.5×10^3 m/s^2; (l) 1.5×10^3 m/s^3
3. −3.15 J
4. (a) 8.0°; (b) more
5. 0.020
6. (a) $(-4.0\,N)\hat{i}$; (b) 0.60 $kg \cdot m^2$
7. (a) 63 rad/s; (b) 4.0 m
8. (a) 37.8 cm; (b) 1.96×10^{-2} N; (c) toward loop's center
9. 4.8 m
10. 7.2×10^{-4} $kg \cdot m^2$
11. (a) 2.0 m; (b) 7.3 m/s
12. 1.34 m/s
13. 0.50
14. 0.25
15. (a) 13 cm/s^2; (b) 4.4 s; (c) 55 cm/s; (d) 18 mJ; (e) 1.4 J; (f) 27 rev/s
16. (a) 0.19 m/s^2; (b) 0.19 m/s^2; (c) 1.1 kN; (d) no; (e) same; (f) greater
17. (a) $(24\,N \cdot m)\hat{j}$; (b) $(-24\,N \cdot m)\hat{j}$; (c) $(12\,N \cdot m)\hat{j}$; (d) $(-12\,N \cdot m)\hat{j}$
18. (a) $(6.0\,N \cdot m)\hat{j} + (8.0\,N \cdot m)\hat{k}$; (b) $(-22\,N \cdot m)\hat{i}$

19. $(-2.0\,\text{N}\cdot\text{m})\hat{\text{i}}$
20. (a) $(6.0\,\text{N}\cdot\text{m})\hat{\text{i}} - (3.0\,\text{N}\cdot\text{m})\hat{\text{j}} - (6.0\,\text{N}\cdot\text{m})\hat{\text{k}}$; (b) $(26\,\text{N}\cdot\text{m})\hat{\text{i}} + (3.0\,\text{N}\cdot\text{m})\hat{\text{j}} - (18\,\text{N}\cdot\text{m})\hat{\text{k}}$; (c) $(32\,\text{N}\cdot\text{m})\hat{\text{i}} - (24\,\text{N}\cdot\text{m})\hat{\text{k}}$; (d) 0
21. (a) $(50\,\text{N}\cdot\text{m})\hat{\text{k}}$; (b) 90°
22. (a) $(-1.5\,\text{N}\cdot\text{m})\hat{\text{i}} - (4.0\,\text{N}\cdot\text{m})\hat{\text{j}} - (1.0\,\text{N}\cdot\text{m})\hat{\text{k}}$; (b) $(-1.5\,\text{N}\cdot\text{m})\hat{\text{i}} - (4.0\,\text{N}\cdot\text{m})\hat{\text{j}} - (1.0\,\text{N}\cdot\text{m})\hat{\text{k}}$
23. $-5.00\,\text{N}$
24. (a) $12\,\text{kg}\cdot\text{m}^2/\text{s}$; (b) positive z direction; (c) $3.0\,\text{N}\cdot\text{m}$; (d) positive z direction
25. (a) $9.8\,\text{kg}\cdot\text{m}^2/\text{s}$; (b) positive z direction
26. (a) $(6.0\times10^2\,\text{kg}\cdot\text{m}^2/\text{s})\hat{\text{k}}$; (b) $(7.2\times10^2\,\text{kg}\cdot\text{m}^2/\text{s})\hat{\text{k}}$
27. (a) 0; (b) $(8.0\,\text{N}\cdot\text{m})\hat{\text{i}} + (8.0\,\text{N}\cdot\text{m})\hat{\text{k}}$
28. (a) 0; (b) $-22\,\text{kg}\cdot\text{m}^2/\text{s}$; (c) $-7.84\,\text{N}\cdot\text{m}$; (d) $-784\,\text{N}\cdot\text{m}$
29. (a) $(3.00\,\text{m/s}^2)\hat{\text{i}} - (4.00\,\text{m/s}^2)\hat{\text{j}} + (2.00\,\text{m/s}^2)\hat{\text{k}}$; (b) $(42.0\,\text{kg}\cdot\text{m}^2/\text{s})\hat{\text{i}} + (24.0\,\text{kg}\cdot\text{m}^2/\text{s})\hat{\text{j}} + (60.0\,\text{kg}\cdot\text{m}^2/\text{s})\hat{\text{k}}$; (c) $(-8.00\,\text{N}\cdot\text{m})\hat{\text{i}} - (26.0\,\text{N}\cdot\text{m})\hat{\text{j}} - (40.0\,\text{N}\cdot\text{m})\hat{\text{k}}$; (d) 127°
30. $(2.0\,\text{N}\cdot\text{m})\hat{\text{i}} - (4.0\,\text{N}\cdot\text{m})\hat{\text{j}}$
31. (a) $(-1.7\times10^2\,\text{kg}\cdot\text{m}^2/\text{s})\hat{\text{k}}$; (b) $(+56\,\text{N}\cdot\text{m})\hat{\text{k}}$; (c) $(+56\,\text{kg}\cdot\text{m}^2/\text{s}^2)\hat{\text{k}}$
32. (a) 0; (b) $(-8.0\,\text{N}\cdot\text{m/s})t\hat{\text{k}}$; (c) $-\dfrac{2.0}{\sqrt{t}}\hat{\text{k}}$ in newton-meters for t in seconds; (d) $8.0t^{-3}\,\hat{\text{k}}$ in newton-meters for t in seconds
33. (a) $(48\,\text{N}\cdot\text{m/s})t\hat{\text{k}}$; (b) increasing
34. (a) $0.53\,\text{kg}\cdot\text{m}^2/\text{s}$; (b) $4.2\times10^3\,\text{rev/min}$
35. (a) $1.47\,\text{N}\cdot\text{m}$; (b) 20.4 rad; (c) $-29.9\,\text{J}$; (d) 19.9 W
36. 1024
37. (a) $4.6\times10^{-3}\,\text{kg}\cdot\text{m}^2$; (b) $1.1\times10^{-3}\,\text{kg}\cdot\text{m}^2/\text{s}$; (c) $3.9\times10^{-3}\,\text{kg}\cdot\text{m}^2/\text{s}$
38. (a) $24\,\text{kg}\cdot\text{m}^2/\text{s}$; (b) $1.5\,\text{k}\cdot\text{m}^2/\text{s}$
39. (a) $1.6\,\text{kg}\cdot\text{m}^2$; (b) $4.0\,\text{kg}\cdot\text{m}^2/\text{s}$
40. 5.0×10^2 rev
41. (a) 3.6 rev/s; (b) 3.0; (c) forces on the bricks from the man transferred energy from the man's internal energy to kinetic energy
42. (a) 750 rev/min; (b) 450 rev/min; (c) clockwise
43. (a) 267 rev/min; (b) 0.667
44. 0.20
45. 0.176 rad/s
46. 3
47. (a) 1.5 m; (b) 0.93 rad/s; (c) 98 J; (d) 8.4 rad/s; (e) $8.8\times10^2\,\text{J}$; (f) internal energy of the skaters
48. (a) 4.2 rad/s; (b) no, because energy is transferred to the cockroach's internal energy
49. 3.4 rad/s
50. (a) 0.180 m; (b) clockwise
51. $1.3\times10^3\,\text{m/s}$
52. (a) 0.347 rad/s; (b) 1.33; (c) energy is transferred from the internal energy of the cockroach to kinetic energy
53. 11.0 m/s
54. 2.6 rad/s
55. (a) 18 rad/s; (b) 0.92
56. (a) $0.24\,\text{kg}\cdot\text{m}^2$; (b) $1.8\times10^3\,\text{m/s}$
57. 1.5 rad/s
58. 0.070 rad/s
59. (a) 0.148 rad/s; (b) 0.0123; (c) 181°
60. 32°
61. (a) 0.33 rev/s; (b) clockwise
62. 0.43 rev/min
63. $(5.55\,\text{kg}\cdot\text{m}^2/\text{s})\hat{\text{k}}$
64. (a) $-(0.11\,\text{m})\omega$; (b) $-2.1\,\text{m/s}^2$; (c) $-47\,\text{rad/s}^2$; (d) 1.2 s; (e) 8.6 m; (f) 6.1 m/s
65. 0.62 J
66. 39.1 J
67. (a) $6.65\times10^{-5}\,\text{kg}\cdot\text{m}^2/\text{s}$; (b) no; (c) 0; (d) yes
68. (a) 0.81 mJ; (b) 0.29; (c) $1.3\times10^{-2}\,\text{N}$
69. $0.47\,\text{kg}\cdot\text{m}^2/\text{s}$
70. (a) 8.0 J; (b) 3.0 m/S; (c) 6.9 J; (d) 1.8 m/s
71. (a) $(-24t^2\,\text{kg}\cdot\text{m}^2\text{s})\hat{\text{k}}$; (b) $(-48t\,\text{N}\cdot\text{m})\hat{\text{k}}$; (c) $(12t^2\,\text{kg}\cdot\text{m}^2/\text{s})\hat{\text{k}}$; (d) $(24t\,\text{N}\cdot\text{m})\hat{\text{k}}$
72. 2.33 m/s
73. 12 s
74. 1.00
75. (a) 0; (b) 0; (c) $(-30t^3\,\text{kg}\cdot\text{m}^2\text{s})\hat{\text{k}}$; (d) $(-90t^2\,\text{N}\cdot\text{m})\hat{\text{k}}$; (e) $(30t^3\,\text{kg}\cdot\text{m}^2\text{s})\hat{\text{k}}$; (f) $90t^2\,\text{N}\cdot\text{m})\hat{\text{k}}$
76. (a) 0.333; (b) 0.111
77. (a) $mvR/(I+MR^2)$; (b) $mvR^2/(I+MR^2)$
78. (a) $(-32\,\text{kg}\cdot\text{m}^2/\text{s})\hat{\text{k}}$; (b) $(-32\,\text{kg}\cdot\text{m}^2/\text{s})\hat{\text{k}}$; (c) $(12\,\text{N}\cdot\text{m})\hat{\text{k}}$; (d) 0
79. $(7.4\,\text{kg}\cdot\text{m}^2\text{s})\hat{\text{k}}$
80. (a) 61.7 J; (b) 3.43 m; (c) no
81. (a) $mR^2/2$; (b) a solid circular cylinder
82. (a) $4.11\,\text{m/s}^2$; (b) $16.4\,\text{rad/s}^2$; (c) $2.55\,\text{N}\cdot\text{m}$

83. (a) 58.8 J; (b) 29.2 J
84. (a) 9.9×10^2 J; (b) 3.0×10^3 J; (c) 1.2×10^5 J
85. (a) $1.6\,\mathrm{m/s^2}$; (b) $16\,\mathrm{rad/s^2}$; (c) $(4.0\,\mathrm{N})\hat{\mathrm{i}}$
86. (a) $(-17.1t^2\,\mathrm{kg\cdot m^2/s})\,\hat{\mathrm{k}}$; (b) $(-34.2t\,\mathrm{kg\cdot m^2/s^2})\,\hat{\mathrm{k}}$; (c) $(-34.2t\,\mathrm{N\cdot m})\,\hat{\mathrm{k}}$
87. (a) 12.7 rad/s; (b) clockwise
88. (a) $(-1.8\,\mathrm{kg\cdot m^2/s})\,\hat{\mathrm{k}}$; (b) $(-3.6\,\mathrm{kg\cdot m^2/s})\,\hat{\mathrm{k}}$; (c) 0; (d) $(-7.3\,\mathrm{N\cdot m})\,\hat{\mathrm{k}}$
89. (a) 0.89 s; (b) 9.4 J; (c) 1.4 m/s; (d) 0.12 J; (e) 4.4×10^2 rad/s; (f) 9.2 J
90. (a) $12.2\,\mathrm{kg\cdot m^2}$; (b) $308\,\mathrm{kg\cdot m^2/s}$
91. (a) $2.9\times10^4\,\mathrm{kg\cdot m^2/s}$; (b) $1.2\times10^6\,\mathrm{N\cdot m}$
92. rotational speed would decrease; day would be about 0.8 s longer
93. (a) $3.14\times10^{43}\,\mathrm{kg\cdot m^2/s}$; (b) 0.614
94. $1.2\times10^8\,\mathrm{kg\cdot m^2/s}$
95. (a) $149\,\mathrm{kg\cdot m^2}$; (b) $158\,\mathrm{kg\cdot m^2/s}$; (c) 0.744 rad/s
97. $2.5\times10^{11}\,\mathrm{kg\cdot m^2/s}$

Chapter 12

1. (a) 2; (b) 7
2. (a) 2; (b) 7
3. 7.92 kN
4. 120°
5. (a) 9.4 N; (b) 4.4 N
6. (a) 8.4×10^2 N; (b) 5.3×10^3 N
7. (a) 1.2 kN; (b) down; (c) 1.7 kN; (d) up; (e) left; (f) right
8. (a) 2.77 kN; (b) 3.89 kN
9. 74.4 g
10. 8.3 kN
11. (a) 2.8×10^2 N; (b) 8.8×10^2 N; (c) 71°
12. 0.536 m
13. (a) 5.0 N; (b) 30 N; (c) 1.3 m
14. (a) 49 N; (b) 28 N; (c) 57 N; (d) 29°
15. (a) 2.7 kN; (b) up; (c) 3.6 kN; (d) down
16. 0.702 m
17. (a) 0.64 m; (b) increased
18. 457 N
19. 13.6 N
20. (a) 1.9 kN; (b) up; (c) 2.1 kN; (d) down
21. (a) 192 N; (b) 96.1 N; (c) 55.5 N
22. (a) 408 N; (b) 245 N; (c) to the right; (d) 163 N; (e) up
23. (a) 6.63 kN; (b) 5.74 kN; (c) 5.96 kN
24. (a) 3.4×10^2 N; (b) 0.88 m; (c) increases; (d) decreases
25. 2.20 m
26. (a) $(-80\,\mathrm{N})\hat{\mathrm{i}}+(1.3\times10^2\,\mathrm{N})\hat{\mathrm{j}}$; (b) $(80\,\mathrm{N})\hat{\mathrm{i}}+(1.3\times10^2\,\mathrm{N})\hat{\mathrm{j}}$
27. (a) 1.50 m; (b) 433 N; (c) 250 N
28. (a) $Wx/(L\sin\theta)$; (b) $Wx/(L\tan\theta)$; (c) $W(1-x/L)$
29. (a) 445 N; (b) 0.50; (c) 315 N
30. (a) 60.0°; (b) 300 N
31. 0.34
32. (a) $(-797\,\mathrm{N})\hat{\mathrm{i}}+(265\,\mathrm{N})\hat{\mathrm{j}}$; (b) $(797\,\mathrm{N})\hat{\mathrm{i}}+(265\,\mathrm{N})\hat{\mathrm{j}}$; (c) $(797\,\mathrm{N})\hat{\mathrm{i}}+(931\,\mathrm{N})\hat{\mathrm{j}}$; (d) $(-797\,\mathrm{N})\hat{\mathrm{i}}+(-265\,\mathrm{N})\hat{\mathrm{j}}$
33. (a) slides; (b) 31°; (c) tips; (d) 34°
34. (a) 30.0°; (b) 51.0 kg; (c) 10.2 kg
35. (a) 211 N; (b) 534 N; (c) 320 N
36. (a) $7.5\times10^{10}\,\mathrm{N/m^2}$; (b) $2.9\times10^8\,\mathrm{N/m^2}$
37. (a) $6.5\times10^6\,\mathrm{N/m^2}$; (b) 1.1×10^{-5} m
38. 56.0 mJ
39. (a) 866 N; (b) 143 N; (c) 0.165
40. (a) 0.80; (b) 0.0; (c) 0.25
41. (a) 1.4×10^9 N; (b) 75
42. (a) 1.2×10^2 N; (b) 68 N
43. 76 N
44. (a) 500 kg; (b) 62.5 kg
45. (a) 8.01 kN; (b) 3.65 kN; (c) 5.66 kN
46. (a) 196 N; (b) 294 N; (c) 441 N; (d) 49.0 N; (e) 0.16 m
47. 71.7 N
48. (a) 50°; (b) $0.77mg$
49. (a) 1.38 kN; (b) 180 N
50. (a) $2mg$; (b) mg; (c) mg; (d) $\sqrt{2}mg$
51. (a) $\mu<0.57$; (b) $\mu>0.57$
52. (a) BC, CD, DA; (b) 535 N; (c) 757 N
53. (a) $L/2$; (b) $L/4$; (c) $L/6$; (d) $L/8$; (e) $25L/24$
54. (a) $3.9\,\mathrm{m/s^2}$; (b) 2.0 kN; (c) 3.5 kN; (d) 0.79 kN; (e) 1.4 kN
55. 0.29
57. 8.7 N
58. 44 N
59. (a) 15 N; (b) 29 N
60. (a) 106 N; (b) 64.0°
61. 60°
62. (a) 200 N; (b) 360 N; (c) 0.35
63. $2.4\times10^9\,\mathrm{N/m^2}$
64. (a) 0.80 mm; (b) 2.3 cm
65. $L/4$
66. $(-1.5\times10^2\,\mathrm{N})\hat{\mathrm{i}}+(2.6\times10^2\,\mathrm{N})\hat{\mathrm{j}}$

67. (a) 88 N; (b) $(30\,N)\hat{i} + (97\,N)\hat{j}$
68. (a) 1.9×10^{-3}; (b) $1.3 \times 10^7\,N/m^2$; (c) $6.9 \times 10^9\,N/m^2$
69. (a) 1.8×10^7 N; (b) 1.4×10^7 N; (c) 16
70. 3.4 m
71. (a) 1.5 kN; (b) 1.9 kN
72. (a) 42 N; (b) 66 N
73. (a) $(1.16\,kN)\hat{j}$; (b) $(1.74\,kN)\hat{j}$
74. (a) $(-671\,N)\hat{j}$; (b) $(400\,N)\hat{i} + (670\,N)\hat{j}$
75. (a) $(35\,N)\hat{i} + (200\,N)\hat{j}$; (b) $(-45\,N)\hat{i} + (200\,N)\hat{j}$; (c) 1.9×10^2 N
76. 3.1 cm
77. 0.19 m
78. (a) 270 N; (b) 72 N; (c) 19°
79. (a) $a_1 = L/2$, $a_2 = 5L/8$, $h = 9L/8$; (b) $b_1 = 2L/3$, $b_2 = L/2$, $h = 7L/6$
80. (a) 51°; (b) $0.64Mg$
81. (a) 2.5 m; (b) 7.3°
82. (a) $6.78\,m^3$; (b) 1.20×10^5 N; (c) $\sigma_0 + (\sigma_m - \sigma_0)r/r_m = (40\,000 + 13r)\,N/m^3$, with r in meters; (d) $2\pi r\,dr$; (e) $[(40\,000 + 13r)\,N/m^3]2\pi r\,dr$, with r in meters; (f) 1.04×10^5 N; (g) -0.13

Chapter 13

1. 19 m
2. 2.16
3. 1/2
4. (a) 2.13×10^{-8} N; (b) 60.6°
5. 2.60×10^5 km
6. $(1.18 \times 10^{-14}\,N)\hat{i} + (1.18 \times 10^{-14}\,N)\hat{j}$
7. $-5.00d$
8. (a) 0.25 kg; (b) 1.0 kg
9. (a) $M = m$; (b) 0
10. (a) -0.20 m; (b) -0.35 m
11. 8.31×10^{-9} N
12. (a) $0.716d$; (b) $-1.07d$
13. (a) $-1.88d$; (b) $-3.90d$; (c) $0.489d$
14. -0.30 N
15. 2.6×10^6 m
16. (a) 17 N; (b) 2.4
17. (a) $7.6\,m/s^2$; (b) $4.2\,m/s^2$
18. (a) $(3.02 \times 10^{43}\,kg \cdot m/s^2)/M_h$; (b) decrease; (c) $9.82\,m/s^2$; (d) $7.30 \times 10^{-15}\,m/s^2$; (e) no
19. 5×10^{24} kg
20. (a) $G(M_1 + M_2)m/a^2$; (b) GM_1m/b^2; (c) 0
21. (a) $(3.0 \times 10^{-7}\,N/kg)m$; (b) $(3.3 \times 10^{-7}\,N/kg)m$; (c) $(6.7 \times 10^{-7}\,N/kg \cdot m)mr$
22. (a) $0.414R$; (b) $0.500R$
23. (a) $9.83\,m/s^2$; (b) $9.84\,m/s^2$; (c) $9.79\,m/s^2$
24. (a) -4.4×10^{-11} J; (b) -2.9×10^{-11} J; (c) 2.9×10^{-11} J
25. (a) 0.74; (b) $3.8\,m/s^2$; (c) 5.0 km/s
26. 1/2
27. (a) 0.0451; (b) 28.5
28. (a) 2.0×10^9 J; (b) $2.5R_s$
29. 5.0×10^9 J
30. (a) 1.33; (b) 2.00; (c) 0
31. (a) 0.50 pJ; (b) -0.5 pJ
32. (a) 2.2×10^7 J; (b) 6.9×10^7 J
33. (a) 1.7 km/s; (b) 2.5×10^5 m; (c) 1.4 km/s
34. (a) -1.7×10^{-8} J; (b) 0.56×10^{-8} J
35. (a) 82 km/s; (b) 1.8×10^4 km/s
36. (a) 0.50 kg; (b) 1.5 kg
37. -4.82×10^{-13} J
38. (a) 5.4×10^4 km/h; (b) 3.8×10^4 km/h
39. 6.5×10^{23} kg
40. 1.87 y
41. 5×10^{10} stars
42. 0.35 lunar months
43. (a) 7.82 km/s; (b) 87.5 min
44. (a) 5.01×10^9 m; (b) 7.20 solar radii
45. (a) 6.64×10^3 km; (b) 0.0136
46. 3.58×10^4 km
47. (a) 1.9×10^{13} m; (b) $3.5R_P$
48. (a) 6×10^{16} kg; (b) $4 \times 10^3\,kg/m^3$
50. 5.8×10^6 m
51. 0.71 y
52. (a) $3.7m_J$; (b) $2.5r_g$
53. $\sqrt{GM/L}$
54. (a) 8.0×10^8 J; (b) 36 N
55. (a) 2.8 y; (b) 1.0×10^{-4}
56. (a) 1/2; (b) 1/2; (c) B; (c) 1.1×10^8 J
57. (a) 3.19×10^3 km; (b) lifting
58. (a) -6.33×10^9 J; (b) -6.33×10^9 J; (c) falling
59. (a) $r^{3/2}$; (b) $1/r$; (c) $\sqrt{r}$; (d) $1/\sqrt{r}$
60. (a) 4.6×10^5 J; (b) 2.6×10^2
61. (a) 7.5 km/s; (b) 97 min; (c) 4.1×10^2 km; (d) 7.7 km/s; (e) 93 min; (f) 3.2×10^{-3} N; (g) no; (h) yes
62. 1.1 s
63. (a) 1.0×10^3 kg; (b) 1.5 km/s
64. (a) $(1 \times 10^2)M_s$; (b) lower
65. 7.2×10^{-9} N
66. (a) 1.4×10^6 m/s; (b) $3 \times 10^6\,m/s^2$

67. $-(0.044\,\mu N\hat{j}$
68. 9.2×10^{-5} rad/s
69. (a) 2.15×10^4 s; (b) 12.3 km/s; (c) 12.0 km/s; (d) 2.17×10^{11} J; (e) -4.53×10^{11} J; (f) -2.35×10^{11} J; (g) 4.04×10^7 m; (h) 1.22×10^3 s; (i) elliptical
70. (a) 2×10^{-5} m/s^2; (b) 2 cm/s
71. (a) 1.3×10^{12} m/s^2; (b) 1.6×10^6 m/s
72. (a) Gm^2/R_i; (b) $Gm^2/2R_i$; (c) $\sqrt{Gm/R_i}$; (d) $2\sqrt{2Gm/R_i}$; (e) Gm^2/R_1; (f) $\sqrt{2Gm/R_i}$; (g) the center-of-mass frame is an inertial frame and in it the conservation of energy principle may be written as in Chapter 8; the reference frame attached to body A is noninertial and the principle cannot be written as in Chapter 8; answer (d) is correct
73. $(0.37\,\mu N)\hat{j}$
74. (a) -1.3×10^{-4} J; (b) less; (c) positive; (d) negative
75. 29 pN
76. (a) $R/3$; (b) $\sqrt{3}R$
77. 2.5×10^4 km
78. (a) 1.9×10^7 m; (b) 7.6×10^8 J; (c) 8.6×10^{24} kg
79. (a) 2.2×10^{-7} rad/s; (b) 89 km/s
80. (a) $(3.4 \times 10^{-3})g$; (b) $(6.1 \times 10^{-4})g$; (c) $(1.4 \times 10^{-11})g$
81. 3.2×10^{-7} N
82. (a) $(2.8 \times 10^4)g$; (b) deadly; (c) $714g$; (d) 1.5 km/s
83. (a) 38.3 MJ; (b) 1.03×10^3 km
84. 2.4×10^4 m/s
85. (a) 0; (b) 1.8×10^{32} J; (c) 1.8×10^{32} J; (d) 0.99 km/s
86. (a) 1.98×10^{30} kg; (b) 2.00×10^{30} kg
87. (a) 1.9×10^{11} m; (b) 4.6×10^4 m/s
88. (a) 5.3×10^{-8} J; (b) $(-6.4 \times 10^{-8}\,N)\hat{i}$
89. 7.9 km/s
90. -1.87 GJ
91. $-Gm(M_E/R + M_m/r)$
92. 3.4×10^5 km
93. 1.1%
94. (a) 120; (b) 4.23×10^4 km; (c) 2.2×10^6 m; (d) smaller; (e) toward
95. $GM_Em/12R_E$
96. (a) $GMmx(x^2 + R^2)^{-3/2}$; (b) $\sqrt{2GM\left(\frac{1}{R} - \frac{1}{\sqrt{R^2+x^2}}\right)}$
98. (b) 1.9 h
100. $2R$
103. $2\pi r^{3/2}/\sqrt{G(M + m/4)}$
104. (a) 1×10^{-8} N; (b) 1×10^{-6} N; (c) 5×10^{-7} N; (d) no

---

Chapter 14

1. 2.9×10^4 N
2. 18 N
3. 1.1×10^5 Pa or 1.1 atm
4. 38 kPa
5. 0.074
6. (a) 1.9×10^2 kPa; (b) 15.9/10.6
7. 26 kN
8. 17 cm
9. 1.90×10^4 Pa
10. 1.08×10^3 atm
11. 7.2×10^5 N
12. -2.6×10^4 Pa
13. 2.80 m
14. 2.0
15. 0.635 J
16. 44 km
17. 4.69×10^5 N
18. (a) 5.0×10^6 N; (b) 5.6×10^6 N
19. (a) 1.88×10^9 N; (b) 2.20×10^{10} N $\cdot$ m; (c) 11.7 m
20. -3.9×10^{-3} atm
21. (a) 7.9 km; (b) 16 km
22. (a) fA/a; (b) 103 N
23. 8.50 kg
24. (a) 35.6 kN; (b) yes, decreases by 0.330 m^3
25. (a) 2.04×10^{-2} m^3; (b) 1.57 kN
26. (a) 37.5 kN; (b) 39.6 kN; (c) 2.23 kN; (d) 2.18 kN
27. five
28. 4.11 kJ
29. (a) 6.7×10^2 kg/m^3; (b) 7.4×10^2 kg/m^3
30. 7.84 cm, down
31. (a) 1.2 kg; (b) 1.3×10^3 kg/m^3
32. (a) 1.5 g/cm^3; (b) 2.7×10^{-3} m^3
33. 57.3 cm
34. 1.8 g/cm^3
35. 0.126 m^3
36. 1.40 m
37. (a) 1.80 m^3; (b) 4.75 m^3

38. (a) 1.84 kg; (b) 2.01 kg
39. (a) 637.8 cm^3; (b) 5.102 m^3; (c) 5.102×10^3 kg
40. 9.7 mm
41. 8.1 m/s
42. 4.0 m
43. 66 W
44. (a) 56 L/min; (b) 1.0
45. (a) 2.5 m/s; (b) 2.6×10^5 Pa
46. 1.7 MPa
47. (a) 3.9 m/s; (b) 88 kPa
48. (a) 2.40 m/s; (b) 245 Pa
49. (a) 1.6×10^{-3} m^3/s; (b) 0.90 m
50. 3.60 cm
51. 1.4×10^5 J
52. (a) 2; (b) 1/2; (c) 3.0 cm
53. (a) 74 N; (b) 1.5×10^2 m^3
54. (a) 6.4 m^3; (b) 5.4 m/s; (c) 9.8×10^4 Pa
55. (a) 35 cm; (b) 30 cm; (c) 20 cm
56. (a) 0.25 m^2; (b) 6.1 m^3/s
57. (a) 0.0776 m^3/s; (b) 69.8 kg/s
58. −2.50 J
59. (b) 2.0×10^{-2} m^3/s
60. (a) 4.1 m/s; (b) 21 m/s; (c) 8.0×10^{-3} m^3/s
61. (b) 63.3 m/s
62. 1.1×10^2 m/s
63. 7.8 cm/h
64. (a) 0.13; (b) 0.96
65. 4.00 cm
66. 44.2 g
67. 45.3 cm^3
68. 9.4%
69. (a) 3.2 m/s; (b) 9.2×10^4 PA; (c) 10.3 m
70. (a) 0.050; (b) 0.41; (c) no
71. (a) 1.8×10^2 kN; (b) 81 kN; (c) 20 kN; (d) 0; (e) 78 kPa; (f) no
72. 5.11×10^{-7} kg
73. (a) 3.1 m/s; (b) 9.5 m/s
74. 6×10^9 capillaries
75. 4.4 mm
76. 3.92 m/s^2
77. 1.00×10^{-2} m/s
78. 1.5 cm
79. (a) 0.38 m; (b) C, D, B, A
80. 1.62×10^6 Pa
81. 1.07×10^3 g
82. 0.412 cm
83. 6.0×10^2 kg/m^3
84. (a) 1.21×10^7 Pa; (b) 1.22×10^7 Pa; (c) 3.82×10^5 N; (d) 5.26 N; (e) 9.04 m/s^2; (f) down
85. 43 cm/s
86. (a) 2; (b) 3; (c) 4/3
87. 60 MPa
88. -1.1×10^3 Pa
89. (a) 42 h; (b) yes
90. 1.5 g/cm^3
91. 1.5 g/cm^3
92. (a) 45 m^2; (b) car should be over the center of the slab if the slab is to be level
93. (a) 6.06×10^9 N; (b) 20 atm; (c) no
94. 0.031 kg
95. 0.50 m
96. 2.79 g/cm^3
97. (a) 2.2; (b) 3.6
98. $0.12(1/\rho - 1/8)$%, with ρ in g/cm^3

Chapter 15

1. (a) 1.0 mm; (b) 0.75 m/s; (c) 5.7×10^2 m/s^2
2. (a) 10 N; (b) 1.2×10^2 N/m
3. (a) 6.28×10^5 rad/s; (b) 1.59 mm
4. 37.8 m/s^2
5. (a) 0.500 s; (b) 2.00 Hz; (c) 12.6 rad/s;
6. (a) 0.75 s; (b) 1.3 Hz; (c) 8.4 rad/s (d) 79.0 N/m; (e) 4.40 m/s; (f) 27.6 N
7. (a) 0.50 s (b) 2.0 Hz; (c) 18 cm
8. (a) 1.29×10^5 N/m; (b) 2.68 Hz
9. (a) 3.0 m; (b) −49 m/s; (c) -2.7×10^2 m/s^2; (d) 20 rad; (e) 1.5 Hz; (f) 0.67 s
10. +1.91 rad (or −4.37 rad)
11. 39.6 Hz
12. −0.927 rad (or +5.36 rad)
13. (a) 498 Hz; (b) greater
14. 2.08 h
15. 3.1 cm
16. 4.00 m
17. (a) 5.58 Hz; (b) 0.325 kg; (c) 0.400 m
18. 1.03 rad (or −5.25 rad)
19. (a) 0.500 m; (b) −0.251 m; (c) 3.06 m/s
20. (a) 25 cm; (b) 2.2 Hz
21. (a) 0.183A; (b) same direction
22. $2\pi/3$ rad
23. (a) 2.2 Hz; (b) 56 cm/s; (c) 0.10 kg; (d) 20.0 cm
24. 54 Hz
25. 22 cm
26. 18.2 Hz

27. (a) 0.525 m; (b) 0.686 s
28. (a) 200 N/m; (b) 1.39 kg; (c) 1.91 Hz
29. 37 mJ
30. (a) yes; (b) 12 cm
31. (a) 0.75; (b) 0.25; (c) $x_m/\sqrt{2}$
32. 8.3×10^2 N/m
33. (a) 2.25 Hz; (b) 125 J; (c) 250 J; (d) 86.6 cm
34. 0.333
35. (a) 3.1 ms; (b) 4.0 m/s; (c) 0.080 J; (d) 80 N; (e) 40 N
36. 2.4 cm
37. (a) 1.1 m/s; (b) 3.3 cm
38. 12 s
39. (a) 39.5 rad/s; (b) 34.2 rad/s; (c) 124 rad/s^2
40. (a) 1.64 s; (b) equal
42. (a) 0.499 m; (b) 0.940 mJ
43. (a) 0.205 kg · m^2; (b) 47.7 cm; (c) 1.50 s
44. 0.366 s
45. 5.6 cm
46. 8.77 s
47. 1.83 s
48. (a) 0.84 m; (b) 0.031 J
49. 0.0653 s
50. (a) 16 cm; (b) circle
51. (a) 0.53 m; (b) 2.1 s
52. 0.18 s
53. (a) 0.845 rad; (b) 0.0602 rad
54. 1.3×10^{-5} kg · m^2
55. (a) 2.26 s; (b) increases; (c) same
56. (a) 2.00 s; (b) 18.5 N · m/rad
57. 0.39
58. 6.0%
59. (a) 14.3 s; (b) 5.27
60. (a) 4.9×10^2 N/cm; (b) 1.1×10^3 kg/s
61. d and e
62. (a) $F_m/b\omega_d$; (b) F_m/b
63. 5.0 cm
64. +1.82 rad or −4.46 rad
65. (a) 1.2 J; (b) 50
66. (a) 11 m/s; (b) 1.7×10^3 m/s^2
67. 1.53 m
68. 65.5%
69. (a) 1.72 ms; (b) 11.2 ms
70. 0.19g
71. (a) 16.6 cm; (b) 1.23%
72. (a) $(r/R)\sqrt{k/m}$; (b) $\sqrt{k/m}$; (c) 0 (no oscillation)
73. (a) 8.11×10^{-5} kg · m^2; (b) 3.14 rad/s
74. (a) 0.015; (b) no
75. (a) 1.23 kN/m; (b) 76.0 N
76. (a) 2.8×10^3 rad/s; (b) 2.1 m/s; (c) 5.7 km/s^2
77. 7.2 m/s
78. (a) 1.6×10^4 m/s^2; (b) 2.5 m/s; (c) 7.9×10^3 m/s^2; (d) 2.2 m/s
79. (a) 1.1 Hz; (b) 5.0 cm
80. (a) 2.1×10^4 N/m; (b) 1.5×10^4 N/m; (c) 3.1×10^2 Hz; (d) 2.6×10^2 Hz
81. (a) 1.3×10^2 N/m; (b) 0.62 s; (c) 1.6 Hz; (d) 5.0 cm; (e) 0.51 m/s
82. (a) 0.735 kg · m^2; (b) 0.0240 N · m; (c) 0.181 rad/s
83. (a) 0.873 s; (b) 6.3 cm
84. 14.0°
85. (a) 0.35 Hz; (b) 0.39 Hz; (c) 0 (no oscillation)
86. 3.5 s
87. (a) 4.0 s; (b) 1.57 rad/s; (c) 0.37 cm; (d) (0.37 cm) cos[(1.57 rad/s)t]; (e) (−0.58 cm/s) sin[(1.57 rad/s)t]; (f) 0.58 cm/s; (g) 0.91 cm/s^3; (h) 0; (i) 0.58 cm/s
88. (a) 7.90 N/m; (b) 1.19 cm; (c) 2.00 Hz
89. (a) 147 N/m; (b) 0.733 s
90. 1.6 kg
91. (a) 1.6 Hz; (b) 1.0 m/s; (c) 0; (d) 10 m/s^2; (e) ±10 cm; (f) (−10 N/m)x
92. (a) 10 N, up; (b) 0.10 m; (c) 0.90 s; (d) 0.50 J
93. (a) 0.20 s; (b) 0.20 kg; (c) −0.20 m; (d) -2.0×10^2 m/s^2; (e) 4.0 J
94. (a) 0.30 m; (b) 0.28 s; (c) 1.5×10^2 m/s^2; (d) 11 J
95. (a) 3.5 m; (b) 0.75 s
96. (a) 62.5 mJ; (b) 31.3 mJ
97. (a) 3.2 Hz; (b) 0.26 m; (c) $x = (0.26\text{ m})\cos[(20\text{ rad/s})t - 1.57\text{ rad}]$
98. (a) 0.21 m; (b) 1.6 Hz; (c) 0.10 m
99. 0.079 kg · m^2
100. (a) 0.20 m; (b) 25; (c) 4.0 J; (d) 2.1 m/s
101. (a) 0.45 s; (b) 0.10 m above and 0.20 m below; (c) 0.15 m; (d) 2.3 J
102. 831.5 mm
103. (a) 0.30 m; (b) 30 m/s^2; (c) 0; (d) 4.4 s
104. (a) 0.44 s; (b) 0.18 m
105. (a) 245 N/m; (b) 0.284 s
106. 7×10^2 N/m
107. (a) 1.0×10^2 N/m; (b) 0.45 s

108. (a) 0.102 kg/s; (b) 0.137 J
109. (a) 8.3 s; (b) no
110. (a) F/m; (b) $2F/mL$; (c) 0
111. 50 cm
112. (a) $y_m = 0.008$ m, $T = 0.18$ s, $\omega = 35$ rad/s; (b) $y_m = 0.07$ m, $T = 0.48$ s, $\omega = 13$ rad/s; (c) $y_m = 0.03$ m, $T = 0.31$ s, $\omega = 20$ rad/s

---

Chapter 16

1. (a) 0.68 s; (b) 1.47 Hz; (c) 2.06 m/s
2. (a) 3.49 m^{-1}; (b) 31.5 m/s
3. 1.1 ms
4. −0.64 rad or 5.64 rad
5. (a) 11.7 cm; (b) π rad
6. (a) 6.0 cm; (b) 1.0×10^2 cm; (c) 2.0 Hz; (d) 2.0×10^2 cm/s; (e) $-x$ direction; (f) 75 cm/s; (g) −2.0 cm
7. (a) 64 Hz; (b) 1.3 m; (c) 4.0 cm; (d) 5.0 m^{-1}; (e) $4.0 \times 10^2\ s^{-1}$; (f) $\pi/2$ rad; (g) negative sign
8. 4.24 m/s
9. (a) 3.0 mm; (b) 16 m^{-1}; (c) $2.4 \times 10^2\ s^{-1}$; (d) negative sign
10. 1.3 cm
11. (a) negative sine functions; (b) 4.0 cm; (c) 0.31 cm^{-1}; (d) 0.63 s^{-1}; (e) π rad; (f) negative sign; (g) 2.0 cm/s; (h) −2.5 cm/s
12. 3.2
13. 129 m/s
14. $\sqrt{2}$
15. (a) 15 m/s; (b) 0.036 N
16. 135 N
17. (0.12 mm); (b) 141 m^{-1}; (c) 628 s^{-1}; (d) positive sign
18. (a) 30 m/s; (b) 17 g/m
19. (a) 5.0 cm ; (b) 40 cm; (c) 12 m/s; (d) 0.033 s; (e) 9.4 m/s; (f) 16 m^{-1}; (g) $1.9 \times 10^2\ s^{-1}$; (h) 0.93 rad; (i) positive sign
20. (a) 0.64 Hz; (b) 63 cm; (c) (5 cm); (d) 0.10 cm^{-1}; (e) 4.0 s^{-1}; (f) negative sign; (g) 0.064 N
21. 2.63 m
22. (a) 28.6 m/s; (b) 22.1 m/s; (c) 188 g; (d) 313 g
24. 198 Hz
25. 3.2 mm
26. 1.75 m/s
27. 0.20 m/s
28. 0.20 m/s
29. $1.4y_m$
30. (a) 82.8°; (b) 1.45 rad; (c) 0.23 wavelength
31. (a) 9.0 mm; (b) 16 m^{-1}; (c) $1.1 \times 10^3\ s^{-1}$; (d) 2.7 rad; (e) negative sign
32. (a) 10 W; (b) 20 W; (c) 40 W; (d) 26 W; (e) 0
33. 5.0 cm
34. 84°
35. (a) π rad; (b) 3.0 mm; (c) 0; (d) 13 mm; (e) 9.4 mm
36. 0
37. (a) 3.29 mm; (b) 1.55 rad; (c) 1.55 rad
38. (a) 4; (b) 8; (c) none
39. 7.91 Hz; (b) 15.8 Hz; (c) 23.7 Hz
40. (a) 66.1 m/s; (b) 26.4 Hz
41. (a) 82.0 m/s; (b) 16.8 m; (c) 4.88 Hz
42. 10 cm
43. (a) 144 m/s; (b) 60.0 cm; (c) 241 Hz
44. (a) $2f_3$; (b) λ_3
45. (a) 105 Hz; (b) 158 m/s
46. 260 Hz
47. (a) 0.25 cm; (b) 1.2×10^2 cm/s; (c) 3.0 cm; (d) zero
48. (a) 4.0 m; (b) 24 m/s; (c) 1.4 kg; (d) 0.11 s
49. (a) 0.50 cm; (b) 3.1 m^{-1}; (c) $3.1 \times 10^2\ s^{-1}$; (d) negative sign
50. (a) 0; (b) 0.20 m; (c) 0.40 m; (d); 50 ms; (e) 8.0 m/s; (f) 0.020 m; (g) 0; (h) 25 ms; (i) 50 ms
51. (a) +4.0 cm; (b) 0; (c) 0; (d) −0.13 m/s
52. 0.25 m
53. (a) 2.00 Hz; (b) 2.00 m; (c) 4.00 m/s; (d) 50.0 cm; (e) 150 cm; (f) 250 cm; (g) 0; (h) 100 cm; (i) 200 cm
54. (a) 4.5 mm; (b) 16 m^{-1}; (c) $5.2 \times 10^2\ s^{-1}$; (d) negative sign
55. (a) 323 Hz; (b) eight
56. 0.845 g/m
57. 2.8 rad or −3.5 rad
58. 2.9 rad or −3.4 rad
59. (a) 5.0 cm/s; (b) $+x$
60. (a) 0.31 m; (b) 1.64 rad; (c) 2.2 mm
61. (a) $0.83y_1$; (b) 37°
62. (a) 3.0 mm; (b) 31 m^{-1}; (c) $7.5 \times 10^2\ s^{-1}$; (d) minus
63. 1.2 rad

65. (a) 0.16 m; (b) 2.4×10^2 N; (c) $y = (0.16\,\mathrm{m}) \sin[(1.57\,\mathrm{m}^{-1})x] \sin[(31.4\,\mathrm{s}^{-1})t]$
66. (a) 3.77 m/s; (b) 12.3 N; (c) 0; (d) 46.4 W; (e) 0; (f) 0; (g) ± 0.50 cm
67. (a) $2\pi y_m/\lambda$; (b) no
68. 300 m/s
69. (a) 1.00 cm; (b) $3.46 \times 10^3\,\mathrm{s}^{-1}$; (c) $10.5\,\mathrm{m}^{-1}$; (d) plus
70. (a) 2.0 cm; (b) $0.63\,\mathrm{cm}^{-1}$; (c) $2.5 \times 10^3\,\mathrm{s}^{-1}$; (d) minus; (e) 50 m/s; (f) 40 m/s
71. (a) 6.7 mm; (b) 45°
72. (a) 1.33 m/s; (b) 1.88 m/s; (c) $16.7\,\mathrm{m/s}^2$; (d) $23.7\,\mathrm{m/s}^3$
73. (a) 75 Hz; (b) 13 ms
74. (a) 880 Hz; (b) 1320 Hz
75. (a) 240 cm; (b) 120 cm; (c) 80 cm
76. (a) -3.9 cm; (b) 0.15 m; (c) $0.79\,\mathrm{m}^{-1}$; (d) $13\,\mathrm{s}^{-1}$; (e) plus; (f) -0.14 m
77. (a) 144 m/s; (b) 3.00 m; (c) 1.50 m; (d) 48.0 Hz; (e) 96.0 Hz
78. (a) $2P_1$; (b) $P_1/4$
79. (a) 2.0 mm; (b) 95 Hz; (c) +30 m/s; (d) 31 cm; (e) 1.2 m/s
80. (a) 5.0 cm/s; (b) 3.2 cm; (c) 0.25 Hz
81. (a) 0.52 m; (b) 40 m/s; (c) 0.40 m
82. 1.3 m; (b) $(2.0\,\mathrm{mm}) \sin[(9.4\,\mathrm{m}^{-1})x] \cos[(3.8 \times 10^3\,\mathrm{s}^{-1})t]$
83. 36 N
84. (a) 0.50 m; (b) 0; (c) 0.25 m; (d) 0.50 s
85. (a) 8.0 cm; (b) 1.0 cm
86. (a) $z = (3.0\,\mathrm{mm}) \sin[(60\,\mathrm{cm}^{-1})y - (31\,\mathrm{s}^{-1})t]$ (b) 9.4 cm/s
87. (a) $\sqrt{k\,\Delta\ell(\ell + \Delta\ell)/m}$
88. (b) $+x$; (c) interchange their amplitudes; (d) $x = \lambda/4 = 6.26$ cm; (e) $x = 0$ and $x = \lambda/2 = 12.5$ cm; (f) the amplitude (4.00 mm) is the sum of the amplitudes of the original waves; (g) the amplitude (1.00 mm) is the difference of the amplitudes of the original waves
89. (a) 4.3×10^{14} Hz to 7.5×10^{14} Hz; (b) 1.0 m to 2.0×10^2 m; (c) 6.0×10^{16} Hz to 3.0×10^{19} Hz
91. (c) 2.0 m/s; (d) $-x$
93. (b) kinetic energy of the transversely moving flat sections of the string
94. (a) $0.5T_A$, $0.75T_A$, $1.75T_A$; (b) T_A, $2T_A$; (c) $0.15T_A$, $1.25T_A$, $1.5T_A$; (d) design b damps out the fundamental oscillations in both; design c does not affect the fundamental oscillation in A but does damp that in B

Chapter 17

1. 1.7×10^2 m
2. (a) 2.6 km; (b) 2.0×10^2
3. (a) 79 m, 41 m; (b) 89 m
4. 0.144 Mpa
5. 40.7 m
6. 44 m
7. 1.9×10^3 km
8. (a) 1.50 Pa; (b) 158 Hz; (c) 2.22 m; (d) 350 m/s
9. (a) 76.2 μm; (b) 0.333 mm
10. 0.23 ms
11. (a) 2.3×10^2 Hz; (b) higher
12. 960 Hz
13. (a) 6.1 nm; (b) $9.2\,\mathrm{m}^{-1}$; (c) $3.1 \times 10^3\,\mathrm{s}^{-1}$; (d) 5.9 nm; (e) $9.8\,\mathrm{m}^{-1}$; (f) $3.1 \times 10^3\,\mathrm{s}^{-1}$
14. 4.12 rad
15. (a) 14; (b) 14
16. 17.5 cm
17. (a) 343 Hz; (b) 3; (c) 5; (d) 686 Hz; (e) 2; (f) 3
18. (a) 0; (b) 0; (c) 4
19. (a) 143 Hz; (b) 3; (c) 5; (d) 286 Hz; (e) 2; (f) 3
20. (a) 0; (b) fully constructive; (c) increase; (d) 128 m; (e) 63.0 m; (f) 41.2 m
21. 15.0 mW
22. (a) $0.080\,\mathrm{W/m}^2$; (b) $0.013\,\mathrm{W/m}^2$
23. 36.8 nm
24. 1.26
25. (a) 1.0×10^3; (b) 32
26. (a) $8.84\,\mathrm{nW/m}^2$; (b) 39.5 dB
27. (a) $10\,\mu\mathrm{W/m}^2$; (b) $0.10\,\mu\mathrm{W/m}^2$; (c) 70 nm; (d) 7.0 nm
28. (a) 0.26 nm; (b) $1.5\,\mathrm{nW/m}^2$
29. (a) 5 dB; (b) 3.2
30. (a) $5.97 \times 10^{-5}\,\mathrm{W/m}^2$; (b) 4.48 nW
31. (a) 0.34 nW; (b) 0.68 nW; (c) 1.4 nW; (d) 0.88 nW; (e) 0
32. (a) 86 Hz; (b) yes, low frequency; (c) higher
33. (a) 2; (b) 1
34. 20 kHz
35. (a) 833 Hz; (b) 0.418 m

36. (a) 57.2 cm; (b) 42.9 cm
37. (a) 405 m/s; (b) 596 N; (c) 44.0 cm; (d) 37.3 cm
38. (a) 4; (b) 0.125 m; (c) 0.375 m
39. (a) 3; (b) 1129 Hz; (c) 1506 Hz
40. (a) 260 Hz; (b) 4; (c) 840 Hz; (d) 7
41. 12.4 m
42. (a) 71.5 Hz; (b) 64.8 N
43. 45.3 N
44. (a) 3; (b) 0.20 m; (c) 0.60 m; (d) 0.60 m; (e) 143 Hz
45. 387 Hz
46. 2.25 ms
47. 0.020
48. (a) 10; (b) 4
49. (a) 526 Hz; (b) 555 Hz
50. zero
51. 4.61 m/s
52. 0.195 MHz
53. 155 Hz
54. (a) 1.02 kHz; (b) 1.04 kHz
55. (a) 1.58 kHz; (b) 0.208 m; (c) 2.16 kHz; (d) 0.152 m
56. 0.236
57. 41 kHz
58. (a) $2v/3$; (b) $2v/3$; (c) $2v/3$; (d) $2v/3$
59. (a) 485.8 Hz; (b) 500.0 Hz; (c) 486.2 Hz; (d) 500.0 Hz
60. (a) 2.0 kHz; (b) 2.0 kHz
61. (a) 598 Hz; (b) 608 Hz; (c) 589 Hz
62. 3.3×10^2 m/s
63. (a) 42°; (b) 11 s
64. 33.0 km
65. 0.250
66. (a) 572 Hz; (b) 1.14 kHz
67. (a) 0; (b) 0.572 m; (c) 1.14 m
68. (a) 221 nm; (b) 35 cm; (c) 24 nm; (d) 35 cm
69. 0
70. 0.33
71. (a) 2; (b) 6; (c) 10
72. (a) 2.10 m; (b) 1.47 m
73. 0.25
74. (a) 0.5; (b) 1.5
75. (a) $L(v_m - v)/v_m v$; (b) 364 m
76. (a) 3.6×10^2 m/s^2; (b) 150 Hz
77. (a) 9.7×10^2 Hz; (b) 1.0 kHz; (c) 60 Hz, no
78. 3.1 m/s
79. (a) 5.5×10^2 m/s; (b) 1.1×10^3 m/s; (c) 1
80. 39.7 μW/m^2; (b) 171 nm; (c) 0.893 Pa
81. (2) 2.00; (b) 1.41; (c) 1.73; (d) 1.85
82. 7.9×10^{10} PA
83. (a) 10 W; (b) 0.032 W/m^2; (c) 99 dB
84. (a) 467 Hz; (b) 494 Hz
85. (a) 7.70 Hz; (b) 7.70 Hz
86. (a) 11 ms; (b) 3.8 m
87. 0.76 μm
88. 2 μW
89. 400 Hz
90. (a) 5.0λ; (b) fully constructive; (c) 5.5λ; (d) fully destructive
91. 3
92. (a) 0.30 cm; (b) 0.26 cm^{-1}; (c) 1.6×10^2 s^{-1}; (d) 6.0 m/s; (e) plus
93. (a) 59.7; (b) 2.81×10^{-4}
94. (b) length
95. (a) 5.2 kHz; (b) 2
96. (a) 5.0×10^3; (b) 71; (c) 71
97. 30°
98. (a) 39.3 Hz; (b) 118 Hz
99. (a) 0.50 m; (b) 0.34 m; (c) 0.66 m
100. (a) 14; (b) 12
101. (a) 88 mW/mn^2; (b) 0.75
102. (a) 3.9×10^2 to 9.2×10^2 GJ; (b) 0.63 to 1.5 W/m^2; (c) 25 to 58 kW/m^2; (d) surface wave
103. 2.1 m
104. (a) 880 Hz; (b) 824 Hz
105. 171 m
106. (b) 0.8 to 1.6 μs
107. (a) rightward; (b) 0.90 m/s; (c) less
108. (a) node; (b) 22 s
109. 1 cm
110. 4.8×10^2 Hz
111. (a) 482 Hz; (b) 660 Hz

Chapter 18

1. 0.05 kPa; (b) nitrogen
2. 1.366
3. 348 K
4. (a) 320°F; (b) −12.3°F
5. (a) −96°F; (b) 56.7°C
6. 1375°X
7. −91.9°X
8. (a) 9.996 cm; (b) 68°C
9. 2.731 cm
10. 1.1 cm
11. 29 cm^3

12. $49.87\,cm^3$
13. $11\,cm^2$
14. $23 \times 10^{-6}/C°$
15. $0.26\,cm^3$
16. (a) −0.69%; (b) aluminum
17. 360°C
18. (a) 0.36%; (b) 0.18%; (c) 0.54%; (d) 0.00%; (e) $1.8 \times 10^{-5}/C°$
19. 0.13 mm
20. 0.217 K/s
21. 7.5 cm
22. 109 g
23. (a) 523 J/kg · K; (b) 26.2 J/mol · K; (c) 0.600 mol
24. 94.6 L
25. 42.7 kJ
26. 0.25 kg
27. 160 s
28. (a) 52 MJ; (b) 0°C
29. 3.0 min
30. (a) 2.03×10^4 cal; (b) 1.11×10^3 cal; (c) 873°C
31. 33 g
32. (a) 68 kJ/kg; (b) 2.3 kJ/kg · K
33. $33\,m^2$
34. 4.0×10^2 J/kg · K
35. 742 kJ
36. (a) 37 W; (b) 2.0 kg; (c) 0.13 kg
37. (a) 5.3°C (b) 0; (c) 0°C; (d) 60 g
38. 82 cal
39. (a) 0°C; (b) 2.5°C
40. 13.5 C°
41. 8.71 g
42. (a) positive; (b) positive; (c) zero; (d) positive; (e) negative; (f) negative; (g) negative; (h) −20 J
43. A: 1.2×10^2 J; (b) 75 J; (c) 30 J
44. (a) −200 J; (b) −293 J; (c) −93 J
45. −30 J
46. (a) +8.0 J; (b)−9.3 J
47. 60 J
48. −5.0 J
49. (a) 6.0 cal; (b) −43 cal; (c) 40 cal; (d) 18 cal; (e) 18 cal
50. (a) 0.13 m; (b) 2.3 km
51. 1.66 kJ/s
52. (a) 8×10^2 W; (b) 2×10^4 J
53. (a) 16 J/s; (b) 0.048 g/s
54. (a) 1.23 kW; (b) 2.28 kW; (c) 1.05 kW W
55. 0.50 min
56. (a) 1.4 W; (b) 3.3
57. (a) $1.7 \times 10^4\,W/m^2$; (b) $18\,W/m^2$
58. (a) 15.8 C°; (b) greater than; (c) 13.8 C°
59. −4.2°C
60. 1.1 m
61. 0.40 cm/h
62. 10%
63. 1.5
64. 0.27 mm
65. (a) 6.61 mm; (b) 3.006 606 m; (c) 6.62 mm; (d) 2.999 985 m; (e) 1.45×10^{-5} m
66. (a) 2.5×10^2 K; (b) 1.5
67. (a) 90 W; (b) 2.3×10^2 W; (c) 3.3×10^2 W
68. 4.5×10^2 J/kg · K
69. (a) 10 000°F; (b) 37.0°C; (c) −57°C; (d) −297°F
70. $0.432\,cm^3$
71. 0.41 kJ/kg · K
72. (a) −45 J; (b) +45 J
73. 1.87×10^4; (b) 10.4 h
74. (a) 84.3°C; (b) 57.6°C
75. 1.7×10^2 km
76. 6.7×10^{12} J
77. (a) $11p_1V_1$; (b) $6p_1V_1$
78. 766°C
79. $4.83 \times 10^{-2}\,cm^3$
80. $35.7\,m^3$
81. 23 J
82. 3.1×10^2 J
83. (a) 80 J; (b) 80 J
84. 4.4×10^{-3} cm
85. 1.17 C°
86. (a) 2.3×10^2 J/s; (b) 15
87. −6.1 nW
88. $0.32\,cm^2$
89. 10.5°C
90. 33.3 kJ
91. 20 MJ
92. −157°C
93. 79.5°C
94. 8.6 J
95. (a) $13 \times 10^{-6}/F°$; (b) 4.2 mm
96. (a) 1.2 W/m · K; (b) 0.70 Btu/ft · F° · h; (c) $5.3 \times 10^{-3}\,m^2 \cdot K/W$
97. $2.16 \times 10^{-5}\,m^2$
98. 45.5°C
99. (a) 1.8 W; (b) 0.024 C°
100. 66°C

101. 333 J
102. 2.5 kJ/kg · K

Chapter 19

1. 0.933 kg
2. (a) 0.0127; (b) 7.64×10^{21} atoms
3. (a) 5.47×10^{-8} mol; (b) 3.29×10^{16} molecules
4. 25 molecules/cm^3
5. (a) 0.0388 mol; (b) 220°C
6. 186 kPa
7. (a) 106 mol; (b) 0.892 m^3
8. (a) 3.14×10^3 J; (b) from
9. 0.2
10. 360 K
11. 207 J
12. (a) 1.5 mol; (b) 1.8×10^3 K; (c) 6.0×10^2 K; (d) 5.0 kJ
13. 5.60 kJ
14. 1.0×10^2 cm^3
15. 2.0×10^5 Pa
16. 442 m/s
17. 1.8×10^2 m/s
18. 2.50 km/s
19. (a) 511 m/s; (b) −200°C; (c) 899°C
20. 9.53×10^6 m/s
21. 1.9 kPa
22. (a) 494 m/s; (b) 27.9 g/mol; (c) N_2
23. 3.3×10^{-20} J
24. (a) 5.65×10^{-21} J; (b) 7.72×10^{-21} J; (c) 3.40 kJ; (d) 4.65 kJ
25. (a) 6.75×10^{-20} J; (b) 10.7
26. 3.7 GHz
27. (a) 6×10^9 km
28. 0.32 nm
29. (a) 3.27×10^{10} molecules/cm^3; (b) 172 m
30. (a) 1.7; (b) 5.0×10^{-5} cm; (c) 7.9×10^{-6} cm
31. (a) 6.5 km/s; (b) 7.1 km/s
32. (a) 3.2 cm/s; (b) 3.4 cm/s; (c) 4.0 cm/s
33. (a) 420 m/s, 458 m/s
34. (a) 2.7×10^2 K; (b) 4.9×10^2 m/s
35. (a) 1.0×10^4 K; (b) 1.6×10^5 K; (c) 4.4×10^2 K; (d) 7.0×10^3 K; (e) no; (f) yes
36. 1.50
37. (a) 7.0 km/s; (b) 2.0×10^{-8} cm; (c) 3.5×10^{10} collisions/s
38. 4.7
39. (a) 0.67; (b) 1.2; (c) 1.3; (d) 0.33
40. 3.4 kJ
41. (a) 0; (b) +374 J; (c) +374 J; (d) $+3.11 \times 10^{-22}$ J
42. (a) +249 J; (b) +623 J; (c) +374 J
43. 15.8 J/mol · K
44. (a) 15.9 J; (b) 34.4 J/mol · K; (c) 26.1 J/mol · K
45. (a) 6.6×10^{-26} kg; (b) 40 g/mol
46. (a) −5.0 kJ; (b) 2.0 kJ; (c) 5.0 kJ
47. (a) 7.72×10^4 J; (b) 5.46×10^4 J; (c) 5.17 J/mol · K; (d) 4.32×10^4 J; (e) 8.86×10^4 J; (f) 8.38 J/mol · K
48. 50 J
49. 8.0 kJ
50. (a) 0.375 mol; (b) 1.09 kJ; (c) 0.714
51. (a) 6.98 kJ; (b) 4.99 kJ; (c) 1.99 kJ; (d) 2.99 kJ
52. 1.5×10^3 $N \cdot m^{2.2}$
53. (a) 14 atm; (b) 6.2×10^2 K
54. (a) 2.46 atm, 336 K; (b) 0.406 L
55. −15 J
56. (a) diatomic; (b) 446 K; (c) 8.10 mol
57. −20 J
58. (a) 0.33; (b) polyatomic (ideal); (c) 1.44
59. (a) 3.74 kJ; (b) 3.74 kJ; (c) 0; (d) 0; (e) −1.81 kJ; (f) 1.81 kJ; (g) −3.22 kJ; (h) −1.93 kJ; (i) −1.29 kJ; (j) 520 J; (k) 0; (l) 520 J; (m) 0.0246 m^3; (n) 2.00 atm; (o) 0.0373 m^3; (p) 1.00 atm
60. (b) 125 J; (c) to
61. (a) 22.4 L
62. 38.8 m
63. 349 K
64. (a) 1.44×10^3 m/s; (b) 5.78×10^{-4}; (c) 71%; (d) 2.03×10^3 m/s; (e) 4.09×10^{-4}; (f) increase; (g) decrease
65. (a) 900 cal; (b) 0; (c) 900 cal; (d) 450 cal; (e) 1200 cal; (f) 300 cal; (g) 900 cal; (h) 450 cal; (i) 0; (j) −900 cal; (k) 900 cal; (l) 450 cal
66. (a) diatomic with rotating molecules; (b) 1.00; (c) 1.90
68. 1.40
69. (a) 1.37; (b) diatomic
70. 9.2×10^{-6}
71. (a) −374 J; (b) 0; (c) +374 J; (d) $+3.11 \times 10^{-22}$ J
72. -1.33×10^4 J

73. (a) −60 J; (b) 90 K
74. 1.52 nm
75. $7.03 \times 10^9\ s^{-1}$
76. (a) 2.00 atm; (b) 333 J; (c) 0.961 atm; (d) 236 J
77. (a) 122 K; (b) 365 K; (c) 0
78. (a) monatomic; (b) 2.7×10^4 K; (c) 4.5×10^4 mol; (d) 3.4 kJ; (e) 3.4×10^2 kJ; (f) 0.010
79. (a) 9.0 atm; (b) 300 K; (c) 4.4 kJ; (d) 3.2 atm; (e) 120 K; (f) 2.9 kJ; (g) 4.6 atm; (h) 170 K; (i) 3.4 kJ
80. 5.0 m^3
81. 653 J
82. (a) 38 L; (b) 71 g
83. 3.11 kJ/kg · K
84. (a) 22.5 L; (b) 2.25; (c) 0.840 μm; (d) 0.840 μm
85. −3.0 J
86. 307°C
87. (a) 1.5; (b) 4.5; (c) 6; (d) 2
88. 4.67 Pa
89. (a) $3/v_0^3$; (b) $0.750v_0$; (c) $0.775v_0$
90. (a) 3.49 kJ; (b) 2.49 kJ; (c) 997 J; (d) 1.50 kJ
91. (a) −2.37 kJ; (b) 2.37 kJ
92. (a) 2.5×10^{25} molecules/m^3; (b) 1.2 kg
93. (a) −45 J; (b) 1.8×10^2 K
94. −6.9 kJ

Chapter 20

1. (a) 9.22 kJ; (b) 23.1 J/K; (c) 0
2. 1.86×10^4 J
3. 14.4 J/K
4. (a) 14.6 J/K; (b) 30.2 J/K
5. (a) 5.79×10^4 J; (b) 173 J/K
6. 2.75 mol
7. (a) 57.0° C; (b) −22.1 J/K; (c) +24.9 J/K; (d) +2.8 J/K
8. 2.75 mol
9. (a) −710 mJ/K; (b) +710 mJ/K; (c) +723 mJ/K; (d) −723 mJ/K; (e) +13 mJ/K; (f) 0
10. 4.5×10^2 J/kg · K
11. (a) 0.333; (b) 0.215; (c) $0.644W$; (d) 1.10; (e) 1.10; (f) 0; (g) 1.10; (h) 0; (i) −0.889; (j) −0.889; (k) −1.10; (l) −0.889; (m) 0; (n) 0.889; (o) 0
12. 0.0368 J/K
13. (a) 320 K; (b) 0; (c) +1.72 J/K
14. (a) 4.5 kJ; (b) −5.0 kJ; (c) 9.5 kJ
15. +0.76 J/K
16. +0.64 J/K
17. (a) −943 J/K; (b) +943 J/K; (c) yes
18. (a) 3.00; (b) 6.00; (c) 0; (d)9.64 J/K; (e) 0
19. (a) 0.693; (b) 4.50; (c) 0.693; (d) 0; (e) 4.50; (f) 23.0 J/K; (g) −0.693; (h) 7.50; (i) −0.693; (j) 3.00; (k) 4.50; (l) 23.0 J/K
20. (a) 1.84 kPa; (b) 441 K; (c) 3.16 kJ; (d) 1.94 J/K
21. (a) 23.6%; (b) 1.49×10^4 J
22. (a) 31%; (b) 16 kJ
23. 97 K
24. 99.999 95%
25. (a) 266 K; (b) 341 K
26. (a) 4.67 kJ/s; (b) 4.17 kJ/s
27. (a) 1.47 kJ; (b) 554 J; (c) 918 J; (d) 62.4%
28. 1.7 kJ
29. (a) 2.27 kJ; (b) 14.8 kJ; (c) 15.4%; (d) 75.0%; (e) greater
31. (a) 33 kJ; (b) 25 kJ; (c) 26 kJ; (d) 18 kJ
32. (a) monatomic; (b) 75%
33. (a) 3.00: (b) 1.98; (c) 0.660; (d) 0.495; (e) 0.165; (f) 34.0%
34. (a) 49 kJ; (b) 7.4 kJ
35. 20 J
36. 13 J
37. 440 W
38. (a) 0.071 J; (b) 0.50 J; (c) 2.0 J; (d) 5.0 J
39. 0.25 hp
40. (a) 167 J; (b) 343 J
41. 2.03
42. 1.08 MJ
44. (a) 1.26×10^{14}; (b) 1.13×10^{15}; (c) 11.1%; (d) 1.01×10^{29}; (e) 1.27×10^{30}; (f) 8.0%; (g) 9.25×10^{58}; (h) 1.61×10^{60}; (i) 5.7%; (j) decrease
45. (a) $W = N!/(n_1!\,n_2!\,n_3!)$; (b) $[(N/2)!\,(N/2)!]/[(N/3)!\,(N/3)!\,(N/3)!]$; (c) 4.2×10^{16}
46. −1.18 J/K
47. (a) 1; (b) 6; (c) 0; (d) 2.47×10^{-23} J/K
48. 13.1%
49. (a) 87 m/s; (b) 1.2×10^2 m/s; (c) 22 J/K
50. (a) 4.45 J/K; (b) no
51. (a) 78%; (b) 82 kg/s
52. (a) 93.8 J; (b) 231 J
53. (a) 66.5°C; (b) 14.6 J/K; (c) 11.0 J/K; (d) −21.2 J/K; (e) 4.39 J/K

54. (a) 40.9°C; (b) −27.1 J/K; (c) 30.5 J/K; (d) 3.4 J/K
55. (a) 6.34 J/K; (b) 6.34 J/K; (c) 6.34 J/K; (d) 6.34 J/K
56. 4.46 J/K
57. 1.18×10^3 J/K
58. (a) 0; (b) 0; (c) −23.0 J/K; (d) 23.0 J/K
59. 75
60. (a) 7.2 kJ; (b) 9.6×10^2 J; (c) 13%
61. (a) 1; (b) 1; (c) 3; (d) 10; (e) 1.5×10^{-23} J/K; (f) 3.2×10^{-23} J/K
62. 25%
63. +3.59 J/K
64. +5.98 J/K
65. (a) 1.95 J/K; (b) 0.650 J/K; (c) 0.217 J/K; (d) 0.072 J/K; (e) decrease
66. (a) 25.5 kJ; (b) 4.73 kJ; (c) 18.5%
67. −40 K
68. (a) −44.2°C; (b) −1.69 J/K; (c) 2.38 J/K; (d) 0.69 J/K
69. (a) 1.26×10^{14}; (b) 4.71×10^{13}; (c) 0.37; (d) 1.01×10^{29}; (e) 1.37×10^{28}; (f) 0.14; (g) 9.05×10^{58}; (h) 1.64×10^{57}; (i) 0.018; (j) decrease
70. 0.141 J/K · s
71. (a) 3.73; (b) 710 J
72. (a) 700 J; (b) 0; (c) 50 J; (d) 700 J; (e) 0.226 m^3; (f) 0.284 m^3; (g) 0; (h) −1.25 kJ; (i) 0; (j) 1.25 kJ
73. (a) 42.6 kJ; (b) 7.61 kJ

Chapter 21

1. (a) 4.9×10^{-7} kg; (b) 7.1×10^{-11} C
2. 2.81 N
3. 1.39 m
4. 0.375
5. 0.500
6. (a) −2.83; (b) no
7. (a) 0.17 N; (b) −0.046 N
8. (a) 9.0; (b) −25
9. (a) −1.00 μC; (b) +3.00 μC
10. −4.00
11. (a) 1.60 N; (b) 2.77 N
12. (a) positive; (b) +9.0
13. (a) 14 cm; (b) 0
14. 1.333
15. (a) 35 N; (b) −10°; (c) −8.4 cm; (d) +2.7 cm
16. (a) −83 μC; (b) 55 μC
17. (a) −0.444; (b) 3.00 cm; (c) 0
18. (a) 1.92 cm; (b) less than
19. 3.8×10^{-8} C
20. (a) 0; (b) 12 cm; (c) 0; (d) 4.9×10^{-26} N
21. (a) 3.2×10^{-19} C; (b) 2
22. 2.89×10^{-9} N
23. 6.3×10^{11}
24. (a) 8.99×10^{-19} N; (b) 625
25. 122 mA
26. 1.3×10^7 C
27. 13e
28. (a) 0.654 rad; (b) 0.889 rad; (c) 0.988 rad
29. (a) 0; (b) 1.9×10^{-9} N
30. (a) positron; (b) electron
31. (a) ^{9}B; (b) ^{13}N; (c) ^{12}C
32. (a) −4; (b) +16
33. 0
34. −11.1 μC
35. (a) $(3.52 \times 10^{-25}\text{ N})\hat{\imath}$; (b) 0
36. +16e
37. 1.31×10^{-22} N
38. (a) −6.05 cm; (b) 6.05 cm
39. (a) $(0.829\text{ N}\hat{\imath}$; (b) $(-0.621\text{ N})\hat{\jmath}$
40. (a) 2.00 cm; (b) 9.21×10^{-24} N
41. (a) 6.16×10^{-24} N; (b) 208°
42. (a) 2.00×10^{10} electrons; (b) 1.33×10^{10} electrons
43. 2.2×10^{-6} kg
44. 9.0 kN
45. 0.707
46. −45 μC
47. 0.19 MC
48. (a) $(L/2)(1 + kqQ/Wh^2)$; (b) $\sqrt{3kqQ/W}$
49. (a) $(89.9\text{ N})\hat{\imath}$; (b) $(-2.50\text{ N})\hat{\imath}$; (c) 68.3 cm; (d) 0
50. (a) 5.1×10^2 N; (b) 7.7×10^{28} m/s^2
51. 3.8 N
52. (a) 3.60 μN; (b) 2.70 μN; (c) 3.60 μN
53. 1.2×10^{-5} C
54. 0.50 C
55. (a) 8.99×10^9 N; (b) 8.99 kN
56. 2.25×10^{20}
57. 4.68×10^{-19} N
58. 0.375
59. 1.7×10^8 N
60. −5.1 m
61. 11.9 cm
62. 10^{18} N

63. (a) 1.25×10^{13} electrons; (b) from you to faucet; (c) positive; (d) from faucet to the cat; (e) stroking the cat transfers electrons from you to the fur, which then induces charge in the cat's body, with surfaces away from the stroked region becoming charged negatively; if you bring your positive hand near the negative nose, electrons can spark across the gap
64. (a) 0.5; (b) 0.15; (c) 0.85
65. (a) 5.7×10^{13} C; (b) cancels out; (c) 6.0×10^5 kg
66. (b) 2.4×10^{-8} C
67. (b) 3.1 cm
68. −2.25
69. (a) Let $J = qQ/4\pi\epsilon_0 d^2$.
For $x < 0$, $F = -J[\alpha^{-2} + (1-\alpha)^{-2}]$;
for $0 < x < d$, $F = J[\alpha^{-2} - (1-\alpha)^{-2}]$;
for $d < x$, $F = J[\alpha^{-2} + (\alpha-1)^{-2}]$
70. -1.32×10^{13} C
71. (a) $1.72L$; (b) 0

Chapter 22

2. (a) 6.4×10^{-18} N; (b) 20 N/C
4. 0.111 nC
5. 56 pC
6. $(-6.39 \times 10^5\ \mathrm{N/C})\hat{\mathrm{i}}$
7. (a) 3.07×10^{21} N/C; (b) outward
8. (a) $2.72L$
9. −30 cm
10. 0
11. $(1.02 \times 10^5\ \mathrm{N/C})\hat{\mathrm{j}}$
12. (a) 160 N/C; (b) 45°
13. (a) 1.38×10^{-10} N/C; (b) 180°
14. (a) 34 cm; (b) 2.2×10^{-8} N/C
15. (a) 3.93×10^{-6} N/C; (b) −76.4°
16. (a) 67.8°; (b) −67.8°
17. 6.88×10^{-28} C · m
18. 0.98
19. (a) $qd/4\pi\epsilon_0 r^3$; (b) −90°
20. $qd^3/4\pi\epsilon_0 z^5$
22. 0.506
23. (a) -1.72×10^{-15} C/m;
(b) $-3.82 \times 10^{-14}\ \mathrm{C/m^2}$;
(c) $-9.56 \times 10^{-15}\ \mathrm{C/m^2}$;
(d) $-1.443 \times 10^{-12}\ \mathrm{C/m^3}$
24. (a) 23.8 N/C; (b) −90°
25. (a) 20.6 N/C; (b) −90°
26. 1.70 cm
27. (a) -5.19×10^{-14} C/m; (b) 1.57×10^{-3} N/C; (c) −180°; (d) 1.52×10^{-8} N/C; (e) 1.52×10^{-8} N/C
28. 1.57
29. (a) 12.4 N/C; (b) 90°
30. 6.3×10^3 N/C
31. 0.346
32. 6.9 cm
33. 28%
34. (a) 2.03×10^{-7} N/C; (b) up
35. $3.51 \times 10^{15}\ \mathrm{m/s^2}$
36. (a) 1.02×10^{-2} N/C; (b) west
37. 6.6×10^{-15} N
38. (a) 4.8×10^{-13} N; (b) 4.8×10^{-13} N
39. (a) 1.5×10^3 N/C; (b) 2.4×10^{-16} N; (c) up; (d) 1.6×10^{-26} N; (e) 1.5×10^{10}
40. (a) 7.12 cm; (b) 28.5 ns; (c) 0.112
41. (a) $1.92 \times 10^{12}\ \mathrm{m/s^2}$; (b) 1.96×10^5 m/s
42. −5e
43. (a) 2.7×10^6 m/s; (b) 1.00 kN/C
44. (a) $(-2.1 \times 10^{13}\ \mathrm{m/s^2})\hat{\mathrm{j}}$;
(b) $(1.5 \times 10^5\ \mathrm{m/s})\hat{\mathrm{i}} - (2.8 \times 10^6\ \mathrm{m/s})\hat{\mathrm{j}}$
45. 27 μm
46. $(1.53 \times 10^6\ \mathrm{m/s})\hat{\mathrm{i}} - (4.34 \times 10^5\ \mathrm{m/s})\hat{\mathrm{j}}$
47. (a) 0.245 N; (b) −11.3°; (c) 108 m; (d) −21.6 m
48. (a) $1.16 \times 10^{16}\ \mathrm{m/s^2}$; (b) $3.94 \times 10^{16}\ \mathrm{m/s^2}$; (c) $3.97 \times 10^{16}\ \mathrm{m/s^2}$; (d) because the net force due to the charged particles near the edge of the disk decreases
49. (a) 27 km/s; (b) 50 μm
50. (a) 0; (b) 8.5×10^{-22} N · m; (c) 0
51. (a) 9.30×10^{-15} C · m; (b) 2.05×10^{-11} J
52. 2.5×10^{-28} C · m
53. 1.22×10^{-23} J
54. (a) −90°; (b) +2.0 μC; (c) −1.6 μC
56. 2.4×10^{-16} C
57. 1.64×10^{-19} C (approx. 2% high)
58. (a) 0; (b) 0; (c) $0.707R$; (d) 3.46×10^7 N/C
59. (a) 14; (b) −4.6
60. 217°
61. $(1.08 \times 10^{-5}\ \mathrm{N/C}\hat{\mathrm{i}}$
62. $-4.19Q$
63. (a) 47 N/C; (b) 27 N/C
64. (a) 1.62×10^6 N/C; (b) −45°
65. (a) 6.0 mm; (b) 180°
66. (a) 3.60×10^{-6} N/C; (b) 2.55×10^{-6} N/C; (c) 3.60×10^{-4} N/C; (d) 7.09×10^{-7} N/C;

(e) As the proton nears the disk, the forces on it from electrons e_s more nearly cancel

67. 5.39 N/C
68. 5.0×10^{-28} C · m
69. (a) 0; (b) 9.96 pN
70. $Q/3\pi\epsilon_0 d^2$
71. (a) −1.0 cm; (b) 0; (c) 10 pC
72. 3.6×10^2 N/C
73. +1.00 μC
74. (a) 0.10 μC; (b) 1.3×10^{17}; (c) 5.0×10^{-6}
75. (a) 8.87×10^{-15} N; (b) 120
76. -3.28×10^{-21} J
77. 61 N/C
78. 38 N/C
79. (a) 2.46×10^{17} m/s^2; (b) 0.122 ns; (c) 1.83 mm
80. (a) $(2q/4\pi\epsilon_0 d^2)\alpha/(1+\alpha^2)^{3/2}$; (c) 0.71; (d) 0.20 and 2.0
81. (a) -1.49×10^{-26} J; (b) $(-1.98 \times 10^{-26}$ N · m) $\hat{k}$; (c) 3.47×10^{-26} J
82. 9:30
83. (a) (−1.80 N/C) $\hat{i}$; (b) (43.2 N/C) $\hat{i}$; (c) (−6.29 N/C) $\hat{i}$
85. (a) −0.029 C; (b) repulsive forces would explode the sphere
86. (a) yes; (b) upper plate, 2.72 cm
87. $(1/2\pi)\sqrt{pE/I}$
88. (a) top row: 4, 8, 12; middle row: 7, 11, 16; (b) 1.63×10^{-19} C
89. (a) 1 kN/C; (b) nonuniform; (c) because the field induces an electric dipole in a grain, the grain then moves toward a region of stronger electric field by moving toward the bee and then toward the stigma; if the grain were positively charged, it would not move to the bee; if it were negatively charged, it would move to bee but not then to the stigma; (d) no, because if it did, the grain would either fall off or be repelled off
91. (a) $2\pi\sqrt{L/|g - qE/m|}$; (b) $2\pi\sqrt{L/(g + qE/m)}$

Chapter 23

1. -0.015 N · m^2/C
2. (a) 0; (b) −3.92 N · m^2/C; (c) 0; (d) 0
3. (a) −72 N · m^2/C; (b) +24 N · m^2/C; (c) −16 N · m^2/C; (d) 0; (e) −48 N · m^2/C
4. -1.1×10^{-4} N · m^2/C
5. 2.0×10^5 N · m^2/C
6. −4.3 nC
7. 3.01 nN · m^2/C
8. (a) -1.3×10^{-8} C/m^3; (b) 8.2×10^{10} charges/m^3
9. 3.54 μC
10. $(-2.8 \times 10^4$ N/C) $\hat{i}$
11. (a) 0; (b) 0.0417
12. (a) +1.8 μC; (b) −5.3 μC; (c) +8.9 μC
13. (a) 8.23 N · m^2/C; (b) 72.9 pC; (c) 8.23 N · m^2/C; (d) 72.8 pC
14. 2.00 N/C · m
15. (a) 37 μC; (b) 4.1×10^6 N · m^2/C
16. 2.0 μC/m^2
17. (a) 4.5×10^{-7} C/m^2; (b) 5.1×10^4 N/C
18. (a) −8.0 μC; (b) +12 μC; (c) −5.3 μC
19. (a) -3.0×10^{-6} C; (b) $+1.3 \times 10^{-5}$ C
20. (a) 0; (b) 5.99×10^3 N/C
21. 5.0 μC/m
22. 2.1×10^{17} m/s^2
23. (a) 0.32 μC; (b) 0.14 μC
24. 8.0 cm
25. (a) 2.3×10^6 N/C; (b) outward; (c) 4.5×10^5 N/C; (d) inward
26. −5.8 nC/m
27. (a) 0.214 N/C; (b) inward; (c) 0.855 N/C; (d) outward; (e) -3.40×10^{-12} C; (f) -3.40×10^{-12} C
28. (a) 0.24 kN/C; (b) −6.4 nC/m^2; (c) +3.2 nC/m^2
29. 3.8×10^{-8} C/m^2
30. (a) 1.9 N/C; (b) 3.6 N/C
31. (a) 5.3×10^7 N/C; (b) 60 N/C
32. (a) $(2.00 \times 10^{-11}$ N/C) $\hat{j}$; (b) 0; (c) $(-2.00 \times 10^{-11}$ N/C) $\hat{j}$
33. −1.5
34. (0.208 N/C) $\hat{k}$
35. (a) 0; (b) 0; (c) $(-7.91 \times 10^{-11}$ N/C) $\hat{i}$
36. 2.9 μC/m^2
37. 0.44 mm
38. 4.9×10^{-10} C
39. 5.0 nC/m^2
40. (a) +69.1 cm; (b) −69.1 cm; (c) +69.1 cm
41. (a) 0; (b) 1.31 μN/C; (c) 3.08 μN/C; (d) 3.08 μN/C
42. 2.2 μC
43. −7.5 nC
44. (a) −750 N · m^2/C; (b) −6.64 nC

45. (a) 2.50×10^4 N/C; (b) 1.35×10^4 N/C
46. −3.3 cm
47. 1.79×10^{-11} C/m^2
48. +6.6 μC
49. (a) 0; (b) 56.2 mN/C; (c) 112 mN/C; (d) 49.9 mN/C; (e) 0; (f) 0; (g) −5.00 fC; (h) 0
50. (a) 0; (b) 0; (c) 0; (d) 7.32 N/C; (e) 12.1 N/C; (f) 1.35 N/c
51. (a) 7.78 fC; (b) 0; (c) 5.58 mN/C; (d) 22.3 mN/C
52. 1.125
53. $6K\epsilon_0 r^3$
54. (a) $E = |\rho|r/2\epsilon_0$; (b) increases; (c) inward; (d) 3×10^6 N/C, at inside pipe surface; (e) yes, along inside pipe surface
55. (a) 3.62 N · m^2/C; (b) 51.1 N · m^2/C
56. (a) 0.125; (b) 0.500
57. (a) 4.9×10^{-22} C/m^2; (b) down
58. +0.213 nC
59. −1.70 nC
60. (a) $0.25R$; (b) $2.0R$
61. (a) +2.0 nC; (b) −1.2 nC; (c) +1.2 nC; (d) +0.80 nC
62. (a) 4.2 kN/C; (b) 2.4 kN/C
63. (a) +4.0 μC; (b) −4.0 μC
64. (a) $0.41R$; (b) $0.50R$
65. (a) 5.4 N/C; (b) 6.8 N/C
66. 7.1 N · m^2/C
67. (a) 0; (b) 2.88×10^4 N/C; (c) 200 N/C
68. (a) 0.50 N · m^2/C; (b) 2.2 pC
69. (a) 15.0 N/C; (b) 25.3 N/C
70. $(5.65 \times 10^4 \text{ N/C})\hat{\text{j}}$
71. 26.6 nC
73. (a) -2.53×10^{-2} N · m^2/C; (b) $+2.53 \times 10^{-2}$ N · m^2/C
74. (a) 4.0×10^6 N/C; (b) 0
75. (a) 0.282 kN/C; (b) 0.621 kN/C
76. 2.00
77. (a) 0; (b) $q_a/4\pi\epsilon_0 r^2$; (c) $(q_a + q_b)/4\pi\epsilon_0 r^2$
78. -4.2×10^{-10} C
79. (a) 0.180 N/C; (b) outward; (c) 0; (d) 4.50 mN/C
80. 0.875
81. −1.04 nC
84. 3.6 nC
85. (b) $\rho R^2/2\epsilon_0 r$
86. (a) 693 kg/s; (b) 693 kg/s; (c) 347 kg/s; (d) 347 kg/s; (e) 575 kg/s
87. (a) $-e/\pi a_0^3$; (b) $5e[\exp(-2)]/4\pi\epsilon_0 a_0^2$, radially outward

Chapter 24

1. (a) 3.0×10^5 C; (b) 3.6×10^6 J
2. 1.2 GeV
3. 8.8 mm
4. (a) 2.46 V; (b) 2.46 V; (c) 0
5. (a) 2.4×10^4 V/m; (b) 2.9 kV
6. (a) 30 V; 40 V; (c) 5.5 m
7. (a) -1.87×10^{-21} J; (b) −11.7 mJ
8. +32.0 V
9. (a) −0.268 mV; (b) −0.681 mV
10. −1.1 nC
11. (a) 3.3 nC; (b) 12 nC/m^2
12. (a) −4.5 kV; (b) −4.5 kV
13. (a) 6.0 cm; (b) −12.0 cm
14. none
15. 0.562 mV
16. 2.21 V
17. (a) 0.54 mm; (b) 790 V
18. $-32e$
19. 16.3 μV
20. 5.6×10^{-37} C · m
21. (a) 24.3 mV; (b) 0
22. −6.20 V
23. (a) −2.30 V; (b) −1.78 V
24. 32.4 mV
25. 47.1 μV
26. 0
27. 13 kV
28. 7.39 mV
29. 19.6 mV
30. 6.7×10^2 V/m
31. $(-12 \text{ V/m})\hat{\text{i}} + (12 \text{ V/m})\hat{\text{j}}$
32. (a) 39 V/m; (b) toward
33. 150 N/C
34. (a) $(2.90 \text{ mV}) \ln[1 + (0.135 \text{ m})/d]$; (b) $(0.392 \text{ nM} \cdot \text{m}^2/\text{C})/[x(x + 0.135 \text{ m})]$; (c) 180°; (d) 32.1 mN/C; (e) 0
35. $(-4.0 \times 10^{-16} \text{ N})\hat{\text{i}} + (1.6 \times 10^{-16} \text{ N})\hat{\text{j}}$
36. (a) 31.6 mV; (b) 0.298 N/C
37. −0.192 pJ
38. 2.1×10^{-25} J
39. (a) 1.15×10^{-19} J; (b) decrease
40. 0
41. (a) $+6.0 \times 10^4$ V; (b) -7.8×10^5 V; (c) 2.5 J; (d) increase; (e) same; (f) same

42. 1.8×10^{-10} J
43. 2.5 km/s
44. 6.63×10^6 m/s
45. 0.32 km/s
46. (a) proton; (b) 65.3 km/s
47. 1.6×10^{-9} m
48. $-5.7\,\mu$C
49. (a) 3.0 J; (b) -8.5 m
50. 4.5×10^{-12} C·m
51. (a) 0; (b) 1.0×10^7 m/s
52. (a) $-12.0\,\mu$C; (b) +0.216 pJ
53. 2.5×10^{-8} C
54. 400 V
55. (a) -1.8×10^2 V; (b) 2.9 kV, -8.9 kV
56. (a) equal; (b) 0.333; (c) 0.667; (d) 2.00
57. (a) 12 kN/C; (b) 1.8 kV; (c) 5.8 cm
58. (a) 1.69 kV/m; (b) 36.7 kV/m; (c) 0; (d) 6.74 kV; (e) 27.0 kV; (f) 34.7 kV; (g) 45.0 kV; (h) 45.0 kV; (i) 45.0 kV
59. 3.71×10^4 V
60. (a) $V = \rho(R^2 - r^2)/4\epsilon_0$; (b) 78 kV
61. 7.0×10^5 m/s
62. (a) 1.7 cm; (b) 20 km/s; (c) 4.8×10^{-17} N; (d) positive; (e) 3.2×10^{-17} N; (f) negative
63. (a) 36 V; (b) 18 V
64. (a) 1.8 cm; (b) 8.4×10^5 m/s; (c) 2.1×10^{-17} N; (d) positive; (e) 1.6×10^{-17} N; (f) negative
65. 10.3 mV
66. -1.7
67. (a) 3.6 kV; (b) 3.6 kV
68. 22 km/s
69. (a) 0.90 J; (b) 4.5 J
70. (a) $+7.19 \times 10^{-10}$ V; (b) $+2.30 \times 10^{-28}$ J; (c) $+2.43 \times 10^{-29}$ J
71. 2.18×10^4 V
72. $(2.9 \times 10^{-2}\,\text{m}^{-2})A$
73. 2.1 days
74. 0.956 V
75. (a) 64 N/C; (b) 2.9 V; (c) 0
76. (a) 12; (b) 2
77. 2.30×10^{-28} J
78. 240 kV
79. 2.30×10^{-22} J
80. (a) -24 J; (b) 0
81. (a) none; (b) 0.41 m
82. (a) 3.6×10^5 V; (b) no
83. 1
84. (a) -6.0 V/m; (b) 0; (c) 3.0 V/m; (d) 3.0 V/m; (e) 15 V/m; (f) 0; (g) -3.0 V/m
85. (a) 1.48 nC; (b) 795 V
86. 0.514 mV
87. -187 V
88. 2.5 kV
89. (a) 2.5 MV; (b) 5.1 J; (c) 6.9 J
90. -1.93 J
91. -1.92 MV
92. 1.48×10^7 m/s
93. (a) 0.225 J; (b) A: 45.0 m/s^2, B: 22.5 m/s^2; (c) A: 7.75 m/s, B: 3.87 m/s
94. $(qQ/8\pi\epsilon_0)(1/r_1 - 1/r_2)$
95. (a) 2.72×10^{-14} J; (b) 3.02×10^{-31} kg, about 1/3 of accepted value
97. 6.4×10^8 V
98. 0.334
100. (a) $q(3R^2 - r^2)/8\pi\epsilon_0 R^3$; (b) $q/8\pi\epsilon_0 R$
101. (a) $Q/4\pi\epsilon_0 r$;
(b) $(\rho/3\epsilon_0)(1.5r_2^2 - 0.50r^2 - r_1^3 r^{-1})$,
$\rho = Q/[(4\pi/3)(r_2^3 - r_1^3)]$;
(c) $(\rho/2\epsilon_0)(r_2^2 - r_1^2)$, with ρ as in (b);
(d) yes
102. 8.8×10^{-14} m
103. (a) -4.8 nm; (b) 8.1 nm; (c) no
105. 2.8×10^5
106. 843 V
107. $p/2\pi\epsilon_0 r^3$
108. 2.90 kV
109. (a) spherical, centered on q, radius 4.5 m; (b) no
110. (a) 0.484 MeV; (b) 0
111. (a) 25 fm; (b) 2.0
112. (a) -0.12 V; (b) 1.8×10^{-8} N/C; (c) inward
113. $-1.2\,\mu$J
117. (a) 38 s; (b) 280 days
118. (c) 4.2 V

Chapter 25

1. 3.0 mC
2. (a) 3.5 pF; (b) 3.5 pF; (c) 57 V
3. (a) 144 pF; (b) 17.3 nC
4. 8.85×10^{-12} m
5. 0.280 pF
6. (a) 84.5 pF; (b) 191 cm^2
7. 9.09×10^3
8. $7.33\,\mu$F
9. $3.16\,\mu$F

10. 315 mC
11. (a) 790 μC; (b) 78.9 V
12. (a) 100 μC; (b) 20.0 μC
13. 43 pF
14. (a) 60 μC; (b) 60 μC
15. (a) 3.00 μF; (b) 60 μC; (c) 10 V; (d) 30.0 μC; (e) 10 V; (f) 20.0 μC; (g) 5.00 V; (h) 20.0 μC
16. 12 μC
17. (a) 50 V; (b) 5.0×10^{-5} C; (c) 1.5×10^{-4} C
18. 3.6 pC
19. (a) 4.0 μF; (b) 2.0 μF
20. (a) 4.5×10^{14}; (b) 1.5×10^{14}; (c) 3.0×10^{14}; (d) 4.5×10^{14}; (e) up; (f) up
21. (a) 32.0 μC; (b) 16.0 μC; (c) 16.0 μC
22. (a) 10 V; (b) 8.0 μF; (c) 2.0 μF
23. (a) 9.00 μC; (b) 16.0 μC; (c) 9.00 μC; (d) 16.0 μC; (e) 8.40 μC; (f) 16.8 μC; (g) 10.8 μC; (h) 14.4 μC
24. 99.6 nJ
25. 72 F
26. (a) 35 pF; (b) 21 nC; (c) 6.3 μJ; (d) 0.60 MV/m; (e) 1.6 J/m^3
27. 0.27 J
28. (a) 750 μC; (b) 50 V; (c) 18.8 mJ; (d) 500 μC; (e) 50.0 V; (f) 12.5 mJ; (g) 250 μC; (h) 450.0 V; (i) 6.25 mJ
29. (a) 9.16×10^{-18} J/m^3; (b) 9.16×10^{-6} J/m^3; (c) 9.16×10^{6} J/m^3; (d) 9.16×10^{18} J/m^3; (e) ∞
30. (a) 400 μC; (b) 100 V; (c) 20.0 mJ; (d) 333 μC; (e) 33.3 V; (f) 5.55 mJ; (g) 333 μC; (h) 66.7 V; (i) 11.1 mJ
31. (a) 16.0 V; (b) 45.1 pJ; (c) 120 pJ; (d) 75.2 pJ
32. 0.11 J/m^3
33. (a) 190 V; (b) 95 mJ
34. 4.0
35. Pyrex
36. (a) 6.2 cm; (b) .28 nF
37. 81 pF/m
38. 1.06 nC
39. 0.63 m^2
40. (a) 0.73 nF; (b) 28 kV
41. 66 μJ
42. 8.41 pF
43. 17.3 pF
44. 45.5 pF
45. (a) 10 kV/m; (b) 5.0 nC; (c) 4.1 nC
46. (a) 13.4 pF; (b) 1.15 nC; (c) 1.13×10^4 N/C; (d) 4.33×10^3 N/C
47. (a) 0.107 nF; (b) 7.79 nC; (c) 7.45 nC
48. (a) 7.2; (b) 0.77 μC
49. (a) 89 pF; (b) 0.12 nF; (c) 11 nC; (d) 11 nC; (e) 10 kV/m; (f) 2.1 kV/m; (g) 88 V; (h) -0.17 μJ
50. (a) 4.9 mJ; (b) no
51. 4
52. (a) 7.20 μC; (b) 18.0 μC; (c) battery supplies charges only to plates to which it is connected; charges on other plates are due to electron transfers between plates, in accord with the new distribution of voltages across the capacitors; so battery does not directly supply charge on capacitor 4
53. (a) 2.0 μF; (b) 6.0 μF
54. (a) 2.0 μF; (b) 0.80 μF
55. (a) 1.1 pm
56. (a) 10 μC; (b) 20 μC
57. (a) five capacitors in series; (b) one possible answer: three rows in parallel, each row containing five capacitors in series
58. (a) 2.0×10^7; (b) away
59. (a) 32 μC; (b) 2.0 V
60. (a) -0.50 μC; (b) 3.6 mJ; (c) no
61. (a) 24.0 μC; (b) 6.00 V
62. 20 μC
63. 45 μC
64. (a) 36 μC; (b) 12 μC
65. (a) 100 μC; (b) 20.0 μC
66. 16 μC
67. 2.28 pF
68. (a) 41 μF; (b) 42 μF
69. (a) 50.0 V; (b) 0
70. 5.3 V
71. (a) 2.40 μF; (b) 0.480 mC; (c) 80 V; (d) 0.480 mC; (e) 120 V
72. (a) 10.0 μF; (b) 1.20 mC; (c) 200 V; (d) 0.800 mC; (e) 200 V
73. 40 μF
75. (a) 200 kV/m; (b) 200 kV/m; (c) 1.77 μC/m^2; (d) 4.60 μC/m^2; (e) -2.83 μC/m^2
76. (a) 0.480 mC; (b) 240 V; (c) 0.480 mC; (d) 60.0 V; (e) 0.192 mC; (f) 96.0 V; (g) 0.768 mC; (h) 96.0 V; (i) 0; (j) 0; (k) 0; (l) 0
77. 4.9%

78. (a) 24 μC; (b) 4 V
79. 6.0 V
80. mica
81. 1.06 nC
82. (a) 0.708 pF; (b) 1.67; (c) −5.44 J; (d) sucked in
83. (a) 0.708 pF; (b) 0.600; (c) 1.02×10^{-9} J; (d) sucked in
84. 32.0 μPC

Chapter 26

1. (a) 1.1 kC; (b) 7.5×10^{21}
2. 6.7 μC/m^2
3. 5.6 ms
4. 14
5. (a) 6.4 A/m^2; (b) north; (c) cross-sectional area
6. (a) 2.4×10^{-5} A/m^2; (b) 1.8×10^{-15} m/s
7. 0.38 mm
8. (a) 0.654 μA/m^2; (b) 83.4 MA
9. 13 min
10. 18.1 μA
11. (a) 1.33 A; (b) 0.666 A; (c) J_a
12. 2.59 mA
13. 2.0×10^6 $(\Omega \cdot \text{m})^{-1}$
14. (a) 5.32×10^5 A/m^2; (b) 1.01 kg/m; (c) 3.27×10^5 A/m^2; (d) 0.495 kg/m
15. 2.0×10^{-8} $\Omega \cdot$ m
16. (a) 1.53 kA; (b) 54.1 MA/m^2; (c) 10.6×10^{-8} $\Omega \cdot$ m; (d) platinum
17. 100 V
18. $2R$
19. 2.4 Ω
20. (a) 1.55 mm; (b) 1.22 mm
21. 54 Ω
22. 3.0 mA
23. 3.0
24. (a) 6.00×10^7 $(\Omega \cdot \text{m})^{-1}$; (b) 7.50×10^6 $(\Omega \cdot \text{m})^{-1}$
25. 1.9×10^{3}°C
26. 3.35×10^{-7} C
27. 8.2×10^{-4} $\Omega \cdot$ m
28. (a) 3.24 pA/m^2; (b) 1.73 cm/s
29. (a) 38.3 mA; (b) 109 A/m^2; (c) 1.28 cm/s; (d) 227 V/m
30. (a) 0.40 Ω
31. (a) 6.00 mA; (b) 1.59×10^{-8} V; (c) 21.2 nΩ
32. 5.44 × 10−9 m/s
33. 981 kΩ
35. (a) 1.0 kW; (b) $0.25 US
36. 14 kC
37. 0.135 W
38. 11.1 Ω
39. (a) 10.9 A; (b) 10.6 Ω; (c) 4.50 MJ
40. (a) 28.8 Ω; (b) 2.60×10^{19} s^{-1}
41. (a) 5.85 m; (b) 10.4 m
42. 12 mW
43. (a) $4.46 US; (b) 144 Ω; (c) 0.833 A
44. 756 kJ
45. (a) 16.9 mV/m; (b) 243 J
46. 0.224 m
47. (a) 5.1 V; (b) 10 V; (c) 10 W; (d) 20 W
48. (a) $i = \rho \pi R^2 v$; (b) 17 μA; (c) no, because current is perpendicular to the radial potential difference; (d) 1.3 W; (e) 0.27 J; (f) exit of the pipe into the silo
49. (a) 64 Ω; (b) 0.25
50. (a) yes; (b) 4.0×10^2 A/m^2
51. 2.1×10^{-6} $\Omega \cdot$ m
52. 3.4×10^4 s
53. (a) upward in the strip; (b) 12 eV; (c) 12 eV
54. 0.536 Ω
55. (a) 2×10^{12}; (b) 5.0×10^3; (c) 10 MV
56. (a) 1.3×10^5 A/m^2; (b) 94 mV
57. 660 W
58. (a) −8.6%; (b) smaller
59. (a) 1.74 A; (b) 2.15 MA/m^2; (c) 36.3 mV/m; (d) 2.09 W
60. (a) silver; (b) 51.6 nΩ
61. 150 s
62. (a) 1.3 mΩ; (b) 4.6 mm
63. (a) 1.37; (b) 0.730
64. 0.20 hp
65. 28.8 kC
66. 95 kJ
67. 146 kJ
68. 13.3 Ω
69. 0.10 V
70. 3.0×10^6 J/kg
71. (a) 0.67 A; (b) toward
72. 2.4 kW
73. (a) 1.5×10^7 A/m^2; (b) toward
74. 57°C
75. (a) 0.81 mm; (b) 1.0 mm
76. (a) 3.1×10^{11}; (b) 25 μA; (c) 1.3 kW; (d) 25 MW
77. 560 W

78. (a) 0.43%; (b) 0.0017%; (c) 0.0034%; (d) R
79. (a) 250°C; (b) yes
81. (a) 0.38 mV; (b) negative; (c) 3 min 58 s
82. 27 cm/s
83. (a) 0.920 mA; (b) 1.08×10^4 A/m^2

Chapter 27

1. (a) \$$3.2 \times 10^2$ US; (b) \$0.048 US
2. 11 kJ
3. 14.4 h
4. (a) 80 J; (b) 67 J; (c) 13 J
5. (a) 0.50 A; (b) 1.0 W; (c) 2.0 W; (d) 6.0 W; (e) 3.0 W; (f) supplied; (g) absorbed
6. −10 V
7. (a) 14 V; (b) 1.0×10^2 W; (c) 6.0×10^2 W; (d) 10 V, 1.0×10^2 W
8. (a) $9.9 \times 10^2\,\Omega$; (b) 9.9×10^{-4} W
9. (a) 50 V; (b) 48 V; (c) negative
10. (a) 12.0 V; (b) 2.15 mV; (c) 24.0 W; (d) 4.30 mW
11. 8.0 Ω
12. (a) 0.20 Ω; (b) 0.30 Ω
13. (a) 0.004 Ω; (b) 1
14. (a) 1.0 kΩ; (b) 0.30 V; (c) 0.23%
15. 5.56 A
16. 4.0 Ω and 12 Ω
17. 4.50 Ω
18. (a) 2.50 Ω; (b) 3.13 Ω
19. (a) 50 mA; (b) 60 mA; (c) 9.0 V
20. 0.25 V
21. 3d
22. (a) 0; (b) 1.25 A
23. 48.3 V
24. (a) same; (b) −2.0 V
25. 1.43 Ω
26. (a) providing; (b) 3.6×10^2 W
27. (a) 0.67 A; (b) down; (c) 0.33 A; (d) up; (e) 0.33 A; (f) up; (g) 3.3 V
28. (a) 6.0 V; (b) 20 Ω; (c) 40 Ω
29. 9
30. (a) 119 Ω; (b) 50.5 mA; (c) 19.0 mA; (d) 19.0 mA; (e) 12.5 mA
31. (a) 0.1‘50 Ω; (b) 240 W
32 (a) 24.0 A; (b) 30.0 A; (c) series; (d) 60.0 A; (e) 48.0 A; (f) parallel
33. (a) 0.709 W; (b) 0.050 W; (c) 0.346 W; (d) 1.26 W; (e) −0.158 W
34. (a) 2.0 kΩ; (b) 4.0 kΩ
35. (a) 1.11 A; (b) 0.893 A; (c) 126 m
36. (a) 19.5 Ω; (b) 0; (c) ∞; (d) 82.3 W; (e) 57.6 W
37. 0.45 A
38. (a) 13.5 kΩ; (b) 1.50 kΩ; (c) 167 Ω; (d) 1.48 kΩ
39. −3.0%
40. (a) 12.5 V; (b) 50.0 A
41. (a) 55.2 mA; (b) 4.86 V; (c) 88.0 Ω; (d) decrease
42. (a) 70.9 mA; (b) 4.70 V; (c) 66.3 Ω; (d) decrease
44. (a) 0.41; (b) 1.1
45. 4.61
46. (a) 2.52 s; (b) 21.6 μC; (c) 3.40 s
47. (a) 2.41 μs; (b) 161 pF
48. 0.208 ms
49. (a) 2.17 s; (b) 39.6 mV
50. 0.72 MΩ
51. (a) 1.0×10^{-3} C; (b) 1.0×10^{-3} A; (c) $(1.0 \times 10^3\text{ V})\,e^{-t}$; (d) $(1.0 \times 10^3\text{ V})\,e^{-t}$; (e) $P = e^{-2t}$ W
52. 2.35 MΩ
53. (a) 1.1 mA; (b) 0.55 mA; (c) 0.55 mA; (d) 0.82 mA; (e) 0.82 mA; (f) 0; (g) 4.0×10^2 V; (h) 6.0×10^2 V
54. 411 μA
55. (a) 0.955 μC/s; (b) 1.08 μW; (c) 2.74 μW; (d) 3.82 μW
56. 162 μs
57. 24.8 Ω; (b) 14.9 kΩ
58. (a) 6.9 km; (b) 20 Ω
59. (a) 6.0 A; (b) 8.0 V; (c) 60 W; (d) 36 W
60. 8
61. (a) 1.32×10^7 A/m^2; (b) 8.90 V; (c) copper; (d) 1.32×10^7 A/m^2; (e) 51.1 V; (f) iron
62. 0.82 mA
63. (a) 80 mA; (b) 0.13 A; (c) 0.40 A
64. (a) 80 Ω; (b) 200 Ω
65. (a) $V_T = \mathcal{E} - ir$; (b) 13.6 V; (c) 0.060 Ω
66. (a) 60.0 mA; (b) down; (c) 180 mA; (d) left; (e) 240 mA; (f) up
69. 2.5 A
70. 2.00 A
71. (a) 12.0 eV; (b) 6.53 W
72. (a) low position connects larger resistance, middle position connects smaller resistance, high position connects filaments in parallel; (b) 72 Ω; (c) 144 Ω

73. the cable
74. (a) 3.00 A; (b) 3.75 A; (c) 3.94 A
75. (a) 7.50 A; (b) left; (c) 10.0 A; (d) left; (e) 87.5 W; (f) supplied
76. 20 Ω
77. (a) 3.0 A; (b) 10 A; (c) 13 A; (d) 1.5 A; (e) 7.5 A
78. (a) 3.00 A; (b) down; (c) 1.60 A; (d) down; (e) supply; (f) 55.2 W; (g) supply; (h) 6.40 W
79. (a) 5.00 A; (b) left; (c) supply; (d) 100 W; (e) supply; (f) 50.0 W; (g) supply; (h) 56.3 W
80. (a) 0.333 A; (b) right; (c) 720 J
81. (a) 85.0 Ω; (b) 915 Ω
82. (a) 4.00 Ω; (b) parallel
83. 7.50 V
84. (a) 38 Ω; (b) 260 Ω
85. (a) −11 V; (b) −9.0 V
86. (a) 1.0 V; (b) 50 mΩ
87. (a) 6.67 Ω; (b) 6.67 Ω; (c) 0
88. 0.143
89. $-13\,\mu C$
90. 13.3 Ω
91. (a) 1.5 mA; (b) 0; (c) 1.0 mA
92. (a) 4.0 A; (b) up; (c) 0.50 A; (d) down; (e) 64 W; (f) 16 W; (g) supplied; (h) absorbed
93. (a) 0; (b) 14.4 W
94. (a) 5.25 V; (b) 1.50 V; (c) 5.25 V; (d) 6.75 V
95. (a) 38.2 mA; (b) down; (c) 10.9 mA; (d) right; (e) 27.3 mA; (f) left; (g) 3.82 V
96. (a) 300 Ω; (b) 2.00 V; (c) 6.67 mA
97. 2.5 V
98. 4.0 V
99. (a) 1.00 A; (b) 24.0 W
100. 0.90%
102. (a) $V_c = \mathcal{E}_0 e^{-t/\tau}$; (b) 12 V; (c) 0.77 s; (d) $3.8\,\mu F$
103. 1.00×10^{-6}
104. (a) 4.0 A; (b) up
105. 3
106. $100\mathcal{E}^2 x^2 R_0^{-2}(100RR_0^{-1} + 10x - x^2)^{-2}$, x in cm
107. (a) 3.41 A; (b) 0.293 A; (c) 0.586 A; (d) 1.71 V
108. (b) yes
109. (a) put each contact roughly in the middle of its range; adjust the current roughly with B; make fine adjustments with A; (b) relatively large percentage changes in A cause only small percentage changes in the equivalent resistance of the parallel combination, thus permitting fine adjustment; any shift in the A contact causes half as much change as any equal shift in the B contact
110. (a) 3.0 kV; (b) 10 s; (c) 11 GΩ
111. $250\,\mu J$
112. (a) $i_2 = i_1+i_4+i_5$, $i_3+i_4+i_5 = 0$, $-16\text{ V}+(7\,\Omega)i_1+(5\Omega)i_2+4\text{ V}=0$, $10\text{ V}+(8\,\Omega)i_3-(9\,\Omega)i_4-4\text{ V}-(5\,\Omega)i_2=0$, $12\text{ V}-(4\,\Omega)i_5+(9\,\Omega)i_4=0$;

$$\text{(b) } [A] = \begin{bmatrix} 1 & -1 & 0 & 1 & 1 \\ 0 & 0 & 1 & 1 & 1 \\ 7 & 5 & 0 & 0 & 0 \\ 0 & -5 & 8 & -9 & 0 \\ 0 & 0 & 0 & 9 & -4 \end{bmatrix} \text{ ohms,}$$

$$[C] = \begin{bmatrix} 0 \\ 0 \\ 12 \\ -6 \\ -12 \end{bmatrix} \text{ volts;}$$

(c) $i_1 = 0.717$ A, $i_2 = 1.40$ A, $i_3 = -0.680$ A, $i_4 = -0.714$ A, $i_5 = 1.39$ A
113. (a) 6.43 V; (b) 3.60 W; (c) 16.7 W; (d) −5.60 W; (e) a
114. (c) it gives the rate with which R dissipates energy

Chapter 28

1. (a) 400 km/s; (b) 835 eV
2. (a) 6.2×10^{-18} N; (b) $9.5 \times 10^8\text{ m/s}^2$; (c) same
3. (a) $(6.2\times10^{-14}\text{ N})\hat{k}$; (b) $(-6.2\times10^{-14}\text{ N})\hat{k}$
4. (a) −3.5 km/s; (b) 7.0 km/s
5. −2.0 T
6. 3.75 km/s
7. −0.267 mT
8. $(-11.4\text{ V/m})\hat{i}-(6.00\text{ V/m})\hat{j}+(4.80\text{ V/m})\hat{k}$
9. 0.068 MV/m
10. (a) 1.25 V/m; (b) $(25.0\text{ mT})\hat{k}$
11. 38.2 cm/s
12. $7.4\,\mu V$
13. (a) $(-600\text{ mV/m})\hat{k}$; (b) 1.20 V
14. (a) 25 cm; (b) 30 cm; (c) 20 cm
15. $21\,\mu T$
16. (a) 1.11×10^7 m/s; (b) 0.316 mm

17. (a) 2.05×10^7 m/s; (b) 467 μT; (c) 13.1 MHz; (d) 76.3 ns
18. (a) 2.60×10^6 m/s; (b) 0.109 μs; (c) 0.140 MeV; (d) 70 kV
19. (a) 0.978 MHz; (b) 96.4 cm
20. (a) 1.0 MeV; (b) 0.5 MeV
21. 65.3 m/s
22. 2.09×10^{-22} J
23. (a) 0.358 ns; (b) 0.166 mm; (c) 1.51 mm
24. (a) 0.252 T; (b) 130 ns
25. (a) 495 mT; (b) 22.7 mA; (c) 8.17 MJ
26. 8.7 ns
27. (a) 5.07 ns
28. (a) 18.3 MHz; (b) 17.2 MeV
29. 2.4×10^2 m
30. (a) 0.787 T; (b) 8.34 MeV; (c) 23.9 MHz; (d) 33.2 MeV
31. (a) 200 eV; (b) 20.0 keV; (c) 0.499%
32. 20.1 N
33. (a) 28.2 N; (b) horizontally west
34. (a) $(-16\,\text{N})\hat{\jmath}$; (b) 0
35. (a) 467 mA; (b) right
36. $(-2.50 \times 10^{-3}\,\text{N})\hat{\jmath} + (0.750 \times 10^{-3}\,\text{N})\hat{k}$
37. (a) 0.10 T; (b) 31°
38. $(-0.35\,\text{N})\hat{k}$
39. $(-4.3 \times 10^{-3}\,\text{N}\cdot\text{m})\hat{\jmath}$
40. (a) 0; (b) 0.138 N; (c) 0.138 N; (d) 0
41. 0.60 μN
42. 6.58×10^{-26} N · m
43. (a) 542 Ω; (b) series; (c) 2.52 Ω; (d) parallel
44. 3.0 mA
45. 2.45 A
46. 2.08 GA
47. (a) 12.7 A; (b) 0.0805 N · m
48. (a) 0.184 A · m^2; (b) 1.45 N · m
49. (a) 0.30 J/T; (b) 0.024 N · m
50. (a) 77°; (b) 77°
51. (a) 2.86 A · m^2; (b) 1.10 A · m^2
52. $(0.150\,\text{A}\cdot\text{m}^2)\hat{\jmath} - (0.300\,\text{A}\cdot\text{m}^2)\hat{k}$
53. (a) $(-9.7 \times 10^{-4}\,\text{N}\cdot\text{m})\hat{\imath} - (7.2 \times 10^{-4}\,\text{N}\cdot\text{m})\hat{\jmath} + (8.0 \times 10^{-4}\,\text{N}\cdot\text{m})\hat{k}$; (b) -6.0×10^{-4} J
54. 4.8×10^{-5} A · m^2
55. (a) 90°; (b) 1; (c) 1.28×10^{-7} N · m
56. 110°
57. 0.53 m
58. 6.7×10^{-2} T
59. 127 u
60. 1.2×10^{-9} kg/C
61. $(-500\,\text{V/m})\hat{\jmath}$
62. $(18.8\,\mu\text{N})\hat{k}$
63. −40 mC
64. (a) $(1.44 \times 10^{-18}\,\text{N})\hat{k}$; (b) $(1.60 \times 10^{-19}\,\text{N})\hat{k}$; (c) $(6.41 \times 10^{-19}\,\text{N})\hat{\imath} + (8.01 \times 10^{-19}\,\text{N})\hat{k}$
65. $-(3.0\,\text{T})\hat{\imath} - (3.0\,\text{T})\hat{\jmath} - (4.0\,\text{T})\hat{k}$
66. (a) 84°; (b) no; (c) no; (d) 5.7 nm
67. (a) −72.0 μJ; (b) $(96.0\,\mu\text{N}\cdot\text{m})\hat{\imath} + (48.0\,\mu\text{N}\cdot\text{m})\hat{k}$
68. (a) 0.67 mm/s; (b) 2.8×10^{29} m^{-3}
69. $(0.80\,\text{mN})\hat{k}$
70. (a) $(12.8\,\text{N})\hat{\imath} + 6.41\,\text{N})\hat{\jmath}$; (b) 90°; (c) 173°
71. (a) 3.34 cm/s; (b) left
72. (a) 3.8 mm; (b) 19 mm; (c) clockwise
73. $(-61\,\text{mT})\hat{k}$
74. (a) 4.99×10^6 m/s; (b) 7.10 mm; (c) 8.93 ns
75. $(-0.600\,\text{N})\hat{k}$
76. (a) 9.56×10^{-14} N; (b) 0; (c) 0.267°
77. (a) 20 min; (b) 5.9×10^{-2} N · m
78. 8.2 mm
79. (a) 0.50; (b) 0.50; (c) 14 cm; (d) 14 cm
80. $(-0.34\,\text{mT})\hat{k}$
81. (a) 2.84×10^{-3}
82. (b) out of the plane of the page
83. $\vec{v} = v_{0x}\,\hat{\imath} + v_{0y}\cos(\omega t)\,\hat{\jmath} - v_{0y}\sin(\omega t)\,\hat{k}$, where $\omega = eB/m$
84. (a) 1.4; (b) 1.0
85. (a) 6.3×10^{14} m/s^2; (c) 3.0 mm
89. $(0.75\,\text{T})\hat{k}$

Chapter 29

1. (a) 3.3 μT; (b) yes
2. 0
3. (a) 16 A; (b) east
4. (a) 1.67 μT; (b) into
5. (a) 0.102 μT; (b) out
6. (a) 0.118 μT; (b) into
7. (a) opposite; (b) 30 A
8. (a) 4.0 cm; (b) unchanged
9. 4.3 A (b) out
10. (a) 0; (b) 3.82 cm
11. $(-7.75 \times 10^{-23}\,\text{N})\hat{\imath}$
12. (a) 30 cm; (b) 2.0 nT; (c) out; (d) into
13. 2 rad
14. (a) −7.0 cm; (b) 7.0 cm
15. $(80\,\mu\text{T})\hat{\jmath}$
16. 14.1
17. 50.3 nT

18. 144°
19. 132 nT
20 2.00 cm
21. (a) $(253\,\text{nT})\hat{k}$; (b) $(192\,\text{nT})\hat{i} + (61.2\,\text{nT})\hat{k}$
22. 1.0 rad
23. $(22.3\,\text{pT})\hat{j}$
24. 2.3 cm
25. (a) 20 μT; (b) into
26. 104°
27. 88.4 pN/m
28. 800 nN/m
29. (a) $(469\,\mu\text{N})\hat{j}$; (b) $(188\,\mu\text{N})\hat{j}$; (c) 0; (d) $(-188\,\mu\text{N})\hat{j}$; (e) $(-469\,\mu\text{N})\hat{j}$
30. $(0.794\,\text{mN/m})\hat{i} - (0.794\,\text{mN/m})\hat{j}$
31. $(-125\,\mu\text{N/m})\hat{i} + (41.7\,\mu\text{N/m})\hat{j}$
32. (a) 0.50 A; (b) out
33. $(3.20\,\text{mN})\hat{j}$
34. $28.3\,\text{nT}\cdot\text{m}$
35. (a) $-2.5\,\mu\text{T}\cdot\text{m}$; (b) 0
36. (a) $-2.5\,\mu\text{T}\cdot\text{m}$; (b) $-16\,\mu\text{T}\cdot\text{m}$
37. (a) 0; (b) 0.850 mT; (c) 1.70 mT; (d) 0.850 mT
38. (a) 3.00 mA; (b) into
39. (a) 0; (b) 0.10 μT; (c) 0.40 μT
40. 5.71 mT
41. 0.30 mT
42. 108 m
43. (a) 533 μT; (b) 400 μT
44. 0.272 A
45. (a) 4.77 cm; (b) 35.5 μT
46. 1.6×10^6 rev
47. $0.47\,\text{A}\cdot\text{m}^2$
48. (a) 4.0; (b) 0.50
49. (a) $2.4\,\text{A}\cdot\text{m}^2$; (b) 46 cm
50. 8.78 μT
51. (a) 0.497 μT; (b) into; (c) $1.06\,\text{mA}\cdot\text{m}^2$; (d) into
52. (a) $(0.060\,\text{A}\cdot\text{m}^2)\hat{j}$; (b) $(96\,\text{pT})\hat{j}$,
53. (a) 79 μT; (b) $1.1 \times 10^{-6}\,\text{N}\cdot\text{m}$
54. (a) 0.90 A; (b) 2.7 A
55. (a) 1.0 mT; (b) out; (c) 0.80 mT; (d) out
56. 1.8 rad
58. 61.3 mA
59. 256 nT
60. 157 nT
61. (a) 15.3 μT
62. (a) −90°; (b) 4.0 A; (c) out; (d) 2.0 A; (e) into
63. 5.3 mm
64. 3.0 μT
65. 32.1 A
66. (a) 15 A; (b) $-z$
67. (a) 4.8 mT; (b) 0.93 mT; (c) 0
68. (a) $(-52.0\,\mu\text{T})\hat{k}$; (b) 8.13 cm; (c) 17,5 cm
69. (a) 1.7 μT; (b) into; (c) 6.7 μT; (d) into
70. (a) $(-400\,\mu\text{T})\hat{i}$; (b) $(400\,\mu\text{T})\hat{j}$
71. 7.7 mT
72. $4.5 \times 10^{-6}\,\text{T}\cdot\text{m}$
73. (a) 5.0 mA; (b) downward
74. (a) 0.17 mN; (b) 0.021 mN
75. (a) $(0.24\,\text{nT})\hat{i}$; (b) 0; (c) $(-43\,\text{pT})\hat{k}$; (c) $(-43\,\text{pT})\hat{k}$; (d) $(0.14\,\text{nT})\hat{k}$
76. (a) 27.5 nT; (b) into
77. 5.0 μT
78. 1.28 mm
79. 4.0 mm
85. $(-0.20\,\text{mT})\hat{k}$
86. (b) 2.3 km/s
87. (a) 3.2×10^{-16} N; (b) 3.2×10^{-16} N; (c) 0
88. $(1.25\,\mu\text{T})\hat{i}$
92. (a) $\mu_0 ir/2\pi c^2$; (b) $\mu_0 i/2\pi r$; (c) $\mu_0 i(a^2 - r^2)/2\pi(a^2 - b^2)r$; (d) 0

Chapter 30

1. 0.198 mV
2. (a) 31 mV; (b) left
3. (a) −11 mV; (b) 0; (c) 11 mV
4. 1.4 T/s
5. 30 mA
6. 0.452 V
7. 0
8. 0
9. (a) 21.7 V; (b) counterclockwise
10. 1.2 mΩ
11. (b) $0.786\,\text{m}^2$
12. (a) 8.0 μA; (b) counterclockwise
13. 29.5 mC
14. 15.5 μC
15. (a) 40 Hz; (b) 3.2 mV
16. (a) 0; (b) none; (c) 6.00 mV; (d) clockwise; (e) 1.00 mV; (f) clockwise; (g) 0; (h) none; (i) 0; (j) none
17. 5.50 kV
18. (a) 24 μV; (b) from c to b
19. (a) $\mu_0 iR^2\pi r^2/2x^3$; (b) $3\mu_0 i\pi R^2 r^2 v/2x^4$; (c) counterclockwise
20. 18 mV

21. (a) 1.26×10^{-4} T; (b) 0; (c) 1.26×10^{-4} T; (d) yes; (e) 5.04×10^{-8} V
22. (a) 0.598 μV; (b) counterclockwise
23. (a) 80 μV; (b) clockwise
24. (a) 14 nWb; (b) 10 μA
25. (a) 13 μWb/m; (b) 17%; (c) 0
26. 750 pJ
27. 3.66 μW
28. 1.0 mΩ
29. (a) 48.1 mV; (b) 2.67 mA; (c) 0.129 mW
30. $v_t = mgR/B^2L^2$
31. (a) 0.60 V; (b) up; (c) 1.5 A; (d) clockwise; (e) 0.90 W; (f) 0.18 N; (g) 0.90 W
32. (a) 85.2 Wb; (b) 56.8 V; (c) 1
33. (a) 240 μV; (b) 0.600 mA; (c) 0.144 μW; (d) 2.87×10^{-8} N; (e) 0.144 μW
34. (a) −1.07 mV; (b) −2.40 mV; (c) 1.33 mV
35. (a) 71.5 μV/m; (b) 143 μV/m
36. 0.030 T/s
37. 0.15 V/m
38. (a) 2.45 mWb; (b) 0.645 mH
39. 0.10 μWb
40. (a) 0.27 μT; (b) 8.0 nH
41. 1.81 μH/m
42. (a) decreasing; (b) 0.68 mH
43. 5.0 A/s
44. (a) 16 kV; (b) 3.1 kV; (c) 23 kV
45. (b) $L_{eq} = \sum_{j=1}^{N} L_j$
46. (b) $\frac{1}{L_{eq}} = \sum_{j=1}^{N} \frac{1}{L_j}$
47. 59.3 mH
48. 12.3 s
49. 6.91
50. (a) 1.00; (b) 0.135; (c) 0.693
51. 46 Ω
52. (a) 3.33 A; (b) 3.33 A; (c) 4.55 A; (d) 2.73 A; (e) 0; (f) −1.82 A (reversed); (g) 0; (h) 0
53. (a) 8.45 ns; (b) 7.37 mA
54. 7.1×10^2 A/s
55. $(42 + 20t)$ V
56. (a) 0.29 mH; (b) 0.29 ms
57. (a) $i(1 - e^{-Rt/L})$; (b) $(L/R)\ln 2$
58. 1.23
59. 25.6 ms
60. (a) 2.4×10^2 W; (b) 1.5×10^2 W; (c) 3.9×10^2 W
61. (a) 97.9 H; (b) 0.196 mJ
62. (a) 18.7 J; (b) 5.10 J; (c) 13.6 J
63. (a) 34.2 J/m^3; (b) 49.4 mJ
64. 5.58 A
65. 1.5×10^8 V/m
66. (a) 1.3 mT; (b) 0.63 J/m^3
67. (a) 1.0 J/m^3; (b) 4.8×10^{-15} J/m^3
68. 13 H
69. (a) 1.67 mH; (b) 6.00 mWb
70. (a) 1.5 μWb; (b) 1.0×10^2 mV; (c) 90 nWb; (d) 12 mV
71. (b) wrap the turns of the two solenoids in opposite directions
72. there is a magnetic field only within the solenoid cross section
73. 13 μH
74. (a) 3.28 ms; (b) 6.45 ms; (c) infinite time; (d) 0; (e) 3.00 ms
75. $(\pi B_0 r^2/\tau)\exp(-t/\tau)$
76. 8.0×10^{-3} T/s
77. (a) 10 μT; (b) out; (c) 3.3 μT; (d) out
78. 1.15 W
79. (a) 1.5 s
80. 2.9 mV
81. 1.54 s
82. (a) 23 mA; (b) 70 mA
83. (a) 400 A/s; (b) 200 A/s; (c) 0.600 A
84. (a) 0.600 mH; (b) 120
85. (a) 0.40 V; (b) 20 A
86. (a) 25 μT/s; (b) 13 μT/s; (c) increasing
87. (a) $(4.4 \times 10^7\ \text{m/s}^2)\hat{\imath}$; (b) 0; (c) $(-4.4 \times 10^7\ \text{m/s}^2)\hat{\imath}$
88. 81.1 μs
89. (a) 2.0 A; (b) 0; (c) 2.0 A; (d) 0; (e) 10 V; (f) 2.0 A/s; (g) 2.0 A; (h) 1.0 A; (i) 3.0 A; (j) 10 V; (k) 0; (l) 0
90. 0.520 ms
91. 95.4 Ω
93. 221 mA
94. 12 A/s
95. (a) 1.0×10^{-3} J/m^3; (b) 8.4×10^{15} J
96. (a) 3.75 mH; (b) 3.75 mH; (c) 100 nWb; (d) 4.24 mV
97. (a) 0.50 mA; (b) counterclockwise; (c) 0.50 mA; (d) counterclockwise; (e) 0; (f) none
98. $L_1\mathcal{E}(L_1 + L_2)^{-1}R^{-1}$
99. (a) 0; (b) 8.0×10^2 A/s; (c) 1.8 mA; (d) 4.4×10^2 A/s; (e) 4.0 mA; (f) 0

100. (a) 10 A; (b) 1.0×10^2 J
101. (a) 51 mV; (b) clockwise
102. 1.0 ns
103. 11 mA
104. (a) 4.7 mH; (b) 2.4 ms
105. (a) 13.9 H; (b) 120 mA
106. (a) 0.10 H/m; (b) 1.3 V/m
107. 45 H
108. (a) 20 A/s; (b) 0.75 A

Chapter 31

1. (a) 1.17 μJ; (b) 5.58 mA
2. 45.2 mA
3. 9.14 nF
4. (a) 6.00 μs; (b) 167 kHz; (c) 3.00 μs
5. (a) 5.00 μs; (b) 2.50 μs; (c) 1.25 μs
6. (a) 89 rad/s; (b) 70 ms; (c) 25 μF
7. (a) 1.25 kg; (b) 372 N/m; (c) 1.75×10^{-4} m; (d) 3.02 mm/s
9. 7.0×10^{-4} s
10. 38 μH
11. (a) 3.0 nC; (b) 1.7 mA; (c) 4.5 nJ
12. (a) 6.0×10^2 Hz; (b) 7.1×10^2 Hz; (c) 1.1 kHz; (d) 1.30 kHz
13. (a) 275 Hz; (b) 364 mA
14. ω
15. (a) 6.0; (b) 36 pF; (c) 0.22 mH
17. (a) 1.98 μJ; (b) 5.56 μC; (c) 12.6 mA; (d) −46.9°; (e) +46.9°
18. (a) 0.500; (b) 0.866
19. (a) 0.180 mC; (b) 70.7 μs; (c) 66.7 W
20. (a) 3.60 mH; (b) 1.33 kHz; (c) 0.188 ms
21. (a) 356 μs; (b) 2.50 mH; (c) 3.20 mJ
22. (a) 46.1 μs; (b) 6.88 n; (c) 6.88 nJ; (d) 1.02×10^3 A/s; (e) 0.938 mW
24. (a) 5.85 μC; (b) 5.52 μC; (c) 1.93 μC
25. 8.66 mΩ
26. $(L/R) \ln 2$
28. (a) 0.283 A; (b) 2.26 A
29. (a) 95.5 mA; (b) 11.9 mA
30. (a) 0.600 A; (b) 0.600 A
31. (a) 0.65 kHz; (b) 24 Ω
32. (a) 5.22 mA; (b) 0; (c) 4.51 mA
33. (a) 6.73 ms; (b) 11.2 ms; (c) inductor; (d) 138 mH
34. (a) 39.1 mA; (b) 0; (c) 33.8 mA
35. (a) 218 Ω; (b) 23.4°; (c) 165 mA
36. (a) 8.0 μF; (b) 2.0 Ω
37. (a) 267 Ω; (b) −41.5°; (c) 135 mA
38. (a) 500 Ω; (b) 40 μF
39. (a) 206 Ω; (b) 13.7°; (c) 175 mA
40. (a) 40 Ω; (b) 60 mH
41. 89 Ω
42. −8.00 V
43. (a) yes; (b) 1.0 kV
44. (a) 100 Ω; (b) 30.6 μF; (c) 301 mH
45. (a) 224 rad/s; (b) 6.00 A; (c) 219 rad/s, (d) 228 rad/s; (e) 0.040
46. (a) 16.6 Ω; (b) 422 Ω; (c) 0.521 A; (d) increase; (e) decrease; (f) increase
48. (b) 318 Hz; (c) +45°; (d) 2.00×10^3 rad/s; (e) 53.0 mA
49. (a) 796 Hz; (b) no change; (c) decreased; (d) increased
50. 100 V
51. 1.84 A
52. 141 V
53. (a) 12.1 Ω; (b) 1.19 kW
55. (a) 0.743; (b) lead; (c) capacitive; (d) no; (e) yes; (f) no; (g) yes; (h) 33.4 W
56. (a) 76.4 mH; (b) yes; (c) 17.8 Ω
57. (a) 117 μF; (b) 0; (c) 90.0 W; (d) 0; (e) 1; (f) 0; (g) −90°; (h) 0
58. (a) 41.4 W; (b) −17.0 W; (c) 44.1 W; (d) 14.4 W; (e) equal
59. (a) 2.59 A; (b) 38.8 V; (c) 159 V; (d) 224 V; (e) 64.2 V; (f) 75.0 V; (g) 100 W; (h) 0; (i) 0
60. 1.0 kV
61. (a) 2.4 V; (b) 3.2 mA; (c) 0.16 A
62. (a) 1.25; (b) 4.00; (c) 5.00; (d) 0.200; (e) 0.250; (f) 0.800
63. (a) 1.9 V; (b) 5.9 W; (c) 19 V; (d) 5.9×10^2 W; (e) 0.19 kV; (f) 59 kW
64. 10
67. (a) 177 Ω; (b) no
68. 1.84 kHz
69. 7.61 A
70. (b) 159 Hz; (c) −45°; (d) 1.00×10^3 rad/s; (e) 170 mA
72. (a) 8.84 kHz; (b) 6.00 Ω
73. (a) 39.1 Ω; (b) 21.7 Ω; (c) capacitive
74. (a) 5.77×10^3 rad/s; (b) 1.09 ms
75. (a) 45.0°; (b) 70.7 Ω
76. 1.59 μF
77. (a) 0.689 μH; (b) 17.9 pJ; (c) 0.110 μC
78. (a) 707 Ω; 32.2 mH; (c) 21.9 nF

79. (a) 6.73 ms; (b) 2.24 ms; (c) capacitor; (d) 59.0 μF
81. (a) 2.41 μH; (b) 21.4 pJ; (c) 82.2 nC
82. (a) 0.588 rad; (b) inductive; (c) 12.0 V
84. (a) 64.0 Ω; (b) 50.9 Ω; (c) capacitive
85. (a) 4.60 kHz; (b) 26.6 nF; (c) 2.60 kΩ; (d) 0.650 kΩ
86. (a) −0.405 rad; (b) 2.76 A; (c) capacitive
87. (a) 165 Ω; (b) 313 mH; (c) 14.9 μF
88. 0.115 A
89. (a) 0.577Q; (b) 0.152
90. (a) 37.0 V; (b) 60.9 V; (c) 113 V; (d) 68.6 W
91. (a) 1.27 μC; (b) 83.1 μs; (c) 5.44 mW
92. (a) 168 Ω; (b) decrease; (c) decrease; (d) decrease
93. (a) +1.22 rad; (b) 0.288 A
94. 69.3 Ω
95. 7.08 mH
96. (a) 79.6 kHz; (b) 4.00 mA; (c) 16.0 nJ; (d) 2.00 kA/s
97. (a) 4.00 μF, 5.00 μF, 5.00 μF, 5.00 μF; (b) 1.78 kHz, 1.59 kHz, 1.59 kHz, 1.59 kHz; (c) 12.0 Ω, 12.0 Ω, 6.00 Ω, 4.00 Ω; (d) 19.8 Ω, 22.4 Ω, 19.9 Ω, 19.4 Ω; (e) 0.605 A, 0.535 A, 0.603 A, 0.619 A
98. (a) 36.0 V; (b) 29.9 V; (c) 11.9 V; (d) −5.85 V
100. (b) 61 Hz; (c) 90 Ω; (d) 61 Hz

Chapter 32

1. (a) 1.1 mWb; (b) inward
2. +3 Wb
3. (a) 47.4 μWb; (b) inward
4. $(\mu_0 iL//\pi)\ln 3$
5. 2.4×10^{13} V/m · s
6. (a) 30 mm; (b) 53 mm; (c) 3.0×10^{-5} T
7. (a) 1.9 pT
8. (a) 1.18×10^{-19} T; (b) 1.06×10^{-19} T
9. (a) 3.54×10^{-17} T; (b) 2.13×10^{-17} T
10. (a) 5.01×10^{-22} T; (b) 4.51×10^{-22} T
11. (a) 3.09×10^{-20} T; (b) 1.67×10^{-20} T
12. 7.5×10^{5} V/s
14. 7.2×10^{12} V/m · s
16. (a) 2.1×10^{-8} A; (b) downward; (c) clockwise
17. (a) 0.63 μT; (b) 2.3×10^{12} V/m · s
18. (a) 1.33 A; (b) 0.300 cm and 4.80 cm
19. (a) 0.71 A; (b) 0; (c) 2.8 A
20. 8.40×10^{-13} T
21. (a) 2.0 A; (b) 2.3×10^{11} V/m · s; (c) 0.50 A; (d) 0.63 μT · m
22. (a) 0.089 mT; (b) 0.18 mT; (c) 0.22 mT; (d) 6.4×10^{-22} T; (e) 6.4×10^{-22} T; (f) 0; (g) out; (h) out
23. (a) 75.4 nT; (b) 67.9 nT
24. (a) 2.22 μT; (b) 2.00 μT
25. (a) 27.9 nT; (b) 15.1 nT
26. (a) 20.0 μT; (b) 12.0 μT
27. 55 μT
28. (a) 13 MWb; (b) outward
29. (a) -9.3×10^{-24} J/T; (b) 1.9×10^{-23} J/T
30. 4.6×10^{-24} J
31. (a) 0; (b) 0; (c) 0; (d) $\pm 3.2 \times 10^{-25}$ J; (e) -3.2×10^{-34} J · s, 2.8×10^{-23} J/T, $+9.7 \times 10^{-25}$ J, $\pm 3.2 \times 10^{-25}$ J
32. 32.3 mT
33. (a) 0; (b) −1, 0, 1; (c) 4.64×10^{-24} J
34. (a) $+x$; (c) clockwise; (d) $+x$
35. $e^2 r^2 B/4m$
36. 0.48 K
37. 20.8 mJ/T
38. (a) 1.5×10^{2} T; (b) 6.0×10^{2} T; (c) no
39. yes
40. 0.30
41. (b) K_i/B; (b) $-z$; (c) 0.31 kA/m
42. 25 km
43. (a) 3.0 μT; (b) 5.6×10^{-10} eV
44. (a) 8.9 A · m^2; (b) 13 N · m
45. 5.15×10^{-24} A · m^2
46. (a) 1.49×10^{-4} N · m; (b) −72.9 μJ
47. (a) 0.14 A; (b) 79 μC
48. 3.19×10^{-9} kg · m^2
49. (a) 1.8×10^{2} km; (b) 2.3×10^{-5}
50. 52 nT · m
51. (a) 16.7 nT; (b) 5.00 mA
52. (b) in the direction of the angular momentum vector
53. (a) $(1.2 \times 10^{-13}\ \text{T})e^{-t/(0.012\ \text{s})}$; (b) 5.9×10^{-15} T
54. (a) 222 μT; (b) 167 μT; (c) 22.7 μT; (d) 1.25 μT; (e) 3.75 μT; (f) 22.7 μT
55. (a) 4 K; (b) 1 K
56. (a) 9; (b) 3.71×10^{-22} J/T; (c) $+9.27 \times 10^{-24}$ J; (d) -9.27×10^{-24} J
57. (a) 0.324 V/m; (b) 2.87×10^{-16} A; (c) 2.87×10^{-18}
58. (a) -2.78×10^{-23} J/T; (b) 3.71×10^{-23} J/T

59. 8.0 A
60. 0.84 kJ/T
61. (a) 7.60 μA; (b) 859 kV · m/s; (c) 3.39 mm; (d) 5.16 pT
62. 0.300 A
63. (a) 7; (b) 7; (c) $3h/2\pi$; (d) $3eh/4\pi m$; (e) $3.5h/2\pi$; (f) 8
64. 3.5×10^{-5} A
65. (b) $-x$; (c) counterclockwise; (d) $-x$
66. (a) 5.3×10^{11} V/m; (b) 20 mT; (c) 6.6×10^2
67. (a) 6.3×10^8 A; (b) yes; (c) no
68. (a) -8.8×10^{15} V/m · s; (b) 5.9×10^{-7} T · m
70. (a) 9.2 mWb; (b) inward
72. (a) 31.0 μT; (b) 0°; (c) 55.9 μT; (d) 73.9°; (e) 62.0 μT; (f) 90.0°
73. (a) 1.66×10^3 km; (b) 383 μT; (c) 61.1 μT; (d) 84.2°
74. (a) 27.5 mm; (b) 110 mm
75. (b) sign is minus;
(c) no, because there is compensating positive flux through the open end nearest to magnet

---

Chapter 33

1. 30 cm
2. (a) 515 nm, 610 nm; (b) 555 nm, 5.41×10^{14} Hz, 1.85×10^{-15} s
3. (a) 4.7×10^{-3} Hz; (b) 3 min 32 s
4. 7.49 GHz
5. 5.0×10^{-21} H
6. 4.7 m
7. 1.07 pT
8. 6.7 nT; (b) y; (c) negative y direction
9. 0.10 MJ
10. 4.8×10^{-29} W/m^2
11. 1.2 MW/m^2
12. (a) 16.7 nT; (b) 33.1 mW/m^2
13. (a) 1.03 kV/m; (b) 3.43 μT
14. (a) 1.4×10^{-22} W; (b) 1.1×10^{15} W
15. (a) 6.7 nT; (b) 5.3 mW/m^2; (c) 6.7 W
16. 3.44×10^6 T/s
17. (a) 87 mV/m; (b) 0.29 nT; (c) 6.3 kW
18. (a) 30.1 nm; (b)345 nm
19. 1.0×10^7 Pa
20. 3.3×10^{-8} Pa
21. 5.9×10^{-8} Pa
22. (a) 6.0×10^8 N; (b) 3.6×10^{22} N
23. (a) 1.0×10^8 Hz; (b) 6.3×10^8 rad/s; (c) 2.1 m^{-1}; (d) 1.0 μT; (e) z; (f) 1.2×10^2 W/m^2; (g) 8.0×10^{-7} N; (h) 4.0×10^{-7} Pa
24. (a) 3.97 GW/m^2; (b) 13.2 Pa; (c) 1.67×10^{-11} N; (d) 3.14×10^3 m/s^2
26. 491 nm
27. 1.9 mm/s
28. 0.96 km^2
29. 0.25 kW
30. (a) 4.68×10^{11} W; (b) any chance disturbance could move the sphere from directly above source, so the two force vectors are no longer along the same axis
31. 4.4 W/m^2
32. 19 W/m^2
33. 3.1%
34. 4.5×10^{-2} %
35. (a) 1.9 V/m; (b) 1.7×10^{-11} Pa
36. 44%
37. 20° or 70°
38. 9.4%
39. 0.67
40. (a) 19.6°; (b) 70.4°
41. (a) 0.16; (b) 0.84
42. 7.38
43. (a) 2 sheets; (b) 5 sheets
44. 180°
45. 1.48
46. (a) greater; (b) greater; (c) 1.4; (d) 1.9
47. 1.26
48. (a) greater; (b) greater; (c) 1.9; (d) 1.4
49. 1.07 m
50. (a) 1.7; (b) 38°
51. (a) 56.9°; (b) 35.3°
52. (a) 3.1°; (b) 0 (no rainbow)
54. 34°
55. 182 cm
56. (a) 26.8°; (b) yes
57. (a) 1.39 (b) 28.1°; (c) no
58. (a) 35.1°; (b) 49.9°; (c) 35.1°; (d) 26.1°; (e) 60.7°; (f) 35.3°
59. 23.2°
60. (a) 3; (b) 2; (c) 40°; (d) none; (e) 2; (f) 3; (g) none; (h) 70°
61. (a) 49°; (b) 29°
62. (a) 35.6°; (b) 53.1°
63. (a) $\sqrt{1 + \sin^2\theta}$; (b) $\sqrt{2}$; (c) yes; (d) no
64. (a) 53°; (b) yes
65. 49.0°

66. 1.0
67. $0.50\,W/m^2$
68. (a) 0; (b) 20°; (c) 0; (d) 20°
69. (a) 15 m/s; (b) 8.7 m/s; (c) higher; (d) 72°
70. (a) 1.6; (b) need more information; (c) 39°
71. (a) 4.56 m; (b) increase
72. (a) 0.33°; (b) 0
73. (a) $-y$; (b) z; (c) $1.91\,kW/m^2$; (d) $E_z = (1.20\,kV/m)\sin[(6.67\times10^6\,m^{-1})y+(2.00\times10^{15}\,s^{-1})t]$; (e) 942 nm; (f) infrared
74. (a) 54.3°; (b)yes; (c) 51.1°; (d) no
75. 1.22
76. $1.5\times10^{-8}\,m/s^2$
77. (c) 139.3°; (d) 137.6°; (e) 1.7°
78. (b) 230.4°; (c) 233.5°; (d) 3.1°; (e) 317.5°; (f) 321.9°; (g) 4.4°
79. (a) 1.60; (b) 58.0°
81. 0.031
83. 22°
84. 1.3
86. (a) 1.91×10^8 Hz; (b) 18.2 V/m; (c) $0.878\,W/m^2$
87. (a) 55.8°; (b) 55.5°
88. 0.024
89. (a) 0.50 ms; (b) 8.4 min; (c) 2.4 h; (d) 5446 B.C.
90. (a) $83\,W/m^2$; (b) 1.7 MW
91. 1.7×10^{-13} N
92. (a) 4.7×10^{-6} Pa; (b) 4.7×10^{-11}
93. (a) $(16.7\,nT)\sin[(1.00\times10^6\,m^{-1})z+(3.00\times10^{14}\,s^{-1})t]$; (b) 6.28 μm; (c) 20.9 fm; (d) $33.2\,mW/m^2$; (e) x; (f) infrared
94. $p_{r1}\cos^2\theta$
95. (a) $3.5\,\mu W/m^2$; (b) 0.78 μW; (c) $1.5\times10^{-17}\,W/m^2$; (d) 1.1×10^{-7} V/m; (e) 0.25 fT
96. (a) $(236\,nT)\sin[(2.51\times10^7\,m^{-1})z+(7.53\times10^{15}\,s^{-1})t]$; (b) 3.83×10^{-20} N
97. 0.21
98. $b_Z = (2.50\times10^{-14}\,t)\sin[(1.40\times10^7\,m^{-1})y+(4.19\times10^{15}\,s^{-1})t]$
99. 0.034
100. 35°
101. (a) 0.33 μT; (b) $-x$
102. 9.2 μN
103. 9.16 μT
104. 9.43×10^{-10} T
106. (b) 5.8×10^{-7} m
107. (a) z axis; (b) 7.5×10^{14} Hz; (c) $1.9\,kW/m^2$
108. 0.125
109. (a) white; (b) white dominated by red end; (c) no refracted light
110. (a) cover the center of each face with an opaque disk of radius 4.5 mm; (b) about 0.63
111. (a) steadily increase; (b) summed discrepancies between the apparent time of eclipse and times observed from x; the radius of Earth's orbit

Chapter 34

1. 40 cm
2. 9.10 m
3. 1.11
4. 1.5 m
5. 351 cm
6. −2.5
7. 10.5 cm
8. +28 cm
9. (a) +24 cm; (b) +36 cm; (c) −2.0; (d) R; (e) I; (f) same
10. (a) +20 cm; (b) +30 cm; (c) −2.0; (d) R; (e) I; (f) same
11. (a) +36 cm; (b) −36 cm; (c) +3.0; (d) V; (e) NI; (f) opposite
12. (a) +72 cm; (b) −72 cm; (c) +3.0; (d) V; (e) NI; (f) opposite
13. (a) −20 cm; (b) −4.4 cm; (c) +0.56; (d) V; (e) NI; (f) opposite
14. (a) −70 cm; (b) −14 cm; (c) +0.61; (d) V; (e) NI; (f) opposite
15. (a) −16 cm; (b) −4.4 cm; (c) +0.44; (d) V; (e) NI; (f) opposite
16. (a) −28 cm; (b) −7.7 cm; (c) +0.45; (d) V; (e) NI; (f) opposite
17. (b) 0.56 cm/s; (c) 11 m/s; (d) 6.7 cm/s
18. +0.32
19. (b) plus; (b) +40 cm; (c) −20 cm; (e) −20 cm; (f) +2.0; (g) V; (h) NI; (i) opposite
20. (a) plane; (b) ∞; (c) ∞; (e) −10 cm; (g) V; (h) NI; (i) opposite
21. (a) concave; (c) +40 cm; (e) +60 cm; (f) −2.0; (g) R; (h) I; (i) same
22. (a) concave; (b) +20 cm; (c) +40 cm; (e) +30 cm; (g) R; (h) I; (i) same
23. (a) convex; (b) −20 cm; (d) +20 cm; (f) +0.50; (g) V; (h) NI; (i) opposite

24. (a) convex; (b) minus; (c) −40 cm; (d) +1.8 m; (e) −18 cm; (g) V; (h) NI; (i) opposite
25. (b) −20 cm; (c) minus; (d) +5.0 cm; (e) minus; (f) +0.80; (g) V; (h) NI; (i) opposite
26. (a) concave; (b) +8.0 cm; (c) +16 cm; (d) +12 cm; (f) minus; (g) R; (i) same
27. (a) convex; (c) −60 cm; (d) +30 cm; (f) +0.50; (g) V; (h) NI; (i) opposite
28. (a) concave; (b) plus; (c) +40 cm; (e) +30 cm; (f) −0.50; (g) R; (h) I
29. (a) concave; (b) +8.6 cm; (c) +17 cm; (e) +12 cm; (f) minus; (g) R; (i) same
30. (a) concave; (b) +16 cm; (c) +32 cm; (e) +28 cm; (g) R; (h) I; (i) same
31. (a) convex; (b) minus; (c) −60 cm; (d) +1.2 m; (e) −24 cm; (g) V; (h) NI; (i) opposite
32. (a) 2.00; (b) none
33. 7.4 cm
34. (d) −18 cm; (e0 V; (f) same
35. (c) −33 cm; (e) V; (f) same
36. (b) +71 cm; (e) R; (f) opposite
37. (c) +30; (e) V; (f) same
38. (b) +10 cm; (e) V; (f) same
39. (d) −26 cm; (e) V; (f) same
40. (a) 1.0; (e) R; (f) opposite
41. 1.86 mm
42. +43 cm
43. (a) 45 mm; (b) 90 mm
44. −16 cm
45. (a) +40 cm; (b) ∞
46. −2.5
47. 5.0 mm
48. +0.30
49. 22 cm
50. (a) +5.3 cm; (b) −0.33; (c) R; (d) I; (e) opposite
51. (a) −48 cm; (b) +4.0; (c) V; (d) NI; (e) same
52. (a) −3.8 cm; (b) +0.38; (c) V; (d) NI; (e) same
53. (a) −4.8 cm; (b) +3.5; (c) V; (d) NI; (e) same
54. (a) −88 cm; (b) +3.5; (c) V; (d) NI; (e) same
55. (a) −8.6 cm; (b) +0.39 (c) V; (d) NI; (e) same
56. (a) −8.7 cm; (b) +0.72; (c) V; (d) NI; (e) same
57. (a) +36 cm; (b) −0.80; (c) R; (d) I; (e) opposite
58. (a) converging; (b) 26.7 cm; (c) 8.89 cm
59. (a) +84 cm; (b) −1.4; (c) R; (d) I; (e) opposite
60. (a) −26 cm; (b) +4.3; (c) V; (d) NI; (e) same
61. (a) −18 cm; (b) +0.76; (c) V; (d) NI; (e) same
62. (a) −9.7 cm; (b) +0.54; (c) V; (d) NI; (e) same
63. (a) −30 cm; (b) +0.86; (c) V; (d) NI; (e) same
64. (a) −63 cm; (b) +2.2; (c) V; (d) NI; (e) same
65. (a) +55 cm; (b) −0.74; (c) R; (d) I; (e) opposite
66. (a) −15 cm; (b) +1.5; (c) V; (d) NI; (e) same
67. (a) +7.5 cm; (b) +0.75; (c) V; (d) NI; (e) same
68. (a) −9.2 cm; (b) +0.92; (c) V; (d) NI; (e) same
69. (a) C; (b) plus; (d) −13 cm; (e) +1.7; (f) V; (g) NI; (h) same
70. (a) D; (b) minus; (d) −5.7 cm; (e) +0.71; (f) V; (h) same
71. (a) D; (b) −5.3 cm; (d) −4.0 cm; (f) V; (g) NI; (h) same
72. (a) C; (b) +3.2 cm; (d) +4.0 cm; (f) R; (g) I; (h) opposite
73. (a) C; (b) +80 cm; (d) −20 cm; (f) V; (g) NI; (h) same
74. (b) plus; (d) +20 cm; (e) −1.0 (f) R; (g) I; (h) opposite
75. (a) C; (d) −10 cm; (f) V; (g) NI; (h) same
76. (a) C; (b) plus; (d) −10 cm; (e) +2.0 cm; (f) V; (g) NI; (h) same
77. (a) D; (b) minus; (d) −3.3 cm; (e) 0.67; (f) V; (g) NI
78. (a) D; (b) −10 cm; (d) +5.0 cm; (f) V; (h) same
79. (a) C; (b) +3.3 cm; (d) +5.0 cm; (f) R; (g) I; (h) opposite
80. (a) +10 cm; (b) −0.75; (c) R; (d) I; (e) opposite
81. (a) +24 cm; (b) +6.0; (c) R; (d) NI; (e) opposite

82. (a) −23 cm; (b) −13; (c) V; (d) I; (e) same
83. (a) +3.1 cm; (b) −0.31; (c) R; (d) I; (e) opposite
84. (a) +9.8 cm; (b) −0.27; (c) R; (d) I; (e) opposite
85. (a) −4.6 cm; (b) +0.69; (c) V; (d) NI; (e) same
86. (a) −3.4 cm; (b) −1.1; (c) V; (d) I; (e) same
87. (a) −5.5 cm; (b) +0.12; (c) VR; (d) NI; (e) same
88. 2.1 mm
89. (a) 13.0 cm; (b) 5.23 cm; (c) −3.25; (d) 3.13; (e) −10.2
90. (a) 3.5; (b) 2.5
91. (a) 2.35 cm; (b) decrease
92. −125
93. (a) 5.3 cm; (b) 3.0 mm
94. (a) 20 cm; (b) 60 cm; (c) 80 cm; (d) 1.0 m
95. (a) 5.8 mm; (b) 1.6 kW/m^2; (c) 4.0 cm
96. −21 cm
97. (a) 3.00 cm; (b) 2.33 cm
98. (a) $2f_1$; (b) −1.0; (c) real; (d) left; (e) inverted
99. 2.2 mm^2
100. (a) 40 cm; (b) real; (c) 80 cm; (d) real; (e) 2.4 m; (f) real; (g) −40 cm; (h) virtual; (i) −80 cm; (j) virtual; (k) −2.4 cm; (l) virtual
101. (a) 20 cm; (b) 15 cm
102. (a) 3; (b) 7; (c) 5; (d) 1; (e) 3
107. (a) 0.15 m; (b) 0.30 mm; (c) no
108. (a) 80 cm; (b) 0 to 12 cm
109. (b) P_n
111. (a) 40 cm; (b) 20 cm; (c) −40 cm; (d) 40 cm
113. (a) +7.5 cm; (b) −0.75; (c) R; (d) I; (e) opposite
114. (a) +10 cm; (b) +0.75; (c) R; (d) NI; (e) opposite
115. (a) +8.6 cm; (b) 2.6; (c) R; (d) NI; (e) opposite
116. (a) −4.0 cm; (b) −1.2; (c) V; (d) I; (e) same
117. (a) +24 cm; (b) −0.58; (c) R; (d) I; (e) opposite
118. (a) −5.2 cm; (b) +0.29; (c) V; (d) NI; (e) same
119. (b) 8.4 mm; (c) 2.5 cm
121. (a) $(0.5)(2-n)x/(n-1)$; (b) right
122. 1.14
123. (a) +36 cm; (b) 1.3 cm; (c) real; (d) inverted
124. (a) −30 cm; (b) not inverted; (c) virtual; (d) 1.0
125. (a) −50 cm; (b) 5.0; (c) virtual; (d) inverted
126. (a) −12 cm
128. +10.0 cm
129. 2.67 cm
130. (a) 8.0 cm; (b) 16 cm; (c) 48 cm
131. (a) convex; (b) 1.60 m
132. 28.0 cm
133. (a) 3.33 cm; (b) left; (c) virtual; (d) not inverted
134. (a) 3.00 cm; (b) left; (c) virtual; (d) not inverted
137. (a) 0.60 m; (b) 0.20; (c) real; (d) left (e) not inverted
138. (a) 1.50 cm; (b) negative; (c) virtual
139. 42 mm
140. (a) $\alpha = 0.500$ rad: 7.799 cm; $\alpha = 0.100$ rad: 8.544 cm; $\alpha = 0.0100$ rad: 8.571 cm; mirror equation: 8.571 cm; (b) $\alpha = 0.500$ rad: −13.56 cm; $\alpha = 0.100$ rad: 12.05 cm; $\alpha = 0.0100$ rad: −12.00 cm; mirror equation: −12.00 cm

Chapter 35

1. 4.55×10^7 m/s
2. (a) 5.09×10^{14} Hz; (b) 388 nm; (c) 1.97×10^8 m/s
3. 1.56
4. 2.0×10^8 m/s
5. (a) 155 nm; (b) 310 nm
6. (a) 0.25; (b) 0.75; (c) 1.25
7. (a) 3.60 μm; (b) intermediate, closer to fully constructive interference
8. (a) 2; (b) 0.03
9. (a) 1.55 μm; (b) 4.65 μm
10. (a) 50°; (b) 0.14 ps
11. (a) 1.70; (b) 1.70; (c) 1.30; (d) all tie
12. (a) 52.50 nm; (b) 157.5 nm
13. (a) 0.833; (b) intermediate, closer to fully constructive interference
14. (a) 0.010 rad; (b) 5.0 mm
15. 648 nm
16. (a) 0.216 rad; (b) 12.4°
17. 16
18. (a) 2.90; (b) 18.2 rad; (c) between $m = 2$ minimum (third minimum from the center) and $m = 3$ maximum (third maximum to one side of center maximum)

19. 2.25 mm
20. 0.15°
21. 72 μm
22. (a) 600 nm to 700 nm; (b) decreased; (c) 0.20 μm
23. 0
24. 7.5
25. 7.88 μm
26. (a) 0; (b) 0; (c) ∞; (d) 6.00; (e) 1.71; (f) intermediate, closer to minimum
27. 6.64 μm
28. 3.5 μm
29. 2.65
30. $17 \sin(\omega t + 13°)$
31. $27 \sin(\omega t + 8.5°)$
32. (a) between central maximum and first minimum ($m = 0$); (b) 0.101
33. $(17.1\,\mu\text{V/m}) \sin[(2.0 \times 10^{14}\,\text{rad/s})t]$
34. (a) 2.33 μV/m; (b) 0.338; (c) between $m =$ 6 maximum (sixth side maximum) and $m =$ 6 minimum (seventh minimum); (d) 1.26×10^{15} rad/s; (e) 39.6 rad
35. (a) 0.117 μm; (b) 0.352 μm
36. (a) 4; (b) 3
37. 70.0 nm
38. (a) 567 nm; (b) 425 nm; (c) longer
39. 120 nm
40. 840 nm
41. 560 nm
42. 608 nm
43. 409 nm
44. 455 nm
45. 509 nm
46. 528 nm
47. 478 nm
48. 339 nm
49. 273 nm
50. 248 nm
51. 161 nm
52. 329 nm
53. 338 nm
54. 673 nm
55. (a) 552 nm; (b) 442 nm
56. 450 nm
57. 608 nm
58. 560 nm
59. 455 nm
60. 409 nm
61. 528 nm
62. 509 nm
63. 339 nm
64. 478 nm
65. 248 nm
66. 273 nm
67. 161 nm
68. 329 nm
69. 140
70. 11
71. 1.89 μm
72. 1.00025
73. 0.012°
74. (a) 10.3 m/s; (b) 1.09 μm
75. $\sqrt{(m + 1/2)\lambda R}$, for $m = 0, 1, 2, \ldots$
76. (a) 34; (b) 46
77. 1.00 m
78. $1.67 \times 10^{-11}\,\text{m}^3/\text{s}$
79. 588 nm
80. 5.2 μm
81. 1.00030
82. 0.345 mm
83. (a) 50.0 nm; (b) 36.2 nm
84. (a) 2.90; (b) intermediate, closer to fully constructive
85. (a) 22°; (b) refraction reduces θ
86. (a) 1.6; (b) 1.4
87. 0.032%
88. (a) ∞; (b) 0; (c) 0; (d) 6.00; (e) 5.80; (f) intermediate, closer to maximum
89. (b) 51.6 ns
90. (a) 1.8; (b) 2.2; (c) 1.25
91. $x = (D/2a)(m + 0.5)\lambda$, for $m = 0, 1, 2, \ldots$
92. (a) 1500 nm; (b) 2250 nm; (c) 0.80
93. 1.95×10^8 m/s
94. 450 nm
95. 0.23°
96. 310.0 nm
97. (a) 1.6 rad; (b) 0.79 rad
98. (a) 411.4°; (b) 51.4°
99. 6.4 m
100. (a) dark; (b) dark; (c) 4
101. (a) 110 nm; (b) 220 nm
102. 8.0 μm
103. (a) 169 nm; (b) they are reflected; (c) blue-violet will be sharply reduced
104. (a) 1; (b) 4.0 fs; (c) 7.5
105. (a) 0.87; (b) intermediate, closer to maximum brightness; (c) 0.37; (d) intermediate, closer to complete darkness

106. (a) 48.6°; (b) away; (c) 1.49 m
107. (a) 1.75 μm; (b) 4.8 mm
108. 33 μm
109. (a) 42.0 ps; (b) 42.3 ps; (c) 43.2 ps; (d) 41.8 ps; (e) 4
110. 492 nm
111. 600 nm
113. 0.20
114. 2.4 μm
115. $I_m \cos^2(2\pi x/\lambda)$
116. 2.1×10^8 m/s
117. (a) 0.253 mm; (b) 2.5λ minimum
118. (a) 39.6; (b) intermediate, closer to complete darkness
119. (a) 88%; (b) 94%
120. (a) 1.80 μm; (b) 9
121. $I = I_0[1 + 8\cos^2(\phi/2)]$, with $\phi = (2\pi d/\lambda)\sin\theta$
122. (a) 0; (b) fully constructive; (c) increase; (d) fully constructive; (e) ∞; (f) fully destructive; (g) 7.88 μm; (h) fully constructive; (i) 3.75 μm; (j) fully destructive; (k) 2.29 μm; (l) fully constructive; (m) 1.50 μm; (n) fully destructive; (o) 0.975 μm

Chapter 36

1. 60.4 μm
2. (a) 0.430°; (b) 0.118 mm
3. (a) 700 nm; (b) 4; (c) 6
4. (a) decrease; (b) 11°; (c) 0.23°
5. (a) 70 cm; (b) 1.0 mm
6. (a) 2.5 mm; (b) 2.2×10^{-4} rad
7. 1.77 mm
8. 24.0 mm
9. 160°
10. (a) 2.33 μm; (b) 6; (c) 15.2°; (d) 51.8°
11. (a) 0.18°; (b) 0.46 rad; (c) 0.93
12. (a) 0.256; (b) between the center and the first minimum
13. (d) 52.5°, 10.1°, 5.06°
15. (b) 0; (c) −0.500; (d) 4.493 rad; (e) 0.930 (f) 7.725 rad; (g) 1.96
16. 31 μm
17. (a) 1.3×10^{-4} rad; (b) 10 km
18. (a) 50 m; (b) no; (c) light pollution on the night side of Earth would be a sure sign
19. 50 m
20. 30 m
21. (a) 1.1×10^4 km; (b) 11 km
22. 53 m
23. (a) 19 cm; (b) larger
24. (a) 17.1 m; (b) 1.37×10^{-10}
25. (a) 0.346°; (b) 0.97°
26. 27 cm
27. (a) 8.8×10^{-7} rad; (b) 8.4×10^7 km; (c) 0.025 mm
28. (a) red; (b) 0.13 nm
29. 5
30. 3
31. (a) 4; (b) every fourth bright fringe is missing
32. $\lambda D/d$
33. (a) 9; (b) 0.255
34. (a) 11.1 μm; (b) 51; (c) 0; (d) 79.0°
35. (a) 5.0 μm; (b) 20 μm
36. (a) 7.43×10^{-3}; (b) between the $m = 6$ minimum (the seventh one) and the $m = 7$ maximum (the seventh side maximum); (c) between the $m = 3$ minimum (the third one) and the $m = 4$ minimum (the fourth one)
37. (a) 62.1°; (b) 45.0°; (c) 32.0°
38. 635 nm
39. 3
40. 2 μm
41. (a) 6.0 μm; (b) 1.5 μm; (c) 9; (d) 7; (e) 6
42. (a) 3; (b) 0.051°
43. 1.09×10^3 rulings/mm
44. 523 nm
45. 470 nm to 560 nm
48. (a) 23 100; (b) 28.7°
49. (a) 0.032°/nm; (b) 4.0×10^4; (c) 0.076°/nm; (d) 8.0×10^4; (e) 0.24°nm; (f) 1.2×10^5
50. 491
51. 3.65×10^3
52. (a) 56 pm; (b) none
53. (a) 10 μm; (b) 3.3 mm
54. (a) $\tan\theta$; (b) 0.89
55. 0.26 nm
56. 6.8°
57. (a) 25 pm; (b) 38 pm
58. 39.8 pm
59. (a) 0.17 nm; (b) 0.13 nm
60. 0.570 nm
61. (a) $0.7071a_0$; (b) $0.4472a_0$; (c) $0.3162a_0$; (d) $0.2774a_0$; (e) $0.2425a_0$

62. 130 pm; (b) 3; (c) 97.2 pm; (d) 4
63. (a) 15.3°; (b) 30.6°; (c) 3.1°; (d) 37.8°
64. 13
65. 41.2 m
66. (a) 1.3×10^{-4} rad; (b) 21 m
67. 4.7 cm
68. 4×10^{-13}
70. 164
71. (a) 80 cm; (b) 1.8 mm
72. (a) 2.4 μm; (b) 0.80 μm; (c) 2
73. (a) 625 nm; (b) 500 nm; (c) 416 nm
74. 1.6×10^3 km
75. 691 nm
76. (a) 2.1°; (b) 21°; (c) 11
77. 106°
78. 500 nm
79. 3.0 mm
80. 9.0
81. (a) fourth; (b) seventh
82. 1.41
83. (a) 6.8°; (b) no
84. 2.9°
85. 2.27 m
86. 11
87. 0.15 nm
88. (a) 32 cm; (b) 2.7 m; (c) The required aperture is too large; the fine-scale resolution is due to "computer enhancement" in which a computer removes much of the blurring due to turbulence.
89. 53.4 cm
90. 2
91. 9
92. 59.5 pm
93. (a) 13; (b) 6
94. 6.1 mm
96. 11
97. 4.9 km
98. 3.3
99. 1.36×10^4
100. 4.84×10^3
101. 36 cm
102. 30.5 μm
106. 2
114. $\theta = 0.143$ rad, $I/I_m = 4.72 \times 10^{-2}$; $\theta = 0.247$ rad, $I/I_m = 1.65 \times 10^{-2}$; $\theta = 0.353$ rad, $I/I_m = 8.35 \times 10^{-3}$

Chapter 37

1. 0.990 50
2. (a) 0.140 370 76; (b) 0.994 987 44; (c) 0.999 950 00; (d) 0.999 999 50
3. 2.68×10^3 y
4. 0.9959
5. 0.446 ps
6. 40 s
7. (a) 0.999 999 50
8. 1.53 cm
9. (a) 87.4 m; (b) 394 ns
10. 0.63 m
11. 1.32 m
12. (a) 0.866; (b) 2.00
13. (a) 26 y; (b) 52 y; (c) 3.7 y
14. 0.25 m
15. (a) 0.999 999 15; (b) 30 ly
16. (a) 0; (b) 2.29 s; (c) 6.54×10^8 m; (d) 3.16 s
17. (a) 138 km; (b) -374 μs
18. (a) 0; (b) -2.5 μs; (c) reverse
19. (a) 25.8 μs; (b) small flash
20. 2.40 μs
21. (a) 1.25; (b) 0.800 μs
22. (a) 09.500 m; (b) 1.00 m; (c) 1.00 m; (d) 19.2 m; (e) 35.5 ns; (f) event 2
23. (a) 0.480c; (b) negative; (c) big flash; (d) 4.39 μs
24. 0.63 μs
25. (a) $\gamma[1.00\,\mu\text{s} - \beta(400\,\text{m})/(2.998 \times 10^8\ \text{m/s})]$; (d) 0.750; (e) $0 < \beta < 0.750$; (f) $0.750 < \beta < 1$; (g) no
26. (a) $\gamma[400\,\text{m} - \beta c(1.00\,\mu\text{s})]$; (d) 0.750; (e) 265 m
27. 0.81c
28. 0.588
29. (a) 0.35; (b) 0.62
30. (a) 0.84$c\hat{\text{i}}$; (b) 1.1$c\hat{\text{i}}$; (c) 0.21$c\hat{\text{i}}$; (d) 0.15$c\hat{\text{i}}$
31. 1.2 μs
32. (a) $-0.36c$; (b) $-c$
33. (a) 1.25 yr; (b) 1.60 yr; (c) 4.00 yr
34. (a) 7000 km/s; (b) away
35. 22.9 MHz
36. (a) 1×10^6 m/s; (b) receding
37. 0.13c
38. 2.97 nm
39. (a) 550 nm; (b) yellow
40. (a) 79.1 keV; (b) 3.11 MeV; (c) 10.9 MeV
41. (a) 196.695; (b) 0.999 987
42. 8.12 MeV

43. (a) 1.0 keV; (b) 1.1 MeV
44. 7.28 MeV
45. (a) 0.222 cm; (b) 701 ps; (c) 7.40 ps
46. (a) 1.2×10^8 N; (b) truck or train; (c) 25 N; (d) backpack
47. $2.83mc$
48. (a) 1.001 957; (b) 0.0624 695 2; (c) 2.956 951; (d) 0.941 079 23; (e) $1.957\,951\,4 \times 10^3$; (f) 0.999 999 87
49. 18 smu/y
50. (a) 20.57; (b) 0.9988; (c) 1.011; (d) 0.1448; (e) 1.003; (f) 0.0731,
51. (a) 0.707; (b) 1.41; (c) 0.414
52. (a) 0.439; (b) 0.866
53. 1.01×10^7 km
54. (c) 207
55. 110 km
56. (a) $mv^2/2 + 3mv^4/8c^2$; (b) 1.0×10^{-16} J; (c) 1.9×10^{-19} J; (d) 2.6×10^{-14} J; (e) 1.3×10^{-14} J; (f) 0.37
57. (a) $\gamma(2\pi m/|q|B)$; (b) no; (c) 4.85 mm; (d) 15.9 mm; (e) 16.3 ps; (f) 0.334 ns
58. (a) 0.948; (b) 226 MeV; (c) 314 MeV/c
59. (a) 2.08 MeV; (b) −1.21 MeV
60. (a) $\gamma[1.00\,\mu\text{s} - \beta(240\text{ m})/(2.998 \times 10^8\text{ m/s})]$; (d) 0.801; (e) 0.599 μs; (f) yes
61. (d) 0.801
62. 0.79 m
63. (a) $vt \sin\theta$; (b) $t[1 - (v/c)\cos\theta]$; (c) $3.24c$
64. (a) $-0.86c$; (b) $-c$
65. (a) 1.93 m; (b) 6.00 m; (c) 13.6 ns; (d) 13.6 ns; (e) 0.379 m; (f) 30.5 m; (g) −101 ns; (h) no; (i) 2; (k) no; (l) both
66. (a) 2.59 μs; (b) 0.572 μs; (c) 2.59 μs; (d) 16.0 μs
68. (a) 1/9; (b) +0.80; (c) $+0.80c$
69. (b) $+0.44c$
70. 0.999 90
71. 6.4 cm
72. 7
73. 55 m
74. (a) 5.71 GeV; (b) 6.65 GeV; (c) 6.58 GeV/c; (d) 3.11 MeV; (e) 3.62 MeV; (f) 3.59 MeV/c
75. 8.7×10^{-3} ly
76. (a) $1/\sqrt{\tau_0(1 - v^2/c^2)}$
77. $0.678c$
78. (a) 2.21×10^{-12}; (b) 5.25 d
79. $0.95c$
80. $0.27c$
81. 2.46 MeV/c
82. (a) 2.24×10^{-13} s; (b) 64.4 μm
83. 189 MeV
84. (a) 1.87×10^4 km/s
85. (a) 2.7×10^{14} J; (b) 1.8×10^7 km; (c) 6.0×10^6
86. (a) 256 kV; (b) $0.745c$
87. (a) 5.4×10^4 km/h; (b) 6.3×10^{-10}
88. 0.75

Chapter 38

1. 2.1 μm, infrared
2. 2.11 eV
3. 1.0×10^{45} photons/s
4. 1.7×10^{21} photons/m$^2 \cdot$ s
5. 2.047 eV
6. 8.6×10^5 m/s
7. 4.7×10^{26} photons
8. 3.6×10^{-17} W
9. (a) infrared lamp; (b) 1.4×10^{21} photons/s
10. (a) 3.61 kW; (b) 1.00×10^{22} photons/s; (c) 60.2 s
11. (a) 2.96×10^{20} photons/s; (b) 4.86×10^7 m; (c) 5.89×10^{18} photons/m$^2 \cdot$ s
12. 3.3×10^{18} photons/s
13. 170 nm
14. barium and lithium
15. 676 km/s
16. 10 eV
17. (a) 1.3 V; (b) 6.8×10^2 km/s
18. 233 nm
19. (a) 2.00 eV; (b) 0; (c) 2.00 V; (d) 295 nm
20. 1.07 eV
21. (a) 382 nm; (b) 1.82 eV item22.(a) 4.12×10^{-15} eV $\cdot$ s; (b) 2.27 eV; (c) 545 nm
23. (a) 3.1 keV; (b) 14 keV
24. 9.68×10^{-20} A
25. (a) 2.73 pm; (b) 6.05 pm
26. (a) 0.511 MeV/c; (b) 2.43 pm; (c) 1.24×10^{20} Hz
27. (a) 8.57×10^{18} Hz; (b) 3.55×10^4 eV; (c) 35.4 keV/c
28. (a) +4.86 pm; (b) −40.6 keV; (c) 40.6 keV; (d) 0°
29. (a) 2.43 pm; (b) 1.32 fm; (c) 0.511 MeV; (d) 939 MeV
30. (a) 2.43 pm; (b) 4.86 pm; (c) 0.255 MeV
31. (a) -8.1×10^{-9} %; (b) -4.9×10^{-4} %;

(c) $-8.8\,\%$; (d) $-66\,\%$
32. 2.64 fm
33. 300 %
34. 1.12 keV
35. (a) 2.43 pm; (b) 4.11×10^{-6}; (c) -8.67×10^{-6} eV; (d) 2.43 pm; (e) 9.78×10^{-2}; (f) -4.45 keV
37. (a) 41.8 keV; (b) 8.2 keV
38. 44°
39. 7.75 pm
40. (a) 0.0388 nm; (b) 1.24 nm; (c) 9.06×10^{-13} nm
41. 4.3 μeV
42. (a) 3.96×10^{6} m/s; (b) 81.7 kV
43. (a) 1.24 μm; (b) 1.22 nm; (c) 1.24 fm; (d) 1.24 fm
44. (a) 3.3×10^{-24} kg · m/s; (b) 3.3×10^{-24} kg · m/s; (c) 38 eV; (d) 6.2 keV
45. (a) 1.9×10^{-21} kg · m/s; (b) 346 fm
46. (a) 1.24 keV; (b) 1.50 eV; (c) 1.24 GeV; (d) 1.24 GeV
47. 0.025 fm; (b) 2.0×10^{2}
48. (a) 5.2 fm; (b) no, the de Broglie wavelength is much less than the distance of closest approach
49. neutron
50. (a) 15 keV; (b) 120 keV
51. 9.76 kV
58. (d) $x = n(\lambda/2)$, where $n = 0, 1, 2, 3, \ldots$
59. 2.1×10^{-24} kg · m/s
60. (a) 124 keV; (b) 40.5 keV
62. 5.1 eV
63. (a) 9.02×10^{-6}; (b) 3.0 MeV; (c) 3.0 MeV; (d) 7.33×10^{-8}; (e) 3.0 MeV; (e) 3.0 MeV
64. (a) 10^{104} years; (b) 2×10^{-19} s
65. (a) $-20\,\%$; (b) $-10\,\%$; (c) $+15\,\%$
66. 4.14×10^{-15} eV · s; (b) 2.31 eV
67. 5.9 μeV
68. (a) no; (b) 544 nm, green
69. (a) 73 pm; (b) 3.4 nm; (c) yes, their average de Broglie wavelength is smaller than their average separation
70. (a) 38.8 meV; (b) 146 pm
72. $T = 10^{-x}$, where $x = 7.2 \times 10^{39}$ (T is very small)
73. 0.19 m
74. (a) no; (b) plane wavefronts of infinite extent, perpendicular to the x axis
75. 1.7×10^{-35} m
81. (a) cesium; (b) both
82. 4.14 eV · fs

Chapter 39

1. 1.41
2. (a) 9.42 eV; (b) 5.13×10^{-3} eV
3. 1.9 GeV
4. 0.020 eV
5. 0.85 nm
6. 90.3 eV
7. 0.65 eV
8. (a) 13; (b) 12
9. 68.7 nm, 25.8 nm, 13.7 nm
10. (a) 11; (b) 10
11. (a) 72.2 eV; (b) 13.7 nm; (c) 17.2 nm; (d) 68.7 nm; (e) 41.2 nm; (g) 68.7 nm; (h) 25.8 nm
12. 350 pm
13. (a) 0.050; (b) 0.10; (c) 0.0095
14. (a) 0.091; (b) 0.091; (c) 0.82
15. 59 eV
16. 280 eV
18. 0.734 eV
19. 3.08 eV
20. (a) 1.25; (b) 2.00; (c) 5.00; (d) 1.00
21. (a) 8; (b) 0.75; (c) 1.00; (d) 1.25; (e) 3.75; (f) 3.00; (g) 2.25
22. (a) 3.00; (b) 9.00; (c) 2.00; (d) 3; (e) 6
23. (a) 7; (b) 1.00; (c) 2.00; (d) 3.00; (e) 9.00; (f) 8.00; (g) 6.00
24. 1.17 eV
25. 4.0
26. 2.6 eV
27. (a) 12.1 eV; (b) 6.45×10^{-27} kg · m/s; (c) 102 nm
28. (a) -3.4 eV; (b) 3.4 eV
30. (a) 0; (b) 10.2 nm^{-1}; (c) 5.54 nm^{-1}
31. (a) 291 nm^{-3}; (b) 10.2 nm^{-1}
32. (a) 12.8 eV; (b) 6; (c) 12.8 eV; (d) 12.1 eV; (e) 10.2 eV; (f) 0.661 eV; (g) 1.89 eV; (h) 2.55 eV
33. (a) 13.6 eV; (b) 3.40 eV
34. 4.1 m/s
35. (a) 13.6 eV; (b) -27.2 eV
36. (a) 31 nm; (b) 8.2×10^{14} Hz; (c) 0.29 μm; (d) 3.7×10^{14} Hz
38. (a) 2; (b) 1; (c) Lyman
39. 0.68

40. (a) 2.6 eV; (b) 4; (c) 2
42. 0.439
43. (a) $(r^4/8a^5)e^{-r/a}\cos^2\theta$;
(b) $(r^4/16a^5)e^{-r/a}\sin^2\theta$
44. (a) 3; (b) 1; (c) Lyman
45. (a) 0.0037; (b) 0.0054
46. 4.3×10^3
48. (c) $(r^2/8a^3)(2 - r/a)^2 e^{-r/a}$
50. (a) 1.3×10^{-19} eV; (b) 1.2×10^{19}; (c) 1.2×10^{13}; (d) yes
51. (a) n; (b) $2\ell + 1$; (c) n^2
52. (b) $\pm(2\pi/h)\sqrt{2mE}$
53. (b) $(2\pi/h)\sqrt{2m(U_0 - E)}$
54. (b) no; (c) no; (d) yes
55. (b) meter$^{-2.5}$
57. (a) 4; (b) 2; (b) Balmer
58. (a) 658 nm; (b) 366 nm

Chapter 40

1. (a) 3; (b) 3
2. (a) 14; (b) 6; (c) 6; (d) 2
3. (a) 3.65×10^{-34} J · s; (b) 3.16×10^{-34} J · s
4. (a) 32; (b) 2; (c) 18; (d) 8
5. 24.1°
6. (a) 3; (b) 2; (c) 14
7. (a) 4; (b) 5; (c) 2
8. 50
9. (a) 3.46; (b) 3.46; (c) 3; (d) 3; (e) −3; (f) 30.0°; (g) 54.5°; (h) 150°
10. (a) 3; (b) 5; (c) 2; (d) 18; (e) 3
12. (a) 58 μeV; (b) 14 GHz; (c) 2.1 cm, short radio wave region
13. (a) 54.7°; (b) 125°
14. (a) 1.5×10^{-21} n; (b) 20 μm
15. 72 km/s^2
16. 51 mT
17. 5.35 cm
18. 19 mT
19. 44
20. 17.25
21. (a) 51; (b) 53; (c) 56
22. (a) 18.00; (b) 18.25; (c) 19.00
23. 42
24. (a) 45; (b) 47; (c) 48
25. (a) 4p; (b) 4; (c) 4p; (d) 5; (e) 4p; (f) 6
26. (a) (1, 0, 0, +1/2); (b) (1, 0, 0, −1/2)
27. (a) (2, 0, 0, +1/2), (2, 0, 0, −1/2);
(b) (2, 1, 1, +1/2), (2, 1, 1, −1/2),
(2, 1, 0, +1/2), (2, 1, 0, −1/2),
(2, 1, −1, +1/2), (2, 1, −1, −1/2)
28. (a) 15; (b) 21
30. 12.4 kV
31. 49.6 pm, 99.2 pm
32. (a) 5.7 keV; (b) 87 pm; (c) 14 keV; (d) 220 pm; (e) 5.7 keV
33. (a) 35.4 pm; (b) 56.5 pm; (c) 49.6 pm
34. 6.44 keV
36. (a) 24.8 pm; (b) same
37. 0.563
38. 2.2 keV
39. (a) 69.5 kV; (b) 17.8 pm; (c) 21.3 pm; (d) 18.5 pm
41. 80.3 pm
42. (a) $(Z-1)^2/(Z'-1)^2$; (b) 57.5; (c) 2.07×10^3
43. (a) −24 %; (b) −15 %; (c) −11 %; (d) −7.9 %; (e) −6.4 %; (f) −4.7 %; (g) −3.5 %; (h) −2.6 %; (i) −2.0 %; (j) −1.5 %
44. 1.3×10^{15} moles
45. (a) 3.60 mm; (b) 5.25×10^{17}
46. 1.0×10^4 K
47. 9.0×10^{-7}
48. -2.75×10^5 K
49. 7.3×10^{17} s^{-1}
50. 4.7 km
51. 2×10^7
52. 2.0×10^{16} s^{-1}
53. (a) 3.03×10^5; (b) 1.43 GHz; (d) 3.31×10^{-6}
54. 1.8 pm
55. (a) 0; (b) 68 J
56. (a) 7.33 μm; (b) 7.07×10^5 W/m^2; (c) 2.49×10^{10} W/m^2
57. (a) 2.13 meV; (b) 18 T
58. (a) 4.3 μm; (b) 10 μm; (c) infrared
59. (a) no; (b) 140 nm
60. (a) 6.9 μeV; (b) radio waves
62. (a) 20 keV; (b) 18 keV; (c) Zr; (d) Nb
63. (a) 6.0; (b) 3.2×10^6 y
64. (a) 2.55 s; (b) 0.50 ns; (c) $(4.5 \times 10^{-4})°$ or 1.6" of arc
67. argon
68. (a) 3×10^{74}; (b) 6×10^{74}; (c) 6×10^{-38} rad
69. $n > 3$; $\ell = 3$; $m_\ell = +3, +2, +1, 0, -1, -2, -3$; $m_s = \pm 1/2$

Chapter 41

1. $8.49 \times 10^{28}\,m^{-3}$
3. (b) $6.81 \times 10^{27}\,m^{-3} \cdot eV^{-3/2}$; (c) $1.52 \times 10^{28}\,m^{-3} \cdot eV^{-1}$
4. $5.90 \times 10^{28}\,m^{-3}$
5. (a) 0; (b) 0.0955
6. $1.9 \times 10^{28}\,m^{-3} \cdot eV^{-1}$
8. 0.91
9. (a) 2.50×10^3 K; (b) 5.30×10^3 K
10. 5.52 eV
11. (a) 6.81 eV; (b) $1.77 \times 10^{28}\,m^{-3} \cdot eV^{-1}$; (c) $1.59 \times 10^{28}\,m^{-3} \cdot eV^{-1}$
12. (a) 90.0%; (b) 12.5%; (c) sodium
13. (a) $1.36 \times 10^{28}\,m^{-3} \cdot eV^{-1}$; (b) $1.68 \times 10^{28}\,m^{-3} \cdot eV^{-1}$; (c) $9.01 \times 10^{27}\,m^{-3} \cdot eV^{-1}$; (d) $9.56 \times 10^{26}\,m^{-3} \cdot eV^{-1}$; (e) $1.71 \times 10^{18}\,m^{-3} \cdot eV^{-1}$
15. (a) 1.0; (b) 0.99; (c) 0.50; (d) 0.014; (e) 2.4 $\times 10^{-17}$; (f) 700 K
16. about 10^{-42}
17. 3
18. (a) $2.7 \times 10^{25}\,m^{-3}$; (b) $8.43 \times 10^{28}\,m^{-3}$; (c) 3.1×10^3; (d) 3.3 nm; (E) 0.23 nm
19. (a) $5.86 \times 10^{28}\,m^{-3}$; (b) 5.49 eV; (c) 1.39 $\times$ 10^3 km/s; (d) 0.522 nm
20. 57 meV
22. (a) $1.31 \times 10^{29}\,m^{-3}$; (b) 9.43 eV; (c) 1.82×10^3 km/s; (d) 0.40 nm
23. 57.1 kJ
24. (a) 0.0055; (b) 0.018
25. 472 K
26. (a) 19.7 kJ; (b) 197 s
27. (a) 226 nm; (b) ultraviolet
28. (a) $+3e$; (b) $+5e$; (c) 2
29. (a) 1.5×10^{-6}; (b) 1.5×10^{-6}
31. 0.22 μg
32. (a) n-type; (b) $5 \times 10^{21}\,m^{-3}$; (c) 2.5×10^5
33. (a) 4.79×10^{-10}; (b) 0.0140; (c) 0.824
34. (a) above; (b) 0.744 eV; (c) 7.13×10^{-7}
35. 6.0×10^5
36. (b) 2.5×10^8
37. 4.20 eV
38. opaque
39. 13 μm
40. (a) 5.0×10^{-17} F; (b) 3.1×10^2
41. (b) $1.8 \times 10^{28}\,m^{-3} \cdot eV^{-1}$
42. 0.03
43. (a) 109.5°; (b) 238 pm
46. (a) $+8 \times 10^{-11}\,\Omega\cdot m/K$; (b) $-2 \times 10^2\,\Omega\cdot m/K$
47. 3.49×10^3 atm

Chapter 42

1. 15.8 fm
2. 1.3×10^{-13} m
3. (a) 0.390 MeV; (b) 4.61 Mev
4. (a) 6; (b) 8
5. (a) $+7.825 \times 10^{-3}$ U; (b) $+7.290\,MeV/c^2$; (c) $+8.664 \times 10^{-3}$ u; (d) $+8.071\,MeV/c^2$; (e) -9.780×10^{-2} u; (f) $-91.10\,MeV/c^2$
6. (a) yttrium; (b) iodine; (c) 50; (d) 74; (e) 19
7. 13 km
8. (a) blow apart; (b) 1.15 GeV; (c) 12.2 MeV/proton; (d) 4.81 MeV/nucleon; (e) strong force is strong
9. (a) $2.3 \times 10^{17}\,kg/m^3$; (b) $2.3 \times 10^{17}\,kg/m^3$; (d) $1.0 \times 10^{25}\,C/m^3$; (e) $8.8 \times 10^{24}\,C/m^3$
10. (b) 0.05%; (c) 0.81%; (d) 0.81%; (e) 0.74%; (f) 0.71%; (g) no
11. (a) 6.2 fm; (b) yes
12. 7.31 MeV
13. (a) 9.303%; (b) 11.71%
14. (a) 19.8 MeV; (b) 6.26 MeV; (c) 2.23 MeV; (d) 28.3 MeV; (e) 7.07 MeV; (f) no
16. 1.6×10^{25} MeV
18. 1.0087 u
19. (b) 7.92 MeV
20. 0.49
21. (a) 0.250; (b) 0.125
22. 280 d
23. (a) $7.5 \times 10^{16}\,s^{-1}$; (b) $4.9 \times 10^{16}\,s^{-1}$
24. 3.0×10^{19}
25. (a) 64.2 h; (b) 0.125; (c) 0.0749
26. (a) 5.04×10^{18}; (b) $4.60 \times 10^6\,s^{-1}$
27. 5.3×10^{22}
28. 1×10^{13} atoms
29. 9.0×10^8 Bq
30. 3.2×10^{12} Bq = 86 Ci
31. (a) 2.0×10^{20}; (b) $2.8 \times 10^9\,s^{-1}$
32. (a) β^- decay; (b) 8.2×10^7; (c) 1.2×10^6
33. 265 mg
34. 209 d
35. 1.12×10^{11} y
36. 87.9 mg
38. 0.66 g

39. (a) $8.88 \times 10^{10}\,s^{-1}$; (b) 1.19×10^{15}; (c) $0.111\,\mu g$
40. (a) 4.25 MeV; (b) −24.1 MeV; (c) 28.3 MeV
41. (a) 1.2×10^{-17}; (b)0
42. (a) −9.50 MeV; (b) 4.66 MeV; (c) −1.30 MeV
43. 4.269 MeV
44. (a) 31.8 MeV; (b) 5.98 MeV; (c) 86 MeV
45. 1.21 MeV
46. (a) 0.90 pm; (b) 6.4 fm; (c) no; (d) yes
47. 0.783, Mev
49. (b) 0.961 MeV
50. (b) 2.7×10^{13} W
51. 78.3 eV
52. 1.61×10^{3} y
53. (a) 1.06×10^{19}; (b) 0.624×10^{19}; (c) 1.68×10^{19}; (d) 2.97×10^{9} y
54. $132\,\mu g$
55. 1.8 mg
56. 4.28×10^{9} y
57. 1.02 mg
58. 145 Bq; (b) 3.92 nCi
59. 13 mJ
60. (a) 18 mJ; (b) 2.9 mSv; (c) 0.29 rem
61. (a) 6.3×10^{18}; (b) 2.5×10^{11}; (c) 0.20 J; (d) 2.3 mGy; (e) 30 mSv
62. 3.87×10^{10} K
63. (a) 6.6 MeV; (b) no
64. (a) ^{18}O, ^{60}Ni, ^{92}Mo, ^{144}Sm, ^{207}Pb; (b) ^{40}K, ^{91}Zr, ^{121}Sb, ^{143}Nd; (c) ^{13}C, ^{40}K, ^{49}Ti, ^{205}Tl, ^{207}Pb
65. (a) 25.4 MeV; (b) 12.8 MeV; (c) 25.0 MeV
66. (b) 1.00; (c) 70.8; (d) 0.0100; (e) 0.708; (f) no
67. (a) 59.5 d; (b) 1.18
68. (a) 7×10^{7} electrons; (b) $(7 \times 10^{7}\ \text{electrons})e^{[-(\ln 2)(D-1996)/T_{1/2}]}$. where D is the current year and $T_{1/2} =$ 30.2 y
69. $730\,cm^2$
70. (a) 3.66^{7} Bq; (b) 3.66×10^{7} Bq; (c) 6.42 ng
71. 600 keV
72. 28.3 MeV
73. 30 MeV
74. 4.9×10^{13} Bq
75. 3.2×10^{4} y
76. (b) $4n + 3$; (c) $4n$; (d) $4n + 2$; (e) $4n + 3$; (f) $4n$; (g) $4n + 1$; (h) $4n + 2$; (i) $4n + 1$; (j) $4n + 1$
77. ^{7}Li
79. ^{225}Ac
80. (a) $4.8 \times 10^{-18}\,s^{-1}$; (b) 4.6×10^{9} y
84. (a) ^{142}Nd, ^{143}Nd, ^{146}Nd, ^{148}Nd, ^{150}Nd; (b) ^{97}Rb, ^{98}Sr, ^{99}Y, ^{100}Zr, ^{100}Sr, ^{101}Nb, ^{102}Mo, ^{103}Tc, ^{105}Rh, ^{109}In, ^{110}Sn, ^{111}Sb, ^{112}Te; (c) ^{60}Zn, ^{60}Cu, ^{60}Ni, ^{60}Co, ^{60}Fe, ^{60}Mn, ^{60}Cr, ^{60}V
85. (a) 11.906 83 u; (b) 236.2025 u
86. 4×10^{-22} s
87. 27

Chapter 43

1. $3.1 \times 10^{10}\,s^{-1}$
2. (a) ^{95}Sr; (b) ^{95}Y; (c) ^{134}Te; (d) 3
3. (a) 2.6×10^{24}; (b) 8.2×10^{13} J; (c) 2.6×10^{4} y
4. 4.54×10^{26} MeV
5. −23.0 MeV
6. (a) +5.00 MeV
7. (a) 16 fissions/day; (b) 4.3×10^{8}
8. 181 MeV
9. (a) ^{153}Nd; (b) 110 MeV; (c) 60 MeV; (d) 1.6×10^{7} m/s; (e) 8.7×10^{6} m/s
10. (a) +25%; (b) 0; (c) −36%
11. (a) 252 MeV; (b) typical fission energy is 200 MeV
12. (a) 10; (b) 226 MeV
13. 462 kg
14. yes
15. 557 W
16. (a) 44 kton
17. (a) 1.2 MeV; (b) 3.2 kg
19. (a) 84 kg; (b) 1.7×10^{25}; (c) 1.3×10^{25}
20. 8.03×10^{3} MW
21. (b) 1.0; (c) 0.89; (d) 0.28; (e) 0.019; (f) 8
22. 1.6×10^{16}
23. 0.99938
24. (a) 75 kW; (b) 5.8×10^{3} kg
25. 3.6×10^{9} y
27. 1.7×10^{9} y
29. 170 keV
30. (a) 170 kV
31. 1.41 MeV
32. 0.151
35. (a) 4.3×10^{9} kg/s; (b) 3.1×10^{-4}
37. (a) $1.8 \times 10^{38}\,s^{-1}$; (b) $8.2 \times 10^{28}\,s^{-1}$
38. (a) 4.0×10^{27} MeV; (b) 5.1×10^{26} MeV

39. (a) 4.1 eV/atom; (b) 9.0 MJ/kg; (c) 1.5×10^3 y
41. 1.6×10^8 y
42. 5×10^9 y
43. (a) 24.9 MeV; (b) 8.65 megaton TNT
45. 14.4 kW
46. (a) 6.3×10^{14} J/kg; (b) 6.2×10^{11} kg/s; (c) 4.3×10^9 kg/s; (e) 1.5×10^{10} y
47. (a) 3.1×10^{31} protons/m^3; (b) 1.2×10^6 times
48. 3.5 MeV; (b) 14.1 MeV
49. $^{238}\text{U} + n \rightarrow {}^{239}\text{U} \rightarrow {}^{239}\text{Np} + \text{e} + \nu$, $^{239}\text{Np} \rightarrow {}^{239}\text{Pu} + \text{e} + \nu$
50. (b) 5.0×10^5 m/s
51. 6×10^2 kg
54. (a) 30 MeV; (b) 6 MeV

Chapter 44

1. 18.4 fm
2. 2.4×10^{-43}
3. 1
4. $\pi^- \longrightarrow \mu^+ + \overline{\nu}$
5. 2.7 cm/s
6. (a) 1.90×10^{-18} kg · m/s; (b) 9.90 m
7. 769 MeV
8. 31 nm
10. (a) 2e^+, e^-, 5ν, $4\overline{\nu}$; (b) boson; (c) meson; (d) 0
11. (a) angular momentum, L_e; (b) charge, L_μ; (c) energy, L_μ
12. b and d
13. (a) 0; (b) −1; (c) 0
14. (a) 605 MeV; (b) −181 MeV
15. (a) energy; (b) strangeness; (c) charge
17. (a) K^+; (b) $\overline{\text{n}}$; (c) K^0
18. 338 MeV
19. (a) 37.7 MeV; (b) 5.35 MeV; (c) 32.4 MeV
20. (a) n; (b) Σ^+; (c) Ξ^-
21. (a) $\overline{\text{uud}}$; (b) $\overline{\text{udd}}$
22. (a) $\overline{\text{sud}}$; (b) $\overline{\text{uss}}$
23. (a) not possible; (b) *uuu*
25. Σ^0, 7.51 km/s
26. 1.4×10^{10} ly
27. 668 nm
28. 2.77×10^8 ly
29. (b) 0.934; (c) 1.28×10^{10} ly
30. (b) 5.7 H atoms/m^3
31. (a) 0.26 μmeV; (b) 4.8 mm
32. $102M_S$
33. (a) 121 m/s; (b) 0.00406; (c) 248 y
34. (b) $2\pi r^{3/2}/\sqrt{GM}$
35. (b) 2.39×10^9 K
36. (a) 2.6 K; (b) 976 nm
37. 1.08×10^{42} J
38. (a) A; (b) J; (c) I; (d) F; (e) G; (f) C; (g) H; (h) D; (i) E
40. 13×10^9 y
41. (a) $0.785c$; (b) $0.993c$; (c) C2; (d) C1; (e) 51 ns; (f) 40 ns
43. (c) $r\alpha/c + (r\alpha/c)^2 + (r\alpha/c)^3 + \ldots$; (d) $r\alpha/c$; (e) $\alpha = H$; (f) 6.5×10^8 ly; (g) 6.9×10^8 y; (h) 6.5×10^8 y; (i) 6.9×10^8 ly; (j) 1.0×10^9 ly; (k) 1.1×10^9 y; (l) 3.9×10^8 ly
44. 6.03×10^{-29} kg

SECTION SEVEN
COMPARISON OF PROBLEMS WITH THE SIXTH EDITION

In the table below, the left column of each group gives the numbers of problems and exercises in the last edition (the sixth) of *Fundamentals of Physics* and for each entry the right column gives the number of the same problem or exercise in this edition (the seventh). The titles (Standard Edition Problems, Enhanced Edition Problems, and Supplement Problems) refer to sixth edition sources.

Standard Edition Problems

Chapter 1

1–1 1–3
1–2 1–32
1–3 1–1
1–4 1–4 revised
1–5 1–5
1–6 1–31
1–7 1–7
1–8 1–34
1–9 1–9
1–10 1–10
1–12 1–58
1–13 1–13
1–14 1–17
1–15 1–47
1–16 1–14
1–17 1–18 revised
1–18 1–16
1–19 1–21
1–20 1–20
1–21 1–19
1–22 1–48
1–23 1–23
1–24 1–33
1–25 1–8
1–26 1–25
1–28 1–53
1–29 1–27

Chapter 2

2–1 2–96
2–2 2–2 revised
2–3 2–3
2–4 106 revised
2–5 2–11
2–6 2–6
2–7 2–5
2–8 2–10
2–9 2–9 revised
2–11 2–12
2–12 2–13
2–13 2–61
2–17 2– revised
2–18 2–18
2–19 2–97
2–20 2–16
2–21 2–19 revised
2–22 2–112
2–23 2–20
2–24 2–110
2–25 2–21
2–26 2–98
2–27 2–23
2–28 2–111
2–29 2–25
2–30 2–26
2–31 2–28
2–32 2–31 revised
2–33 2–27
2–34 2–64 revised
2–35 2–68
2–36 2–63 revised
2–37 2–95
2–38 2–37
2–39 2–29
2–40 2–40
2–41 2–39
2–42 2–38
2–43 2–41
2–44 2–116
2–45 2–70
2–46 2–117
2–47 2–42
2–48 2–91
2–49 2–47
2–50 2–92
2–51 2–49 revised
2–53 2–51 revised

2–54 2–56
2–55 2–53 revised
2–56 2–58
2–57 2–94
2–58 2–93
2–59 2–43
2–60 2–52
2–61 4–117
2–62 2–89
2–63 2–90
2–64 2–105
2–65 2–55

Chapter 3

3–1 3–74
3–2 3–75
3–3 3–1
3–4 3–2
3–5 3–3
3–6 3–4 revised
3–7 3–39
3–8 3–37 revised
3–9 3–7
3–10 3–10
3–11 3–51
3–12 3–8
3–13 3–11
3–14 3–12
3–15 3–62
3–16 3–14
3–17 3–15
3–18 3–49
3–19 3–17
3–20 3–20
3–21 3–19 revised
3–22 3–18
3–24 3–48
3–26 3–59
3–27 3–25 revised
3–28 3–46
3–29 3–45
3–30 3–79 revised
3–31 3–31
3–32 3–78 revised
3–33 3–77
3–34 3–32
3–35 3–76
3–36 3–34
3–37 3–35
3–38 3–38

Chapter 4

4–1 4–2
4–2 4–1
4–3 4–92
4–4 4–85
4–5 4–5
4–6 4–7
4–7 4–6
4–8 4–81
4–9 4–9
4–10 4–95
4–11 4–11
4–12 4–97
4–13 4–15
4–14 4–14
4–15 4–99
4–16 4–16
4–17 4–87
4–18 4–20
4–19 4–32
4–20 4–21
4–21 4-104
4–22 4–18
4–23 4–24
4–24 4–34
4–25 4–31
4–26 4–106
4–28 4–30
4–29 4–19
4–30 4–42
4–31 4–108
4–32 4–25
4–33 4–23
4–34 4–29
4–35 4–41
4–36 4–28
4–37 4–27
4–38 4–110
4–39 4–37
4–40 4–26
4–41 4–39 revised
4–42 4–44
4–43 4–112
4–44 4–48
4–45 4–115
4–46 4–47 revised
4–47 4–116
4–48 4–88
4–49 4–45 revised
4–50 4–82
4–51 4–53
4–52 4–96
4–53 4–55 revised
4–54 4–56 revised
4–55 4–118
4–56 4–58
4–57 4–59
4–58 4–86
4–59 4–63
4–60 4–65
4–61 4–61
4–62 4–136
4–63 4–66
4–64 4–68

Chapter 5

5–1 5–2 sig. figs.
5–2 5–3
5–3 5–1
5–4 5–7
5–5 5–62
5–6 5–4
5–7 5–5 sig. figs.
5–8 5–61
5–9 5–9
5–10 5–10 revised
5–11 5–81
5–12 5–80
5–13 5–64
5–15 5–13
5–16 5–88
5–17 5–19
5–18 5–18
5–19 5–17
5–20 5–22 revised
5–21 5–23
5–22 5–20
5–23 5–25
5–24 5–32
5–25 5–27
5–26 5–89
5–27 5–21 revised
5–28 5–87

5–29 5–31
5–30 5–28
5–31 5–43
5–32 5–86
5–33 5–33
5–34 5–42 revised
5–35 5–85
5–36 5–45
5–37 5–72
5–38 5–63
5–39 5–83
5–40 5–48
5–41 5–39 revised
5–43 5–51
5–44 5–54
5–45 5–29
5–46 5–82
5–47 5–49
5–48 5–40 revised
5–49 5–41
5–50 5–50 revised
5–51 5–60
5–52 5–24 revised
5–53 5–53
5–54 5–56 revised
5–55 5–35 revised
5–56 5–34

Chapter 6

6–1 6–3 revised
6–2 6–1
6–3 6–2
6–4 6–4 revised
6–5 6–5
6–6 6–81
6–7 6–104
6–8 6–102
6–9 6–15 sig. figs.
6–10 6–10 revised
6–12 6–11 revised
6–13 6–13 revised
6–14 6–8
6–15 6–28
6–16 6–7
6–17 6–14 revised
6–18 6–20 revised
6–19 6–21 revised
6–20 6–19 revised
6–21 6–27 revised
6–22 6–26
6–23 6–25 (c) removed
6–24 6–22 revised
6–25 6–29 revised
6–26 6–68
6–27 6–30 revised
6–28 6–76
6–29 6–67
6–30 6–17 revised
6–31 6–31
6–32 6–35 revised
6–33 6–100
6–34 6–32
6–35 6–33
6–36 6–38
6–37 6–36
6–38 6–40 revised
6–39 6–37
6–40 6–44 revised
6–41 6–49 revised
6–42 6–99
6–43 6–45 revised
6–44 6–46
6–45 6–47 revised
6–46 6–98
6–47 6–51 revised
6–48 6–53

Chapter 7

7–2 7–2
7–4 7–5
7–5 7–3
7–6 7–8
7–7 7–52
7–8 7–9 revised
7–9 7–11
7–10 7–35
7–11 7–13
7–12 7–6
7–13 7–14
7–14 7–16
7–15 7–63
7–16 7–51
7–17 7–17
7–18 7–22
7–19 7–19
7–20 7–24
7–22 7–50
7–23 7–29
7–24 7–32
7–25 7–31
7–26 7–49
7–27 7–33
7–28 7–36
7–29 7–65
7–30 7–40
7–31 7–41
7–32 7–44
7–33 7–43
7–34 7–42
7–35 7–45
7–36 7–46
7–38 7–67
7–39 7–77
7–40 7–64

Chapter 8

8–1 8–1
8–2 8–2
8–3 8–3
8–4 8–5 revised
8–5 8–4 revised
8–6 8–7 revised
8–7 8–6
8–8 8–8
8–9 8–9
8–10 8–10
8–11 8–12
8–12 8–11
8–13 8–13
8–14 8–20
8–15 8–14
8–16 8–28
8–17 8–17
8–18 8–26 revised
8–19 8–31 revised
8–20 8–21 revised
8–21 8–29 revised
8–22 8–24
8–23 8–23 sig. figs.
8–24 8–25
8–25 8–18
8–26 8–22
8–27 8–43 revised
8–28 8–16

8–29 8–64 revised
8–30 8–76
8–31 8–69 revised
8–32 8–30
8–33 8–70 revised
8–35 8–36 revised
8–36 8–71 revised
8–37 8–38 revised
8–38 8–39 revised
8–39 8–40
8–40 8–105
8–41 8–43
8–42 8–41
8–43 8–45
8–44 8–68
8–45 8–44
8–46 8–46
8–47 8–47
8–48 8–131
8–49 8–48
8–50 8–52
8–51 8–51 revised
8–52 8–49 revised
8–53 8–72
8–54 8–54
8–55 8–106 revised
8–56 8–55
8–57 8–57 revised
8–58 8–73
8–59 8–50
8–60 8–56
8–61 8–58 revised
8–62 8–65 revised
8–63 8–59 revised
8–64 8–62
8–65 8–66
8–66 8–117

Chapter 9

9–1 9–133 revised
9–2 9–102 revised
9–3 9–1 revised
9–4 9–3 revised
9–5 9–82 revised
9–6 9–6 revised
9–7 9–5 revised
9–8 9–7 revised
9–9 9–8 revised
9–10 9–10
9–11 9–104
9–12 9–89 revised
9–13 9–9
9–14 9–11
9–15 9–15 revised
9–16 9–12
9–17 9–81 revised
9–18 9–16
9–19 9–17 revised
9–22 9–18
9–23 9–19
9–24 9–20 revised
9–25 9–87
9–26 9–22
9–27 9–25 revised
9–28 9–106 revised
9–29 9–108 revised
9–30 9–36
9–31 9–37 revised
9–32 9–80 revised
9–33 9–85
9–34 9–39
9–35 9–110 revised
9–36 9–44
9–37 9–43 revised
9–38 9–40
9–39 9–41 revised
9–40 9–93 revised
9–41 9–72
9–42 9–112
9–43 9–71
9–44 9–70
9–45 9–114
9–46 9–79
9–47 9–73
9–48 9–135
9–49 8–110
9–50 8–112
9–51 8–113
9–52 8–103
9–53 8–107
9–54 8–108
9–56 8–84
9–57 8–85
9–58 8–75

Chapter 10

10–1 5–
10–2 5–
10–3 5–
10–4 5–
10–4 9–24 revised
10–5 9–23 revised
10–6 9–25
10–7 9–26
10–8 9–27
10–9 9–101
10–10 9–29 revised
10–11 5–
10–13 9–99
10–14 9–32
10–15 9–100
10–16 9–103
10–17 9–33 revised
10–19 9–105
10–20 9–45
10–22 9–46
10–25 9–48 revised
10–26 9–49
10–28 9–51
10–29 9–107
10–30 9–50
10–31 9–109 revised
10–33 9–53
10–34 9–54
10–35 9–56
10–36 9–111
10–37 9–55
10–40 9–60
10–41 9–59
10–42 9–88
10–43 9–57
10–44 9–113 revised
10–45 9–63
10–46 9–66
10–47 9–65
10–48 9–67
10–49 9–132
10–50 9–68
10–51 9–69
10–52 9–131 revised
10–53 9–130 revised
10–56 9–30
10–57 9–129

Chapter 11

11–1 10–110
11–2 10–1
11–3 10–106
11–4 10–4
11–5 10–6
11–6 10–7
11–7 10–5
11–8 10–9
11–9 10–107
11–10 10–10
11–11 119 sig. figs.
11–12 10–11
11–13 10–13
11–14 10–18
11–15 10–17 revised
11–16 10–16
11–17 10–15 revised
11–20 10–121
11–21 10–122
11–22 10–21
11–23 10–22
11–24 10–23
11–25 10–29
11–26 10–30 revised
11–27 10–27
11–28 10–26
11–29 10–28
11–30 10–25
11–31 10–32 revised
11–32 10–31
11–33 10–33
11–34 10–123
11–35 10–35
11–36 10–120
11–37 10–39 revised
11–39 10–37
11–40 10–42
11–41 10–41
11–42 10–43
11–43 10–85
11–44 10–44
11–45 10–45
11–46 10–46
11–47 10–47(a) deleted
11–48 10–48
11–49 10–49
11–50 10–50
11–51 10–108
11–52 10–53
11–54 10–83
11–55 10–55
11–56 10–57
11–57 10–54
11–58 10–59
11–59 10–58
11–60 10–60
11–61 10–61 revised
11–62 10–124
11–63 10–63
11–64 10–64
11–65 10–84 revised
11–66 10–67
11–67 10–66
11–68 10–73
11–69 10–111
11–70 10–75

Chapter 12

12–1 11–1
12–2 11–2 revised
12–3 11–3
12–4 11–74
12–5 11–5
12–6 11–81
12–7 11–4
12–8 11–6 revised
12–9 11–9
12–10 11–68 revised
12–11 11–7 revised
12–12 11–8 revised
12–13 11–70
12–14 11–64
12–15 11–15
12–16 11–89
12–17 11–98
12–18 11–17 revised
12–19 11–18 revised
12–20 11–22
12–21 11–21
12–22 11–20
12–23 11–25 revised
12–24 11–24 revised
12–25 11–27
12–26 11–26
12–27 11–67 revised
12–28 11–78
12–29 11–31
12–30 11–30 revised
12–31 11–32 revised
12–32 11–71 revised
12–33 11–35
12–34 11–34
12–35 11–37 revised
12–36 11–100
12–37 11–101
12–38 11–39
12–39 11–41
12–40 11–40
12–41 11–43
12–42 11–42 revised
12–43 11–95
12–44 11–46
12–45 11–45 revised
12–46 11–47 revised
12–47 11–48 revised
12–48 11–77 revised
12–49 11–49
12–50 11–51 revised
12–51 11–59
12–52 11–52 revised
12–53 11–92
12–54 11–54
12–55 11–55
12–56 11–58 revised
12–57 11–56
12–58 11–57
12–59 11–60 revised

Chapter 13

13–1 12–2
13–2 12–81
13–4 12–4
13–5 12–3
13–6 12–6
13–7 12–5 revised
13–8 12–8
13–9 12–7 revised
13–10 12–10
13–11 12–9 sig. figs.
13–12 12–12
13–13 12–11 revised
13–14 12–15 revised
13–15 12–45 revised
13–16 12–14 revised

13–17 12–47
13–18 12–20 revised
13–19 12–13
13–20 12–22 revised
13–21 12–19 revised
13–22 12–24 revised
13–23 12–21 revised
13–24 12–53
13–25 12–23 revised
13–26 12–26 revised
13–27 12–25
13–28 12–28 revised
13–29 12–31
13–30 12–27 revised
13–31 12–35
13–32 12–32 revised
13–33 12–29
13–34 12–79
13–35 12–33 revised
13–36 12–36
13–37 12–37
13–38 12–40
13–39 12–39
13–40 12–41 revised
13–41 12–80
13–42 82

Chapter 14

14–1 13–1
14–2 13–104
14–3 13–75
14–4 13–2
14–5 13–3
14–6 13–92
14–7 13–5
14–8 13–4 revised
14–9 13–6 revised
14–10 13–9
14–11 13–81
14–12 13–73
14–13 13–11 revised
14–14 13–14 revised
14–15 13–15
14–16 13–16
14–17 13–98
14–18 13–17
14–20 13–18
14–21 13–19
14–22 13–20
14–23 13–21
14–24 13–76
14–25 13–23
14–26 13–24
14–27 13–74
14–28 13–26
14–29 13–25
14–30 13–27
14–31 13–31 revised
14–32 13–32
14–34 13–84
14–35 13–33
14–36 13–83
14–37 13–35 sig. figs.
14–38 13–34
14–39 13–77
14–40 13–40
14–41 13–39
14–43 13–41
14–44 13–42
14–45 13–43
14–46 13–44
14–47 13–45
14–48 13–46
14–49 13–47
14–50 13–48
14–51 13–49
14–52 13–50
14–53 13–51
14–54 13–103
14–55 13–53
14–56 13–58 revised
14–57 13–55
14–58 13–56 revised
14–61 13–57
14–62 13–60
14–63 13–61
14–64 13–62

Chapter 15

15–1 14–3
15–2 14–2
15–3 14–1
15–4 14–6
15–5 14–5
15–6 14–4
15–7 14–7
15–8 14–9
15–12 14–93
15–13 14–11 slight change
15–14 14–14 revised
15–15 14–15 revised
15–16 14–16 revised
15–18 14–18
15–19 14–19 revised
15–20 14–20
15–21 14–21
15–22 14–22
15–23 14–75
15–24 14–24
15–25 14–25
15–26 14–26
15–27 14–29
15–29 14–31
15–30 14–91
15–31 14–33
15–32 14–38 sig. fig.
15–33 14–35
15–34 14–98
15–35 14–92
15–36 14–27
15–38 14–37
15–39 14–41
15–40 14–42
15–41 14–43
15–42 14–44
15–43 14–45
15–44 14–48
15–45 14–47
15–46 14–46 revised
15–47 14–49
15–50 14–52 revised
15–51 14–54
15–52 14–73
15–53 14–53 revised
15–54 14–55 revised
15–55 14–59
15–56 14–60
15–57 14–61
15–58 14–62
15–59 14–71
15–60 14–64
15–61 14–70

Chapter 16

16–1 15–7
16–2 15–6
16–3 15–5
16–4 15–4
16–5 15–13 revised
16–6 15–75
16–7 15–3
16–8 15–2
16–9 15–1
16–10 15–76
16–11 15–8
16–12 15–9
16–13 15–77
16–15 15–14
16–16 15–25
16–17 15–15
16–18 15–20
16–19 15–17
16–20 15–19
16–21 15–23 revised
16–22 15–22
16–23 15–21
16–24 15–11 revised
16–25 15–24 revised
16–26 15–78
16–27 15–26 revised
16–28 15–27 revised
16–29 15–80 revised
16–30 15–79 revised
16–31 15–29
16–32 15–28
16–33 15–33
16–35 15–81
16–36 15–37 revised
16–37 15–31
16–38 15–35 revised
16–39 15–71 sig. figs.
16–40 15–82 sig. figs
16–41 15–39
16–42 15–109
16–44 15–46
16–45 15–45
16–46 15–44 revised
16–47 15–55 revised
16–48 15–83
16–49 15–43 revised
16–50 15–40
16–51 15–41
16–52 15–51 revised
16–53 15–49
16–54 15–86
16–55 15–85
16–56 15–84
16–58 15–72
16–59 15–57
16–60 15–58
16–61 15–59
16–62 15–60
16–63 15–62 revised
16–64 15–63

Chapter 17

17–1 16–2
17–2 16–89
17–3 16–1
17–4 16–69 revised
17–6 16–6
17–7 16–70 revised
17–8 16–11 revised
17–9 16–5
17–10 16–12
17–11 16–13
17–12 16–14
17–13 16–15
17–14 16–18
17–15 16–17 revised
17–16 16–68 sig. figs.
17–17 16–67
17–18 16–20 revised
17–19 16–19 revised
17–20 16–22
17–21 16–21 revised
17–22 16–87
17–23 16–23
17–24 16–24
17–25 16–66
17–26 16–30
17–27 16–29 sig. figs.
17–28 16–60
17–29 16–33
17–30 16–34
17–31 16–61
17–32 16–44
17–33 16–43 sig. figs.
17–34 16–42
17–35 16–41
17–36 16–40
17–37 16–39 revised
17–38 16–38 revised
17–39 16–45
17–40 16–93
17–41 16–47
17–42 16–84 revised
17–43 16–49
17–44 16–83
17–45 16–82
17–46 16–48
17–47 16–53 sig. figs, revised
17–48 16–50 revised
17–49 16–92
17–50 16–51 sig. figs.
17–51 16–55
17–52 16–85

Chapter 18

18–1 17–91
18–3 17–3 revised
18–4 17–1
18–5 17–7
18–6 17–75 revised
18–7 17–5
18–9 17–9
18–10 17–92 revised
18–11 17–8
18–13 17–17 revised
18–14 17–14
18–15 17–19 revised
18–16 17–16
18–17 17–21
18–18 17–22
18–19 17–23
18–20 17–24
18–21 17–25 sig. figs.
18–22 17–26
18–23 17–93
18–26 17–96
18–27 17–94
18–28 17–30
18–29 17–95
18–30 17–35
18–31 17–36
18–32 17–38 revised
18–33 17–37
18–35 17–39 revised

18–37 17–41
18–38 17–42
18–39 17–108
18–40 17–44 revised
18–41 17–43
18–42 17–46
18–43 17–45
18–44 17–48
18–45 17–47 revised
18–46 17–50
18–48 17–51
18–49 17–49
18–50 17–52
18–51 17–54
18–52 17–55
18–53 17–53
18–54 17–57
18–55 17–59
18–56 17–60
18–57 17–61
18–58 17–97
18–59 17–63
18–60 17–64

Chapter 19

19–1 18–1
19–2 18–2
19–3 18–3
19–4 18–4
19–6 18–5
19–9 18–7 note added
19–10 18–10
19–12 18–8
19–13 18–9
19–14 18–12 sig. figs
19–15 18–11
19–17 18–15
19–18 18–14
19–19 18–17 sig. figs.
19–20 18–16
19–22 18–18
19–24 18–20
19–25 18–21
19–26 18–24
19–27 18–23
19–28 18–22
19–29 18–25
19–32 18–26
19–34 18–28
19–35 18–27
19–36 18–30
19–38 18–29
19–39 18–35
19–41 18–38
19–42 18–33
19–43 18–31
19–44 18–37 revised
19–45 18–39
19–46 18–40
19–47 18–41
19–48 18–44
19–49 18–43
19–50 18–42 revised
19–51 18–45
19–52 18–48
19–53 18–49
19–54 18–78
19–55 18–50
19–56 18–86
19–57 18–51
19–58 18–52
19–59 18–53
19–61 18–55
19–62 18–54
19–63 18–57
19–64 18–59
19–65 18–61
19–66 18–60
19–67 18–67

Chapter 20

20–1 19–1
20–2 19–2
20–5 19–3
20–6 19–4
20–7 19–5
20–8 19–6
20–9 19–7
20–10 19–81
20–11 19–11 revised
20–12 19–9
20–13 19–13 sig. figs
20–14 19–12
20–15 19–14
20–17 19–15
20–18 19–18
20–19 19–17
20–20 19–16
20–21 19–20
20–22 19–19
20–23 19–21
20–24 19–22
20–25 19–23
20–26 19–24
20–27 19–25
20–30 19–28
20–31 19–27
20–32 19–26
20–34 19–30
20–35 19–29
20–36 19–32
20–37 19–31
20–38 19–33 revised
20–39 19–35 revised
20–40 19–36
20–41 19–37
20–42 19–38 sig. figs.
20–43 19–39
20–44 19–40
20–46 19–44
20–47 19–43 revised
20–48 19–46
20–49 19–45
20–50 19–48
20–51 19–49 revised
20–52 19–50
20–53 19–51
20–54 19–54 sig. figs.
20–55 19–53
20–56 19–52
20–58 19–67
20–59 19–68
20–60 19–58 revised
20–61 19–59

Chapter 21

21–1 20–3
21–2 20–2
21–3 20–1 sig. figs.
21–4 20–6
21–5 20–5
21–6 20–6
21–7 20–4
21–8 20–14

21–9 20–7
21–10 20–9
21–12 20–11
21–13 20–13
21–14 20–19 revised
21–15 20–15
21–16 20–16
21–17 20–17
21–19 20–18
21–20 20–20
21–21 20–22
21–22 20–23
21–23 20–21
21–24 20–24
21–25 20–25 revised
21–26 20–26
21–27 20–27
21–29 20–29 revised
21–30 20–30
21–31 20–51
21–32 20–32
21–33 20–33 revised
21–35 20–35
21–36 20–36
21–37 20–37
21–38 20–38
21–39 20–39
21–40 20–42
21–41 20–41 revised
21–42 20–43
21–44 20–44
21–45 20–45

Chapter 22

22–1 21–3
22–2 21–2
22–3 21–1
22–4 21–4 revised
22–5 21–7 revised
22–6 21–8 revised
22–7 21–9 revised
22–8 21–10 revised
22–9 21–17 revised
22–10 21–13 revised
22–11 21–65 revised
22–12 21–15 revised
22–13 21–5 revised
22–14 21–6 revised
22–15 21–66 revised
22–16 21–67 revised
22–17 21–48
22–18 21–22
22–19 21–70
22–20 21–47
22–21 21–21
22–22 21–24
22–23 21–23
22–24 21–60 revised
22–25 21–25
22–26 21–26
22–27 21–29
22–28 21–59
22–29 21–31
22–30 21–64 revised

Chapter 23

23–1 22–2
23–2 22–1
23–3 22–3
23–4 22–4
23–5 22–5
23–6 22–6 revised
23–7 22–7 revised
23–9 22–9 revised
23–10 22–8 sig. figs.
23–11 22–10 revised
23–12 22–12 revised
23–13 22–11 revised
23–14 22–90
23–15 22–17
23–16 22–19 revised
23–17 22–21
23–18 22–22 sig. figs.
23–19 22–84
23–20 22–24 revised
23–21 22–25 revised
23–22 22–26 revised
23–23 22–27 revised
23–24 22–29 revised
23–25 22–55
23–26 22–30
23–27 22–31 revised
23–28 22–36 revised
23–29 22–35
23–30 22–34
23–31 22–37
23–32 22–38
23–33 22–39 revised
23–34 22–85
23–35 22–41
23–36 22–40
23–37 22–42
23–38 22–57
23–39 22–43
23–40 22–44
23–41 22–45
23–42 22–47
23–43 22–86
23–44 22–51
23–45 22–50
23–46 22–53 revised
23–47 22–87
23–48 22–89

Chapter 24

24–1 23–86
24–2 23–1 revised
24–3 23–2
24–5 23–5
24–6 23–4 revised
24–7 23–13 revised
24–8 23–8
24–9 23–9
24–10 23–6
24–11 23–11 revised
24–12 23–16
24–13 23–15
24–14 23–17
24–15 23–19
24–16 23–23
24–17 23–21
24–18 23–20 revised
24–19 23–27 revised
24–20 23–29
24–21 23–25 revised
24–22 23–30
24–23 23–84
24–24 23–28
24–25 23–85
24–26 23–32 revised
24–27 23–31
24–28 23–34 revised
24–29 23–39
24–30 23–35 revised

24–31	23–37 revised
24–32	23–38
24–33	23–41 revised
24–34	23–44
24–35	23–43
24–36	23–45
24–37	23–83
24–38	23–82
24–39	23–81
24–41	23–51 revised
24–42	23–87
24–43	23–49 revised
24–44	23–50 revised
24–45	23–47 revised
24–46	23–72
24–47	23–53
24–48	23–54

Chapter 25

25–1	24–1
25–2	24–2
25–4	24–4
25–5	24–3
25–6	24–5
25–8	24–9 revised
25–9	24–100 revised
25–10	24–7 revised
25–11	24–101
25–12	24–10
25–13	24–12
25–14	24–99
25–15	24–13 revised
25–16	24–14
25–17	24–17
25–18	24–11
25–19	24–97
25–20	24–98 revised
25–21	24–15 revised
25–22	24–19
25–23	24–96
25–24	24–21 revised
25–25	24–23 revised
25–26	24–22 revised
25–27	24–25 revised
25–28	24–28 revised
25–29	24–29 revised
25–30	24–30
25–31	24–31 revised
25–32	24–32 revised
25–33	24–114
25–34	24–36 revised
25–35	24–34 revised
25–36	24–39
25–37	24–37 revised
25–38	24–113
25–39	24–41
25–40	24–40 revised
25–41	24–94
25–43	24–43
25–44	24–42
25–45	24–93
25–46	24–44 revised
25–47	24–45
25–48	24–92
25–49	24–47
25–50	24–54
25–51	24–53
25–52	24–56 revised
25–53	24–55 revised
25–54	24–57
25–55	24–112 revised
25–56	24–58
25–57	24–60

Chapter 26

26–2	25–2
26–3	25–1
26–5	25–3 sig. figs.
26–6	25–4 revised
26–7	25–5 revised
26–8	25–6
26–10	25–8
26–11	25–7
26–12	25–10
26–13	25–9
26–14	25–11 sig. figs.
26–16	25–12 sig. figs.
26–17	25–13
26–18	25–15 revised
26–19	25–17
26–20	25–23 revised
26–21	25–21 revised
26–22	25–24
26–23	25–25
26–24	25–26
26–25	25–27
26–28	25–28 revised
26–29	25–31 revised
26–30	25–30 sig. figs.
26–31	25–74
26–32	25–32
26–34	25–34
26–35	25–35
26–36	25–36
26–37	25–37
26–38	25–40
26–39	25–39
26–40	25–42 revised
26–41	25–43 revised
26–42	25–44 revised
26–43	25–45
26–44	25–46
26–45	25–47 revised
26–46	25–48
26–48	25–50

Chapter 27

27–1	26–1 revised
27–2	26–2
27–3	26–3
27–4	26–6
27–5	26–5
27–6	26–4
27–7	26–7
27–8	26–8
27–9	26–55
27–10	26–11 revised
27–11	26–9
27–12	26–13
27–13	26–15
27–14	26–54
27–15	26–17
27–16	26–16 revised
27–17	26–19
27–18	26–79
27–19	26–21
27–20	26–18
27–21	26–23
27–22	26–31
27–23	26–27
27–24	26–29
27–25	26–25
27–26	26–28 revised
27–27	26–78 revised

27–28 26–30
27–29 26–33 revised
27–30 26–34
27–31 26–77 sig. figs.
27–32 26–36
27–33 26–35 revised
27–34 26–38
27–35 26–37
27–36 26–40
27–37 26–39
27–38 26–41
27–39 26–57
27–40 26–43 revised
27–41 26–76 revised
27–42 26–56
27–43 26–45 sig. figs.
27–44 26–48

Chapter 28

28–1 27–1
28–2 27–2
28–3 27–3 sig. figs.
28–5 27–5 revised
28–6 27–4 revised
28–7 27–7
28–8 27–114
28–9 27–9
28–10 27–6 revised
28–11 27–97
28–12 27–8
28–13 27–11
28–14 27–73
28–15 27–13 revised
28–16 27–14 revised
28–17 27–68
28–18 27–16 revised
28–19 27–15
28–20 27–17 revised
28–21 27–19 revised
28–22 27–74 revised
28–24 27–20 revised
28–25 27–21
28–26 27–18
28–27 27–29
28–28 27–30 sig. figs.
28–29 27–31 revised
28–30 27–32 revised
28–31 27–33 revised
28–32 27–36 revised
28–33 27–27 revised
28–34 27–25 revised
28–35 27–35 revised
28–36 27–38 sig. figs.
28–37 27–37
28–38 27–40 sig. figs.
28–39 27–39
28–40, 42 27–42 revised
28–41, 42 27–41 revised
28–43 27–43
28–44 27–44 revised
28–45 27–45 sig. figs.
28–46 27–46
28–47 27–47
28–48 27–50
28–49 27–55
28–50 27–67
28–51 27–49
28–52 27–52
28–53 27–51
28–54 27–57 revised
28–55 27–53 revised

Chapter 29

29–1 28–2 revised
29–2 28–76
29–3 28–1
29–4 28–3
29–5 28–85 revised
29–6 28–64 revised
29–7 28–80 revised
29–8 28–6 revised
29–9 28–7 sig. figs.
29–10 28–8
29–11 28–9 revised
29–12 28–12 revised
29–13 28–81
29–14 28–11
29–15 28–15 sigfigs
29–16 28–16
29–17 28–17
29–18 28–59
29–19 28–19
29–20 28–18
29–21 28–82
29–22 28–57
29–23 28–20
29–24 28–84 revised
29–25 28–25
29–26 28–83
29–27 28–23 sig. figs.
29–28 28–24
29–29 28–27 revised
29–30 28–28 sig. figs.
29–31 28–29 sig. figs.
29–32 28–30 revised
29–33 28–33 revised
29–34 28–32
29–35 28–35 revised
29–36 28–36 sig. figs.
29–37 28–37 revised
29–38 28–96
29–39 28–39 revised
29–40 28–40 revised
29–41 28–55 revised
29–42 28–86
29–43 28–41 revised
29–44 28–87
29–45 28–43
29–46 28–42 revised
29–47 28–45
29–48 28–46
29–49 28–47
29–50 28–48
29–51 28–49
29–52 28–77
29–53 28–51 revised
29–54 28–52 revised
29–55 28–53
29–56 28–62 revised
29–58 28–50
29–59 28–5

Chapter 30

30–1 29–1
30–3 29–3
30–4 29–79
30–5 29–11 revised
30–6 29–2
30–7 29–13 sig. figs.
30–8 29–4 revised
30–9 29–5 revised
30–10 29–6 revised
30–11 29–17 revised
30–12 29–81

30–13 29–19 revised
30–14 29–82
30–15 29–83
30–17 29–84
30–18 29–25 revised
30–19 29–23 revised
30–20 29–85 revised
30–21 29–7 revised
30–22 29–8 revised
30–23 29–9 revised
30–24 29–29 revised
30–25 29–15 revised
30–26 29–30 revised
30–27 29–31 revised
30–28 29–86
30–29 29–33 revised
30–30 29–34 revised
30–31 29–35 revised
30–32 29–37 revised
30–33 29–93
30–34 29–36 revised
30–35 29–39 revised
30–37 29–61 revised
30–38 29–38 revised
30–39 29–89
30–40 29–40
30–41 29–41
30–42 29–42
30–43 29–43
30–46 29–44
30–47 29–45
30–48 29–48
30–49 29–47
30–50 29–50 revised
30–51 29–49
30–54 29–52 revised
30–55 29–80
30–56 29–51 revised
30–57 29–53
30–58 29–21 revised

Chapter 31

31–2 30–1 revised
31–3 30–2 revised
31–4 30–3
31–5 30–4 revised
31–7 30–5
31–8 30–6
31–9 30–19
31–10 30–7
31–12 30–21 revised
31–13 30–13
31–14 30–14
31–15 30–9
31–16 30–18
31–17 30–11 revised
31–18 30–15 revised
31–19 30–17
31–20 30–8
31–21 30–23 revised
31–22 30–22
31–23 30–25
31–24 30–24 revised
31–25 30–27
31–26 30–26 revised
31–27 30–29
31–28 30–30
31–29 30–31 revised
31–30 30–32
31–31 30–33
31–32 30–34 revised
31–33 30–35
31–34 30–37
31–36 30–38
31–37 30–39
31–38 30–40 revised
31–39 30–41 revised
31–40 30–42
31–41 30–43
31–42 30–44
31–43 30–45 revised
31–44 30–46 revised
31–45 30–49 revised
31–46 30–48
31–47 30–51
31–48 30–50 revised
31–49 30–53
31–50 30–55
31–52 30–56
31–53 30–52 revised
31–54 30–89 revised
31–55 30–57 revised
31–56 30–58 revised
31–57 30–59
31–58 30–60 sig. figs.
31–59 30–61
31–60 30–62
31–62 30–64
31–63 30–63
31–65 30–65
31–66 30–66
31–67 30–67
31–68 30–70 revised
31–69 30–69
31–70 30–68
31–71 30–71
31–72 30–72
31–74 30–92
31–75 30–73 revised

Chapter 32

32–2 32–2
32–3 32–3 revised
32–4 32–28 revised
32–5 32–27
32–6 32–71
32–7 32–72 revised
32–8 32–73 revised
32–9 32–29
32–10 32–30
32–11 32–31 revised
32–12 32–34 revised
32–13 32–35
32–14 32–36
32–15 32–37
32–16 32–65 revised
32–17 32–39
32–18 32–38
32–19 32–41 revised
32–20 32–42
32–21 32–43
32–22 32–44
32–23 32–45
32–25 32–49
32–26 32–74 revised
32–27 32–5
32–28 32–7
32–29 32–13
32–30 32–12
32–31 32–15
32–32 32–14
32–33 32–17
32–34 32–16 revised
32–35 32–19 revised
32–37 32–21 revised

32–38 32–18 revised

Chapter 33

33–1 31–3
33–2 31–2
33–3 31–1
33–4 31–4
33–5 31–5 revised
33–6 31–6
33–7 31–7
33–8 31–10
33–9 31–9
33–10 31–8
33–11 31–11
33–12 31–20
33–13 31–13
33–14 31–12
33–15 31–15 revised
33–16 31–18 revised
33–17 31–17
33–18 31–16
33–19 31–19 revised
33–20 31–14
33–21 31–21
33–24 31–99
33–25 31–25
33–26 31–24 revised
33–27 31–26
33–29 31–27
33–30 31–28
33–31 31–29
33–32 31–30
33–33 31–31
33–34 31–32
33–35 31–33
33–36 31–34
33–37 31–35 revised
33–38 31–37 revised
33–39 31–39 revised
33–40 31–98 revised
33–41 31–43 revised
33–42 31–98
33–43 31–41
33–44 31–46
33–45 31–45 revised
33–46 31–112 revised
33–47 31–47
33–48 31–49
33–49 31–51
33–50 31–50
33–51 31–52
33–54 31–66
33–55 31–53
33–56 31–58 revised
33–57 31–55 revised
33–58 31–54
33–59 31–57 revised
33–60 31–56 revised
33–61 31–59 revised
33–62 31–60
33–63 31–61 revised
33–64 31–62 revised
33–65 31–64 revised

Chapter 34

34–1 33–89
34–2 33–3
34–3 33–2 revised
34–4 33–4
34–5 33–111
34–6 33–6
34–7 33–5
34–8 33–7
34–9 33–8 revised
34–11 33–9
34–12 33–10
34–14 33–11
34–15 33–12
34–16 33–13 revised
34–17 33–15
34–18 33–14
34–19 33–17 revised
34–20 33–20
34–21 33–19
34–22 33–22
34–23 33–21
34–24 33–24
34–25 33–23 revised
34–26 33–26 revised
34–27 33–25
34–29 33–27
34–30 33–28
34–31 33–106
34–33 33–35
34–34 33–34
34–35 33–33
34–36 33–37
34–37 33–31
34–38 33–32
34–39 33–39
34–40 33–41
34–41 33–43
34–42 33–44
34–43 33–45
34–44 33–84 revised
34–45 33–47 revised
34–46 33–51
34–47 33–49 revised
34–49 33–85
34–50 33–82
34–51 33–53
34–52 33–54
34–53 33–75 revised
34–54 33–55
34–55 33–61
34–58 33–62
34–57 33–110
34–58 33–62
34–59 33–63 revised
34–60 33–64
34–61 33–65
34–62 33–87 revised
34–63 33–69
34–65 33–66
34–66 33–1

Chapter 35

35–1 34–1
35–2 34–2
35–3, 35–4 34–102 revised
35–5 34–120
35–6 34–4
35–7 34–3 revised
35–8 34–5
35–9 34–7
35–10(a) 34–19 revised
35–10(b) 34–20 revised
35–10(c) 34–21 revised
35–10(d) 34–22 revised
35–10(e) 34–23 revised
35–10(f) 34–24 revised
35–10(g) 34–25 revised
35–10(h) 34–26 revised
35–11 34–110

35–12 34–17
35–13 34–32
35–14(a) 34–34 revised
35–14(b) 34–35 revised
35–14(c) 34–36 revised
35–14(e) 34–37 revised
35–14(f) 34–38 revised
35–14(g) 34–39 revised
35–14(h) 34–40 revised
35–15 34–112
35–16 34–139
35–17 34–126
35–18 34–41
35–19 34–43 revised
35–20 34–45
35–21 34–103
35–22 34–47
35–23 34–49
35–24(a) 34–74 revised
35–24(b) 34–75 revised
35–24(c) 34–76 revised
35–24(d) 34–77 revised
35–24(e) 34–66 revised
35–24(f) 34–67 revised
35–24(g) 34–68 revised
35–24(h) 34–78 revised
35–24(i) 34–79 revised
35–25 34–104
35–26 34–123 revised
35–27 34–123 revised
35–28 34–125 revised
35–29 34–105
35–30 34–58
35–31 34–106
35–32 34–88
35–33 34–89
35–34 34–90 revised
35–35 34–91
35–36 34–92
35–37 34–93

Chapter 36

36–1 35–2
36–2 35–1
36–3 35–3
36–4 35–116
36–5 35–85
36–6 35–8 revised
36–7 35–7
36–8 35–11 revised
36–9 35–13
36–10 35–9
36–11 35–16
36–12 35–133
36–13 35–19
36–14 35–14
36–15 35–15
36–16 35–20
36–17 35–17 revised
36–18 35–23 revised
36–19 35–21
36–20 35–25 revised
36–21 35–27
36–22 35–117 revised
36–23 35–29
36–24 35–30
36–25 35–31
36–26 35–131
36–27 35–130 revised
36–28 35–129
36–29 35–121 revised
36–30 35–5 revised
36–31 35–134
36–32 35–6 revised
36–33 35–35 revised
36–34 35–113 revised
36–35 35–37
36–36 35–132
36–37 35–39
36–39 35–55
36–40 35–54
36–41 35–112
36–42 35–53
36–43 35–69
36–45 35–71
36–46 35–40
36–47 35–114
36–48 35–72
36–49 35–75
36–50 35–76
36–51 35–77
36–52 35–123
36–53 35–91
36–54 35–80
36–55 35–79
36–56 35–82
36–57 35–81
36–58 35–115 revised
36–59 35–122 revised

Chapter 37

37–1 36–1
37–2 36–2
37–3 36–3 revised
37–4 36–6
37–5 36–5
37–6 36–65 revised
37–7 36–7
37–8 36–9
37–9 36–105
37–10 36–11
37–11 36–13 revised
37–12 36–14
37–13 36–15 revised
37–14 36–16 revised
37–15 36–17
37–16 36–66
37–17 36–19
37–18 36–20
37–19 36–21
37–20 36–22
37–21 36–26
37–22 36–67
37–23 36–25 revised
37–24 36–28
37–25 36–27
37–26 36–68 revised
37–27 36–29
37–28 36–30
37–29 36–31
37–30 36–32
37–31 36–33
37–32 36–35
37–33 36–37
37–34 36–38 revised
37–35 36–39
37–36 36–40
37–37 36–41 revised
37–38 36–42
37–39 36–43
37–40 36–44
37–41 36–69
37–42 36–108
37–43 36–107
37–44 36–45

37–45 36–47
37–46 36–50
37–47 36–51
37–48 36–52
37–49 36–103
37–50 36–53
37–51 36–109
37–52 36–54 revised
37–53 36–55
37–54 36–57 revised
37–55 36–58
37–56 36–59
37–57 36–110
37–58 36–60
37–59 36–61 revised
37–60 36–62 revised
37–61 36–63 revised
37–62 36–4
37–63 36–18

Chapter 38

38–3 37–1 sig. figs.
38–4 37–2 revised
38–5 37–5
38–6 37–7 sig. figs.
38–7 37–11
38–8 37–8
38–9 37–10
38–10 37–12 revised
38–11 37–9
38–12 37–15 revised
38–13 37–13
38–14 37–16 revised
38–15 37–17 revised
38–16 37–18 revised
38–17 37–19
38–18 37–23 revised
38–19 37–21
38–20 37–20 revised
38–21 37–27 revised
38–22 37–30 revised
38–23 37–29 revised
38–24 37–28 revised
38–25 37–31
38–26 37–33
38–27 37–35
38–28 37–34
38–29 37–36 revised
38–30 37–37
38–31 37–39 revised
38–32 37–40
38–33 37–48 revised
38–34 37–50 revised
38–35 37–41
38–36 37–42
38–37 37–49
38–38 37–43
38–39 37–51
38–40 37–52 revised
38–41 37–47 sig. figs.
38–42 37–54 revised
38–43 37–53
38–44 37–58 revised
38–45 37–55
38–51 37–59
38–52 37–65
38–53 37–63

Chapter 39

39–1 38–82
39–2 38–1
39–3 38–83
39–4 38–2
39–5 38–3
39–6 38–4
39–7 38–67
39–8 38–6
39–9 38–5
39–10 38–8
39–11 38–7
39–12 38–12
39–13 38–9
39–14 38–10
39–15 38–11
39–16 38–68 revised
39–17 38–14
39–18 38–81 revised
39–19 38–13
39–20 38–16
39–21 38–15
39–22 38–17
39–23 38–19
39–24 38–20
39–25 38–18
39–26 38–22
39–27 38–21
39–28 38–66
39–29 38–24
39–30 38–23
39–31 38–25
39–32 38–26 revised
39–33 38–27 revised
39–34 38–28 revised
39–35 38–79
39–36 38–30
39–37 38–29
39–38 38–32
39–39 38–33
39–40 38–31
39–41 38–80
39–42 38–78
39–43 38–37
39–44 38–35 revised
39–45 38–34
39–46 38–71
39–47 38–38
39–48 38–36
39–49 38–76
39–50 38–75
39–51 38–39
39–52 38–40
39–53 38–41
39–54 38–42
39–55 38–70
39–56 38–44 revised
39–57 38–43 revised
39–58 38–69
39–59 38–45
39–60 38–46 revised
39–61 38–47 revised
39–62 38–48 revised
39–63 38–49
39–64 38–50
39–65 38–51 revised
39–66 38–52
39–67 38–53
39–68 38–54
39–69 38–55
39–70 38–77
39–71 38–57
39–72 38–56
39–73 38–58 revised
39–74 38–74
39–75 38–73
39–76 38–59 revised

39–77 38–61
39–78 38–60 revised
39–79 38–63 revised
39–80 38–62
39–81 38–65
39–82 38–64
39–83 38–72

Chapter 40

40–1 39–2 revised
40–2 39–1 revised
40–3 39–3
40–4 39–5
40–5 39–4
40–6 39–7
40–7 39–6
40–8 39–8 revised
40–9 39–49
40–10 39–10 revised
40–11 39–9 revised
40–12 39–11 revised
40–13 39–50
40–14 39–
40–15 39–54
40–16 39–14
40–17 39–13
40–18 39–55
40–19 39–15 revised
40–20 39–16
40–21 39–53 revised
40–22 39–52 revised
40–23 39–17
40–24 39–18
40–25 39–19
40–26 39–20
40–27 39–21 revised
40–28 39–22
40–29 39–23 revised
40–30 39–56
40–31 39–26
40–32 39–24
40–33 39–25
40–34 39–28
40–35 39–27
40–36 39–58
40–37 39–29
40–38 39–31
40–39 39–30
40–40 39–32 revised
40–41 39–33
40–42 39–34
40–43 39–57 revised
40–44 39–36 revised
40–45 39–35
40–46 39–38 revised
40–47 39–40 revised
40–48 39–59
40–49 39–39
40–50 39–44 revised
40–51 39–37
40–52 39–42
40–53 39–41
40–54 39–51
40–55 39–45
40–56 39–46
40–57 39–47
40–58 39–48
40–59 39–43 revised

Chapter 41

41–1 40–70
41–2 40–2
41–3 40–1
41–4 40–3
41–5 40–4
41–6 40–6 revised
41–7 40–5 revised
41–8 40–7 revised
41–9 40–69
41–10 40–8
41–11 40–9 revised
41–12 40–10
41–13 40–11
41–14 40–68
41–15 40–13 revised
41–16 40–12 revised
41–17 40–15
41–18 40–14
41–19 40–17
41–20 40–16 revised
41–21 40–57 revised
41–22 40–18 revised
41–23 40–19 revised
41–24 40–20 revised
41–25 40–21 revised
41–26 40–22 revised
41–27 40–23 revised
41–28 40–24 revised
41–29 40–66
41–30 40–25 revised
41–31 40–67
41–32 40–26 revised
41–33 40–27 revised
41–34 40–28 revised
41–35 40–29
41–36 40–30
41–37 40–65
41–38 40–61
41–39 40–31 revised
41–40 40–32 revised
41–41 40–35
41–42 40–36 revised
41–43 40–33
41–44 40–34
41–45 40–37
41–46 40–38
41–47 40–40
41–48 40–62 revised
41–49 40–39 revised
41–50 40–41
41–51 40–42
41–52 40–43 revised
41–53 40–63
41–54 40–44
41–55 40–47
41–56 40–64
41–57 40–46
41–58 40–48
41–59 40–45
41–60 40–49
41–61 40–50
41–62 40–51
41–63 40–52
41–64 40–54
41–65 40–53
41–66 40–55
41–67 40–59
41–68 40–56
41–69 40–58
41–70 40–60

Chapter 42

42–1 41–1
42–2 41–49

42–3 41–47
42–4 41–2
42–5 41–46
42–6 41–4
42–7 41–3
42–8 41–48
42–9 41–5
42–10 41–6
42–11 41–7
42–12 41–10
42–13 41–8
42–14 41–11 sig. figs.
42–15 41–9
42–16 41–44
42–17 41–12 sig. figs.
42–18 41–13 revised
42–19 41–18 revised
42–20 41–16 revised
42–21 41–15 revised
42–22 41–14
42–23 41–17
42–24 41–20
42–25 41–19
42–26 41–22
42–27 41–41 revised
42–28 41–45
42–29 41–21
42–30 41–23
42–31 41–26
42–32 41–24
42–33 41–25 sig. figs.
42–34 41–42
42–35 41–27 sig. figs.
42–36 41–28 revised
42–37 41–43
42–38 41–29
42–39 41–30
42–40 41–32
42–41 41–31
42–42 41–34 revised
42–43 41–33 revised
42–44 41–36
42–45 41–35
42–46 41–38
42–47 41–37
42–48 41–39
42–49 41–40 revised

Chapter 43

43–1 42–72
43–2 42–1
43–3 42–3
43–4 42–87
43–5 42–82
43–6 42–7
43–7 42–4
43–8 42–84
43–9 42–83
43–10 42–6 revised
43–11 42–8 revised
43–12 42–9 revised
43–13 42–15 revised
43–14 42–10 revised
43–15 42–11
43–16 42–16
43–17 42–73 revised
43–18 42–18 revised
43–19 42–17
43–20 42–14 revised
43–21 42–13 revised
43–22 42–5 revised
43–23 42–16
43–24 42–18 revised
43–25 42–19 revised
43–26 42–21 sig. figs.
43–27 42–22
43–28 42–24
43–29 42–23
43–30 42–80
43–31 42–25
43–32 42–26
43–33 42–27
43–34 42–33
43–35 42–31
43–36 42–67
43–37 42–34
43–38 42–36
43–39 42–35
43–40 42–78
43–41 42–37
43–42 42–38
43–43 42–39 revised
43–44 42–70 revised
43–45 42–69
43–46 42–81
43–47 42–41 revised
43–48 42–40
43–49 42–43
43–50 42–42
43–51 42–44 revised
43–52 42–52 revised
43–53 42–77
43–54 42–46
43–55 42–45
43–56 42–48
43–57 42–47
43–58 42–58
43–59 42–49
43–60 42–50
43–61 42–51
43–62 42–52
43–63 42–53 revised
43–64 42–54
43–65 42–55
43–66 42–58 revised
43–67 42–57
43–69 42–59
43–70 42–60 revised
43–71 42–61
43–72 42–62
43–73 42–63
43–74 42–64
43–75 42–65
43–76 42–66
43–77 42–20
43–78 42–29
43–79 42–32
43–80 42–68
43–81 42–30
43–82 42–79
43–83 42–56
43–84 42–28
43–85 42–2
43–86 42–12
43–87 42–75
43–88 42–74

Chapter 44

44–1 43–3
44–2 43–2 revised
44–3 43–1
44–4 43–4
44–5 43–53
44–6 43–6
44–7 43–5

44–8 43–7
44–9 43–8
44–10 43–52
44–11 43–9 revised
44–12 43–12
44–13 43–11
44–14 43–10
44–15 43–13
44–16 43–51
44–17 43–14
44–18 43–16
44–19 43–15
44–20 43–17
44–21 43–49
44–22 43–18
44–23 43–19 revised
44–24 43–22
44–25 43–23 sig. figs.
44–26 43–20
44–27 43–21 revised
44–28 43–25
44–29 43–24
44–30 43–26
44–31 43–27
44–32 43–54
44–33 43–29
44–34 43–28
44–35 43–30
44–36 43–31
44–37 43–32
44–38 43–50
44–41 43–47
44–42 43–36
44–43 43–35
44–44 43–38
44–45 43–37
44–46 43–40
44–47 43–39
44–48 43–42
44–49 43–41
44–50 43–46
44–51 43–43
44–52 43–44
44–53 43–45
44–54 43–48

Chapter 45

45–1 44–44
45–2 44–2
45–3 44–1
45–4 44–4
45–5 44–37
45–6 44–3
45–7 44–5
45–8 44–8
45–9 44–7
45–10 44–6
45–11 44–9
45–12 44–42
45–13 44–11
45–14 44–10 revised
45–15 44–13
45–16 44–12
45–17 44–15
45–18 44–14
45–19 44–18
45–20 44–16
45–21 44–17 revised
45–22 44–19 revised
45–23 44–21
45–24 44–20
45–25 44–23
45–26 44–22
45–27 44–24
45–28 44–39
45–29 44–25 revised
45–30 44–26
45–31 44–27
45–32 44–28
45–33 44–30
45–34 44–29
45–35 44–31
45–36 44–32 revised
45–37 44–33
45–38 44–34
45–39 44–35
45–40 44–36
45–41 44–41
45–42 44–38 revised
45–43 44–43
45–44 44–40

Enhanced Edition Problems

Chapter 1

1–N1 1–51
1–N2 1–26
1–N3 1–50
1–N4 1–24
1–N5 1–49
1–N6 1–29
1–N7 1–46
1–N8 1–11
1–N9 1–41
1–N10 1–2
1–N11 1–52
1–N12 1–45 revised
1–N13 1–44
1–N14 1–6 revised
1–N15 1–54
1–N16 1–15
1–N17 1–43
1–N18 1–28 revised
1–N19 1–55

Chapter 2

2–N1 2–30
2–N2 2–65
2–N3 2–44
2–N4 2–45
2–N5 2–100
2–N6 2–48
2–N7 2–1
2–N8 2–46
2–N10 2–8
2–N11 2–66
2–N12 2–35
2–N13 2–107
2–N14 2–75
2–N15 2–69
2–N16 2–33
2–N17 2–115
2–N18 2–99
2–N19 2–24
2–N20 2–32
2–N21 2–67
2–N23 2–17
2–N24 2–108
2–N29 2–109

2–N32 2–36

Chapter 3

3–N1 3–24
3–N2 3–40 revised
3–N3 3–21
3–N4 3–22
3–N5 3–41
3–N6 3–42
3–N7 3–43
3–N8 3–44 revised
3–N9 3–27
3–N10 3–33
3–N11 3–36
3–N12 3–52
3–N13 3–73
3–N14 3–50
3–N15 3–53
3–N16 3–56
3–N17 3–57
3–N18 3–58
3–N19 3–47
3–N20 3–23
3–N21 3–30
3–N22 3–55
3–N23 3–54
3–N24 3–60
3–N25 3–16
3–N26 3–61

Chapter 4

4–N1 4–67
4–N2 4–75
4–N3 4–122
4–N4 4–54
4–N5 4–121
4–N6 4–64
4–N8 4–52 revised
4–N9 4–120
4–N10 4–38
4–N11 4–102
4–N12 4–78
4–N13 4–62
4–N14 4–40
4–N15 4–125
4–N16 4–71 revised
4–N17 4–60 revised
4–N18 4–51
4–N19 4–127
4–N20 4–46
4–N21 4–113
4–N22 4–57
4–N23 4–50 revised
4–N24 4–70
4–N25 4–84
4–N26 4–76
4–N27 4–72 revised
4–N28 4–73
4–N29 4–124
4–N30 4–74
4–N31 4–114
4–N32 4–35
4–N33 4–49
4–N34 4–83 revised
4–N35 4–11
4–N36 4–13
4–N37 4–12 revised

Chapter 5

5–N1 5–57
5–N2 5–79 revised
5–N3 5–65 revised
5–N4 5–44
5–N5 5–58
5–N6 5–36
5–N7 5–6
5–N8 5–59
5–N9 5–30
5–N10 5–15
5–N11 5–26
5–N12 5–11
5–N13 5–90
5–N14 5–38
5–N16 5–46
5–N17 5–66
5–N18 5–8
5–N19 5–91
5–N20 5–52
5–N21 5–92
5–N22 5–55
5–N23 5–104
5–N22 5–55

Chapter 6

6–N1 6–55 revised
6–N2 6–54
6–N3 6–57 revised
6–N4 6–56
6–N5 6–6
6–N6 6–58
6–N7 6–59
6–N8 6–60
6–N9 6–61 revised
6–N10 6–18
6–N11 6–62 revised
6–N12 6–63
6–N13 6–64
6–N14 6–24
6–N15 6–65 revised
6–N16 6–48
6–N17 6–43
6–N18 6–41
6–N19 6–52
6–N20 6–66 revised
6–N21 6–84
6–M22 6–42
6–N23 6–103 revised
6–N24 6–69 revised
6–N25 6–16
6–N26 6–70
6–N27 6–71
6–N28 6–23 revised
6–N29 6–72
6–N30 6–73
6–N31 6–39
6–N32 6–9
6–N33 5–74 revised
6–N34 5–75

Chapter 7

7–N1 7–37
7–N2 7–54
7–N3 7–76
7–N4 7–30
7–N5 7–7
7–N6 7–55
7–N7 7–56
7–N8 7–80
7–N9 7–47
7–N10 7–21

7–N11 7–57
7–N12 7–18
7–N13 7–75
7–N14 7–12
7–N15 7–74
7–N16 7–58
7–N18 7–59
7–N19 7–73
7–N20 7–60

Chapter 8

8–N1 8–60
8–N2 8–53
8–N3 8–104
8–N4 8–27
8–N5 8–102
8–N6 8–82
8–N7 8–130
8–N8 8–80 revised
8–N9 8–81
8–N10 8–101
8–N11 8–79
8–N12 8–100
8–N13 8–99 revised
8–N14 8–134
8–N15 8–98
8–N16 8–42
8–N17 8–63 revised
8–N18 8–78
8–N19 8–109
8–N20 8–97
8–N21 8–96
8–N22 8–129
8–N23 8–95
8–N24 8–128
8–N25 8–132
8–N26 8–77
8–N27 8–135
8–N28 8–67
8–N29 8–61 revised

Chapter 9

9–N1 9–42
9–N2 9–77
9–N3 9–91 revised
9–N4 9–13
9–N5 9–78
9–N6 9–116
9–N8 9–138 revised
9–N10 9–38
9–N11 9–118
9–N13 9–2 revised
9–N14 9–21
9–N15 9–120 revised
9–N16 9–14
9–N17 9–83
9–N18 9–95
9–N19 9–122
9–N20 9–140
9–N22 9–4

Chapter 10

10–N1 9–62 revised
10–N2 9–64
10–N3 9–34
10–N4 9–75
10–N6 9–76
10–N8 9–58 revised
10–N9 9–115
10–N10 9–117 revised
10–N11 9–94
10–N12 9–47 revised
10–N13 9–86
10–N14 9–92
10–N15 9–28
10–N16 9–31 revised
10–N17 9–134
10–N19 9–119 revised
10–N20 9–90
10–N21 9–84
10–N22 9–74
10–N23 9–52
10–N24 9–61

Chapter 11

11–N1 10–38
11–N2 10–72
11–N3 10–56
11–N4 10–52
11–N5 10–36
11–N6 10–3 revised
11–N7 10–14
11–N8 10–8
11–N9 10–40 revised
11–N10 10–20 revised
11–N11 10–62
11–N12 10–65
11–N13 10–105
11–N14 10–12
11–N15 10–34
11–N16 10–51
11–N17 10–71
11–N18 10–100
11–N19 10–82
11–N20 10–68 revised
11–N21 10–103 revised
11–N22 10–102 revised

Chapter 12

12–N1 11–73
12–N2 11–44
12–N3 11–33
12–N4 11–13
12–N5 11–63 sig. figs.
12–N6 11–14
12–N7 11–38
12–N8 11–12
12–N9 11–11
12–N10 11–66 sig. figs.
12–N11 11–91
12–N12 11–10
12–N13 11–65
12–N14 11–29
12–N15 11–23
12–N16 11–36 sig. figs.
12–N17 11–99
12–N18 11–53
12–N19 11–72 sig. figs.
12–N20 11–50
12–N21 11–28

Chapter 13

13–N1 12–34
13–N2 12–46
13–N4 12–30
13–N5 12–43 revised
13–N6 12–48 revised
13–N7 12–78

13–N8 12–17
13–N9 12–70
13–N10 12–1
13–N11 12–62 revised
13–N12 12–44
13–N13 12–77 revised
13–N14 12–16 sig. figs.
13–N15 12–49
13–N16 12–61
13–N17 12–38
13–N18 12–18 revised
13–N19 12–58
13–N20 12–60
13–N21 12–54 revised

Chapter 14

14–N1 13–63
14–N2 13–28
14–N3 13–65
14–N4 13–37 revised
14–N6 13–29
14–N7 13–7 revised
14–N8 13–12 revised
14–N9 13–13 revised
14–N10 13–10 revised
14–N11 13–93
14–N12 13–95
14–N13 13–91
14–N14 13–8
14–N16 13–64
14–N17 13–68
14–N18 13–36
14–N19 13–22
14–N20 13–30 revised
14–N21 13–96
14–N22 13–90

Chapter 15

15–N1 14–56
15–N2 14–58 revised
15–N4 14–50
15–N5 14–72
15–N6 14–30
15–N7 14–65
15–N8 14–32
15–N9 14–8
15–N10 14–36
15–N11 14–17
15–N12 14–13
15–N13 14–40
15–N14 14–67
15–N15 14–90
15–N16 14–23
15–N18 14–89
15–N19 14–85
15–N20 14–34
15–N22 14–68
15–N23 14–88
15–N24 14–28
15–N25 14–66

Chapter 16

16–N1 15–10
16–N2 15–12
16–N3 15–64
16–N4 15–34
16–N5 15–61 revised
16–N6 15–18
16–N7 15–32
16–N8 15–52
16–N9 15–30 revised
16–N10 15–65
16–N11 15–42 sig. figs.
16–N12 15–50 revised
16–N13 15–66
16–N14 15–54 sig. figs.
16–N15 15–74
16–N16 15–73
16–N17 15–67
16–N18 15–68
16–N19 15–112
16–N20 15–16
16–N21 15–56
16–N22 15–36
16–N24 15–53
16–25N 15–69 sig. figs.
16–N26 15–111

Chapter 17

17–N1 16–9 revised
17–N2 16–10
17–N3 16–57
17–N4 16–4
17–N5 16–58
17–N6 16–63
17–N7 16–8
17–N8 16–25
17–N9 16–3
17–N10 16–35
17–N11 16–65
17–N12 16–62 revised
17–N13 16–37
17–N14 16–36
17–N15 16–64
17–N16 16–52
17–N17 16–32
17–N18 16–31 revised
17–N19 16–56
17–N20 16–54
17–N21 16–46

Chapter 18

18–N2 17–12 sig. figs.
18–N3 17–13 revised
18–N4 17–27 revised
18–N5 17–56 revised
18–N6 17–65 revised
18–N7 17–29
18–N8 17–58
18–N9 17–68
18–N10 17–10
18–N11 17–15
18–N12 17–100
18–N13 17–31
18–N14 17–99
18–N16 17–18 revised
18–N17 17–98 revised
18–N18 17–40
18–N19 17–34
18–N20 17–69
18–N22 17–74
18–N24 17–70 revised
18–N25 17–71 revised
18–N26 17–72 sig. figs.
18–N28 17–90
18–N30 17–62
18–N31 17–89 sig. figs.
18–N32 17–33

Chapter 19

19–N2 18–34
19–N4 18–68
19–N5 18–101
19–N6 18–32
19–N8 18–66
19–N10 18–58 sig. figs.
19–N12 18–36
19–N14 18–65
19–N17 18–64
19–N18 18–93
19–N19 18–100
19–N20 18–47
19–N22 18–63 revised
19–N23 18–79
19–N24 18–46
19–N25 18–89

Chapter 20

20–N2 19–10
20–N4 19–34
20–N5 19–60
20–N6 19–77
20–N7 19–70
20–N8 19–71 revised
20–N9 19–63
20–N10 19–8
20–N11 19–64
20–N12 19–72
20–N14 19–73
20–N16 19–57
20–N18 19–74
20–N19 19–56
20–N20 19–75
20–N21 19–47 revised
20–N22 19–76 sig. figs.

Chapter 21

21–N1 20–50
21–N2 20–8
21–N3 20–12
21–N4 20–28
21–N6 20–49 revised
21–N8 20–10
21–N10 20–40
21–N12 20–55
21–N13 20–74
21–N14 20–73
21–N16 20–56 revised
21–N17 20–53
21–N18 20–72
21–N19 20–71
21–N20 20–48
21–N21 20–70
21–N22 20–57
21–N23 20–47

Chapter 22

22–N1 21–38
22–N2 21–37 revised
22–N3 21–27
22–N4 21–35 revised
22–N5 21–12 revised
22–N6 21–20 revised
22–N7 21–44
22–N8 21–36
22–N9 21–71 revised
22–N10 21–14
22–N11 21–34
22–N12 21–33 revised
22–N13 21–45 revised
22–N14 21–32
22–N15 21–28 revised
22–N16 21–18 revised
22–N17 21–30
22–N18 21–40
22–N19 21–69
22–N20 21–56

Chapter 23

23–N1 22–33
23–N2 22–54
23–N3 22–23
23–N4 22–14
23–N5 22–56
23–N6 22–58 revised
23–N7 22–67 revised
23–N8 22–62
23–N9 22–60
23–N10 22–28
23–N11 22–88
23–N12 22–18
23–N13 22–83 revised
23–N14 22–16
23–N15 22–69 revised
23–N16 22–20
23–N17 22–63
23–N18 22–52
23–N19 22–70
23–N20 22–48
23–N21 22–49
23–N22 22–46
23–N24 22–66 sig. figs.
23–N25 22–13
23–N26 22–64
23–N27 22–61 revised
23–N28 22–32
23–N30 22–68

Chapter 24

24–N1 23–12
24–N2 23–18
24–N3 23–40 revised
24–N4 23–10 revised
24–N5 23–26
24–N6 23–48
24–N7 23–61
24–N8 23–60 revised
24–N9 23–62
24–N10 23–58
24–N11 23–59 revised
24–N12 23–46
24–N13 23–56 revised
24–N14 23–24
24–N15 23–66
24–N16 23–42
24–N17 23–80 revised
24–N18 23–33
24–N19 23–52
24–N20 23–64 revised
24–N21 23–22
24–N22 23–36

Chapter 25

25–N1 24–49
25–N2 24–18 revised
25–N3 24–16 revised
25–N4 24–48
25–N5 24–67

25–N6 24–20
25–N7 24–69
25–N8 24–50
25–N9 24–68
25–N10 24–26
25–N11 24–70 revised
25–N12 24–71 sig. figs.
25–N13 24–77
25–N14 24–78
25–N15 24–79
25–N16 24–38
25–N17 24–46 revised
25–N18 24–59
25–N19 24–95
25–N20 24–62 revised
25–N21 24–118
25–N22 24–64 revised
25–N23 24–102
25–N24 24–35 revised
25–N25 24–72
25–N26 24–66
25–N27 24–73
25–N28 24–51
25–N29 24–103
25–N30 24–52
25–N31 24–80
25–N32 24–27 revised
25–N33 24–81
25–N34 24–61
25–N35 24–117 revised
25–N37 24–104

Chapter 26

26–N1 25–51
26–N2 25–52 sig. figs.
26–N3 25–14 revised
26–N4 25–22
26–N5 25–18
26–N6 25–19
26–N7 25–53 revised
26–N8 25–20 revised
26–N9 25–38 revised
26–N10 25–54
26–N11 25–61 sig. figs.
26–N12 25–55 revised
26–N13 25–62
26–N14 25–16
26–N15 25–63
26–N16 25–56 revised
26–N17 25–64
26–N18 25–58
26–N19 25–81
26–N20 25–65 sig. figs.
26–N22 25–66

Chapter 27

27–N2 26–32 revised
27–N4 26–22
27–N5 26–72
27–N6 26–53
27–N7 26–71 revised
27–N8 26–20
27–N9 26–70
27–N10 26–24
27–N12 26–46 sig. figs.
27–N14 26–69
27–N16 26–10
27–N18 26–26
27–N19 26–68
27–N20 26–42
27–N21 26–44
27–N22 26–50
27–N23 26–67 sig. figs.
27–N24 26–66
27–N25 26–75
27–N26 26–65

Chapter 28

28–N1 27–22 revised
28–N2 27–34 revised
28–N3 27–103
28–N4 27–24
28–N6 27–10 revised
28–N8 27–48
28–N10 27–69
28–N11 27–65
28–N12 27–12
28–N13 27–75 revised
28–N14 27–60
28–N15 27–76
28–N16 27–64 sig. figs.
28–N17 27–77 revised
28–N18 27–28
28–N19 27–78 revised
28–N20 27–79 revised
28–N21 27–80 revised
28–N22 27–62 revised
28–N23 27–26 revised
28–N24 27–112
28–N25 27–111 revised
28–N26 27–56
28–N27 27–113
28–N28 27–102 sig. figs.

Chapter 29

29–N1 28–4
29–N2 28–10 revised
29–N3 28–54
29–N4 28–44
29–N6 28–60
29–N8 28–22
29–N9 28–89
29–N10 28–21
29–N12 28–58
29–N14 28–31
29–N15 28–67
29–N16 28–66
29–N17 28–26
29–N18 28–56
29–N20 28–14

Chapter 30

30–N1 29–16 revised
30–N2 29–10
30–N3 29–64
30–N4 29–26 revised
30–N5 29–22
30–N6 29–62 revised
30–N7 29–12
30–N8 29–56
30–N9 29–78
30–N10 29–18
30–N11 29–77
30–N12 29–54
30–N13 29–59
30–N14 29–14
30–N15 29–76 revised
30–N16 29–24
30–N17 29–60 revised
30–N18 29–27

30–N20 29–32 revised
30–N22 29–20
30–N24 29–46
30–N25 29–58

Chapter 31

31–N1 30–74 revised
31–N2 30–76
31–N3 30–10
31–N4 30–78 sig. figs.
31–N5 30–20
31–N6 30–54 sig. figs.
31–N7 30–80
31–N8 30–28
31–N9 30–12 revised
31–N10 30–82
31–N11 30–84 sig. figs.
31–N12 30–86
31–N13 30–94
31–N14 30–96 sig. figs.
31–N15 30–88 sig. figs.
31–N16 30–36
31–N17 30–91 sig. figs.
31–N18 30–90
31–N19 30–75
31–N20 30–16 revised
31–N21 30–77 revised

Chapter 32

32–N1 32–1
32–N2 32–22 revised
32–N3 32–32
32–N4 32–51
32–N5 32–50
32–N6 32–48
32–N7 32–8
32–N8 32–9
32–N9 32–10 sig. figs
32–N10 32–11
32–N11 32–20
32–N12 32–46
32–N13 32–54
32–N15 32–56 revised
32–N16 32–40
32–N17 32–23
32–N18 32–24
32–N19 32–25
32–N20 32–26

Chapter 33

33–N1 31–76
33–N2 31–36
33–N3 31–74
33–N4 31–38
33–N5 31–72
33–N6 31–70
33–N7 31–68
33–N8 31–48
33–N10 31–40
33–N11 31–78
33–N12 31–42
33–N13 31–80
33–N14 31–22
33–N15 31–23
33–N16 31–81
33–N20 31–84
33–N21 31–100
33–N22 31–86
33–N23 31–97
33–N24 31–44
33–N25 31–88
33–N27 31–90

Chapter 34

34–N2 33–46
34–N3 33–95
34–N4 33–48
34–N6 33–50
34–N8 33–70
34–N9 33–94
34–N10 33–29
34–N11 33–100
34–N12 33–36
34–N13 33–97
34–N14 33–60
34–N16 33–38
34–N17 33–81 revised
34–N18 33–42
34–N19 33–40
34–N20 33–18
34–N21 33–56
34–N22 33–52
34–N23 33–58
34–N24 33–86
34–N25 33–16
34–N26 33–30
34–N27 33–72
34–N28 33–98
34–N29 33–67
34–N30 33–102
34–N31 33–107

Chapter 35

35–N1 34–6
35–N2 34–8
35–N3 34–9 revised
35–N4 34–11 revised
35–N5 34–13 revised
35–N6 34–15 revised
35–N7 34–12 revised
35–N8 34–16 revised
35–N9 34–14 revised
35–N10 34–10 revised
35–N12 34–96
35–N13 34–29 revised
35–N14 34–31 revised
35–N15 34–27 revised
35–N16 34–30 revised
35–N17 34–28 revised
35–N18 34–18
35–N24 34–42
35–N25 34–50
35–N26 34–51
35–N27 34–52
35–N28 34–53
35–N29 34–54
35–N30 34–55
35–N31 34–56
35–N32 34–57
35–N33 34–59
35–N34 34–60
35–N35 34–61
35–N36 34–62
35–N37 34–63
35–N38 34–64
35–N39 34–65
35–N40 34–44
35–N41 34–46
35–N42 34–48
35–N43 34–69 revised

35–N44 34–70 revised
35–N45 34–71 revised
35–N46 34–72 revised
35–N47 34–73 revised
35–N48 34–80
35–N49 34–81
35–N50 34–82
35–N51 34–83
35–N52 34–84
35–N53 34–85
35–N54 34–86
35–N55 34–87
35–N56 34–109
35–N57 34–121 revised
35–N58 34–122
35–N59 34–127
35–N60 34–136
35–N61 34–140
35–N62 34–113
35–N63 34–114
35–N64 34–115
35–N65 34–116
35–N66 34–117
35–N67 34–118

Chapter 36

36–N1 35–36
36–N2 35–78
36–N3 35–135
36–N4 35–86
36–N5 35–127
36–N6 35–26
36–N7 35–126
36–N8 35–88
36–N9 35–124
36–N10 35–34
36–N12 35–32
36–N14 35–4
36–N15 35–98
36–N16 35–22 revised
36–N17 35–70
36–N18 35–74
36–N19 35–38 revised
36–N20 35–18
36–N21 35–100
36–N22 35–24
36–N23 35–12
36–N24 35–96
36–N25 35–94
36–N26 35–56
36–N27 35–90
36–N28 35–92
36–N29 35–10 revised
36–N30 35–84
36–N31 35–28

Chapter 37

37–N1 36–102
37–N2 36–12
37–N3 36–24
37–N4 36–10
37–N5 36–101
37–N6 36–36 sig. figs.
37–N7 36–106
37–N8 36–34
37–N9 36–104
37–N10 36–100
37–N11 36–99
37–N12 36–46
37–N13 36–113
37–N14 36–98
37–N15 36–112
37–N16 36–97
37–N17 36–95
37–N18 36–96
37–N19 36–56
37–N20 36–94
37–N21 36–93
37–N22 36–64
37–N23 36–92
37–N24 36–91
37–N25 36–111
37–N26 36–90
37–N27 36–89
37–N28 36–114

Chapter 38

38–N1 37–25 revised
38–N2 37–26 revised
38–N3 37–70
38–N4 (part) 37–60 revised
38–4 (part) 37–61 revised
38–N5 37–46
38–N6 37–22
38–N8 37–62
38–N9 37–56 revised
38–N10 37–4 revised
38–N11 37–67
38–N12 37–68
38–N13 37–69
38–N14 37–24
38–N15 37–32
38–N16 37–14
38–N17 37–6
38–N18 37–64

Supplement Problems

Chapter 1

1–31 1–12
1–32 1–42
1–33 1–40
1–34 1–22
1–35 1–59
1–36 1–39 revised
1–37 1–38
1–39 1–37
1–40 1–57 revised
1–41 1–30 revised
1–43 1–56
1–44 1–35 revised

Chapter 2

2–66 2–81
2–67 2–74
2–68 2–113
2–69 2–76
2–70 2–77
2–71 2–114
2–72 2–78
2–73 2–79
2–74 2–54
2–75 2–80
2–76 2–101
2–77 2–14
2–78 2–82 revised
2–79 2–34
2–80 2–4
2–81 2–73 revised
2–82 2–83

2–83 2–72
2–85 2–84
2–86 2–50
2–87 2–71
2–89 2–85 revised
2–91 2–7
2–94 2–86
2–95 2–22
2–96 2–87
2–97 2–102
2–98 2–103
2–100 2–104
2–103 2–57
2–104 2–88

Chapter 3

3–39 3–71
3–40 3–66
3–42 3–28
3–43 3–29
3–44 3–65
3–47 3–67
3–48 3–64
3–49 3–13
3–50 3–9
3–51 3–63
3–52 3–68
3–53 3–26 revised
3–55 3–69
3–56 3–5
3–57 3–6 revised
3–58 3–70
3–59 3–72

Chapter 4

4–65 4–109
4–67 4–134
4–68 4–3
4–69 4–10
4–70 4–107
4–71 4–105 revised
4–72 4–133
4–75 4–103
4–76 4–100
4–77 4–98
4–78 4–4 revised
4–79 4–119
4–80 4–94 revised
4–81 4–131
4–82 4–80
4–83 4–22 revised
4–84 4–93
4–85 4–130
4–86 4–17
4–87 4–91
4–88 4–89
4–89 4–129
4–90 4–128
4–91 4–126 revised
4–92 4–90 revised
4–93 4–101
4–94 4–8 revised
4–95 4–135 revised
4–97 4–132 revised
4–98 4–79 revised
4–99 4–77
4–100 4–123
4–104 4–36
4–105 4–33 revised
4–106 4–43 revised
4–107 4–69 revised

Chapter 5

5–57 5–37 revised
5–58 5–67 revised
5–59 5–12 revised
5–60 5–93
5–61 5–94 revised
5–62 5–68
5–63 5–95 revised
5–64 5–16 revised
5–65 5–69
5–66 5–96
5–67 5–70 revised
5–68 5–71
5–69 5–101
5–70 5–73
5–71 5–74 revised
5–72 5–75 revised
5–73 5–97
5–74 5–98
5–75 5–14 revised
5–77 5–76 revised
5–78 5–78
5–79 5–100 revised
5–80 5–77 revised
5–81 5–99 revised
5–82 5–84

Chapter 6

6–49 6–77 revised
6–50 6–108
6–51 6–106
6–52 6–12
6–53 6–34
6–54 6–78
6–55 6–79
6–56 6–80 revised
6–57 6–101
6–58 6–82
6–59 6–83
6–60 6–85
6–61 6–86
6–62 6–87 revised
6–63 6–88
6–64 6–89 revised
6–65 6–90 revised
6–66 6–91
6–67 6–92
6–68 6–107
6–70 6–50
6–73 6–94
6–74 6–95
6–75 6–105
6–76 6–96 revised
6–77 6–97 revised

Chapter 7

7–41 7–66
7–42 7–34
7–43 7–61
7–44 7–62
7–45 7–78
7–46 7–72
7–47 7–1
7–48 7–71
7–49 7–70
7–50 7–69
7–51 7–68
7–53 7–79

7-54 7-48

Chapter 8

8-68 8-94
8-69 8-136
8-70 8-126
8-71 8-125
8-72 8-93
8-73 8-133
8-74 8-124 revised
8-75 8-123 revised
8-76 8-122
8-77 8-127
8-78 8-33
8-79 8-116
8-80 8-92
8-81 8-115
8-82 8-118
8-83 8-121
8-84 8-111
8-85 8-114
8-86 8-32 revised
8-87 8-91
8-88 8-90 sig. figs.
8-89 8-89 revised
8-90 8-88 revised
8-91 8-87
8-92 8-19 revised
8-94 8-86
8-95 8-83
8-96 8-119
8-97 8-120

Chapter 9

9-60 9-96
9-61 9-125
9-62 9-127

Chapter 10

10-60 9-136
10-61 9-121
10-62 9-137
10-63 9-139 revised
10-67 9-124 revised
10-68 9-126 revised
10-80 9-98 revised
10-81 9-128

Chapter 11

11-71 10-80 sig. figs.
11-72 10-79 revised
11-73 10-19
11-74 10-70
11-75 10-86
11-76 10-78 sig. figs
11-77 10-101
11-78 10-104
11-79 10-69 revised
11-80 10-93
11-81 10-77 revised
11-82 10-98 revised
11-84 10-97
11-86 10-96
11-87 10-95
11-88 10-91
11-89 10-109
11-90 10-90
11-91 10-89 revised
11-92 10-87
11-93 10-88 revised
11-94 10-113
11-96 10-99
11-98 10-117
11-99 10-24
11-100 10-115
11-101 10-114
11-102 10-118
11-103 10-2

Chapter 12

12-61 11-16 revised
12-62 11-87 revised
12-63 11-76 revised
12-64 11-86
12-65 11-88 revised
12-66 11-69 revised
12-67 11-93
12-68 11-19
12-69 11-85 revised
12-70 11-84
12-71 11-103
12-72 11-90
12-73 11-75
12-74 11-79
12-75 11-97
12-76 11-94
12-77 11-96
12-78 11-82
12-79 11-80

Chapter 13

13-43 12-68
13-44 12-64
13-45 12-76
13-46 12-59 revised
13-47 12-75 revised
13-48 12-74 revised
13-49 12-69 revised
13-50 12-67 revised
13-51 12-73 revised
13-52 12-63
13-53 12-51 revised
13-54 12-66 revised
13-55 12-42 revised
13-56 12-57 revised
13-57 12-55 revised
13-58 12-72
13-59 12-56
13-60 12-50 revised
13-61 12-71 revised
13-62 12-52 revised
13-63 12-65

Chapter 14

14-65 13-70
14-66 13-38
14-67 13-66
14-68 13-94
14-69 13-72
14-70 13-52
14-71 13-82 revised
14-72 13-89
14-73 13-97
14-74 13-99
14-75 13-88 revised
14-76 13-100

14–78 13–80
14–81 13–101
14–82 13–87
14–83 13–69
14–84 13–71
14–85 13–67 revised
14–87 13–86
14–88 13–79
14–89 13–54
14–90 13–78
14–91 13–85
14–92 13–59 revised
14–94 13–102

Chapter 15

15–64 14–39
15–66 14–87
15–67 14–10
15–68 14–69
15–69 14–86
15–70 14–12
15–71 14–76
15–72 14–51
15–73 14–84
15–74 14–83
15–75 14–96
15–76 14–82
15–77 14–97
15–79 14–81
15–83 14–80
15–84 14–94
15–86 14–95
15–87 14–78
15–88 14–77
15–89 14–57

Chapter 16

16–65 15–113
16–66 15–110
16–67 15–70
16–69 15–108
16–73 15–93
16–74 15–92
16–75 15–91 revised
16–77 15–90
16–78 15–89
16–79 15–88
16–80 15–107
16–81 15–106 sig. figs.
16–82 15–105
16–83 15–104
16–85 15–103
16–89 15–47
16–90 15–102
16–91 15–101
16–92 15–100
16–93 15–48
16–94 15–38
16–95 15–99
16–96 15–98
16–98 15–97
16–99 15–96 sig. figs.
16–101 15–95
16–102 15–94
16–103 15–87

Chapter 17

17–53 16–94
17–54 16–80
17–55 16–81
17–57 16–91 revised
17–58 16–79 revised
17–59 16–59
17–60 16–78
17–61 16–16
17–62 16–77 revised
17–63 16–76 revised
17–64 16–86
17–65 16–75 revised
17–66 16–90
17–67 16–74 revised
17–68 16–73
17–69 16–72 sig. figs.
17–70 16–7 revised
17–71 16–71 revised
17–72 16–88

Chapter 18

18–61 17–88
18–62 17–32
18–63 17–28
18–64 17–87
18–65 17–79
18–66 17–103
18–67 17–105
18–68 17–102
18–69 17–86
18–70 17–107
18–71 17–2
18–72 17–73 revised
18–76 17–85
18–77 17–84
18–80 17–11
18–81 17–110
18–82 17–20 revised
18–83 17–82
18–84 17–6
18–86 17–81 revised
18–87 17–80
18–88 17–109
18–89 17–4
18–91 17–67
18–92 17–11
18–93 17–101
18–96 17–78
18–97 17–77
18–98 17–76
18–99 17–104

Chapter 19

19–69 18–74 revised
19–70 18–91
19–71 18–94
19–72 18–62
19–73 18–72 revised
19–74 18–77
19–75 18–90
19–76 18–85
19–77 18–84
19–79 18–13
19–81 18–69 revised
19–82 18–76
19–83 18–75
19–84 18–88
19–85 18–96
19–86 18–70
19–87 18–87
19–88 18–92
19–91 18–81
19–93 18–56

19–96 18–82
19–97 18–83
19–98 18–98
19–99 18–95
19–100 18–19
19–103 18–71
19–106 18–80
19–107 18–73
19–108 18–6
19–109 18–99
19–110 18–97

Chapter 20

20–62 19–65 revised
20–63 19–78 revised
20–64 19–79 revised
20–65 19–80
20–67 19–69
20–70 19–82
20–71 19–66 revised
20–73 19–83
20–74 19–84
20–75 19–55
20–76 19–85
20–77 19–86
20–78 19–87 revised
20–79 19–61
20–80 19–88 sig. figs.
20–81 19–89
20–82 19–90
20–84 19–62
20–86 19–91 sig. figs.
20–87 19–92
20–89 19–93
20–91 19–94

Chapter 21

21–46 20–7
21–47 20–31
21–50 20–69 revised
21–53 20–54
21–54 20–68
21–55 20–67
21–56 20–66
21–57 20–65 revised
21–58 20–63 revised
21–59 20–64
21–61 20–58
21–62 20–62
21–63 20–61
21–64 20–60
21–66 20–46
21–67 20–59
21–68 20–34
21–69 20–52

Chapter 22

22–31 21–55
22–32 21–63
22–33 21–54
22–34 21–53 revised
22–35 21–11 revised
22–36 21–52 revised
22–37 21–62 revised
22–38 21–61
22–39 21–51
22–40 21–50
22–41 21–49 revised
22–42 21–39 revised
22–43 21–16 revised
22–44 21–46
22–45 21–43
22–46 21–19
22–47 21–42
22–48 21–57
22–49 21–58
22–50 21–68 revised
22–51 21–41 revised

Chapter 23

23–49 22–82
23–50 22–81
23–52 22–80
23–53 22–79
23–54 22–78
23–55 22–77
23–58 22–15 revised
23–59 22–76
23–60 22–91 revised
23–61 22–75
23–62 22–74
23–63 22–73
23–65 22–72
23–66 22–65
23–67 22–59 revised
23–68 22–71 revised

Chapter 24

24–49 23–79 revised
24–50 23–76 revised
24–52 23–73 revised
24–53 23–71
24–54 23–3 sig. figs.
24–55 23–78
24–56 23–70 revised
24–57 23–69 revised
24–58 23–68
24–59 23–55 sig. figs.
24–61 23–67 sig. figs.
24–61 23–75 sig. figs.
24–63 23–7 revised

Chapter 25

25–58 24–65 revised
25–60 24–82
25–61 24–105
25–62 24–83 revised
25–63 24–106
25–64 24–84 revised
25–65 24–107
25–66 24–85
25–67 24–116
25–68 24–108
25–69 24–24 revised
25–70 24–115
25–71 24–74 revised
25–72 24–75
25–73 24–86 revised
25–74 24–63
25–75 24–109 sig. figs.
25–76 24–87 sig. figs.
25–77 24–6
25–78 24–88
25–79 24–8 sig. figs.
25–80 24–89
25–81 24–110
25–82 24–111 revised
25–83 24–90 sig. figs.

25–84 24–91 sig. figs.
25–85 24–33 sig. figs.
25–86 24–76 revised

Chapter 26

26–50 25–60
26–52 25–59
26–53 25–41
26–56 25–68
26–57 25–69 revised
26–58 25–67 revised
26–59 25–70
26–60 25–71 revised
26–61 25–72 revised
26–62 25–73
26–63 25–57 revised
26–64 25–29 revised
26–65 25–33
26–66 25–75
26–67 25–76 revised
26–68 25–77
26–69 25–78
26–71 25–79
26–72 25–80
26–74 25–82 revised
26–75 25–83 revised
26–76 25–84 sig. figs.
26–78 25–49 revised

Chapter 27

27–49 26–63 revised
27–50 26–62
27–51 26–74
27–52 26–80
27–53 26–81
27–54 26–82
27–55 26–61 sig. figs.
27–56 26–52
27–57 26–83
27–58 26–60
27–59 26–64
27–60 26–59
27–61 26–14 revised
27–62 26–58
27–63 26–51
27–65 26–73 revised
27–66 26–12 revised
27–67 26–49
27–68 26–47

Chapter 28

28–58 27–110
28–59 27–63
28–60 27–61 revised
28–61 27–81 revised
28–62 27–98
28–63 27–82 revised
28–64 27–83 revised
28–65 27–101
28–66 27–109 revised
28–67 27–84 revised
28–68 27–108
28–69 27–23 revised
28–70 27–54 revised
28–71 27–99 revised
28–72 27–70 revised
28–73 27–85
28–74 27–59 revised
28–75 27–71 revised
28–76 27–58
28–77 27–86
28–78 27–107 revised
28–80 27–87
28–81 27–105
28–82 27–106
28–83 27–72 revised
28–84 27–88 revised
28–85 27–100
28–86 27–89 revised
28–87 27–90 revised
28–88 27–91 revised
28–89 27–92 revised
28–91 27–93 revised
28–92 27–104 revised
28–93 27–96 revised
28–94 27–66 revised
28–95 27–95 revised
28–96 27–94

Chapter 29

29–60 28–75
29–61 28–13 revised
29–62 28–74
29–63 28–72
29–64 28–73
29–65 28–63
29–66 28–79
29–68 28–71 revised
29–69 28–68
29–70 28–61 revised
29–71 28–70
29–72 28–34 revised
29–73 28–69
29–74 28–65
29–76 28–78 revised
29–77 28–88
29–78 28–38

Chapter 30

30–59 29–91
30–60 29–92
30–61 29–55 revised
30–62 29–75
30–63 29–88 revised
30–65 29–74
30–66 29–73 revised
30–67 29–72
30–68 29–57
30–69 29–71
30–70 29–70 revised
30–71 29–69 revised
30–72 29–68 revised
30–73 29–67
30–74 29–90
30–75 29–87 revised
30–76 29–65
30–78 29–66 revised
30–79 29–63
30–80 29–28 revised

Chapter 31

31–76 30–98
31–78 30–81
31–81 30–102
31–84 30–85
31–85 30–83
31–86 30–105
31–87 30–106

31–88 30–107
31–90 30–87 revised
31–91 30–108
31–92 30–109
31–94 30–93
31–95 30–79
31–97 30–95
31–101 30–97 revised
31–103 30–99
31–104 30–101 revised
31–105 30–103

Chapter 32

32–39 32–57
32–40 32–6 revised
32–41 32–61
32–42 32–4
32–43 32–55
32–44 32–63 revised
32–45 32–53
32–46 32–70 revised
32–47 32–62
32–48 32–64
32–49 32–69
32–50 32–66
32–51 32–60
32–52 32–59
32–53 32–58
32–55 32–67
32–56 32–52
32–57 32–33 sig. figs.
32–58 32–47
32–60 32–68

Chapter 33

33–66 31–67
33–67 31–69
33–69 31–71
33–70 31–63 revised
33–71 31–73
33–72 31–75
33–73 31–77
33–74 31–79
33–76 31–82
33–79 31–83
33–81 31–85 revised
33–82 31–65
33–83 31–87 revised
33–85 31–89
33–86 31–91 revised
33–87 31–92 sig. figs.
33–88 31–93 sig. figs.
33–89 31–94
33–91 31–95
33–92 31–96

Chapter 34

34–67 33–77
34–68 33–78
34–69 33–71
34–71 33–96
34–75 33–80
34–77 33–92
34–78 33–68
34–79 33–83
34–81 33–99
34–82 33–93
34–83 33–76
34–84 33–91
34–85 33–104
34–86 33–105
34–87 33–101 revised
34–88 33–90
34–89 33–108
34–91 33–79
34–92 33–57
34–93 33–88 revised
34–94 33–73
34–95 33–103
34–98 33–59
34–99 35–89
34–100 33–74

Chapter 35

35–39 34–119
35–40 34–97
35–41 34–111 revised
35–42 34–1012
35–43 34–138
35–45 34–100 revised
35–46 34–135
35–47 34–134 revised
35–48 34–133
35–49 34–132
35–50 34–108
35–51 34–131
35–52 34–98
35–54 34–99
35–55 34–107 revised
35–56 34–130
35–57 34–129
35–58 34–33
35–59 34–128
35–60 34–137
35–61, 35–62 34–95 revised
35–64 34–94 revised

Chapter 36

36–61 34–120
36–62 34–119 revised
36–63 34–119
36–64 34–87
36–65 34–93
36–66 35–33
36–67 34–118
36–68 34–136
36–70 34–111 revised
36–71 34–110
36–72 34–109 revised
36–73 34–108
36–74 35–73
36–75 34–107 revised
36–76 34–106 revised
36–77 34–105 revised
36–78 34–104 revised
36–79 34–128
36–80 34–103
36–81 34–102
36–83 34–101
36–84 34–95
36–85 34–99
36–86 34–83
36–87 34–98

Chapter 37

37–65 36–88
37–66 36–87
37–67 36–86

37–73 36–81
37–74 36–80
37–75 36–79
37–76 36–49 revised
37–77 36–78 sig. figs.
37–78 36–23
37–79 36–8 revised
37–80 36–77
37–81 36–76
37–82 36–70
37–83 36–71
37–84 36–72
37–85 36–73 revised
37–86 36–74
37–87 36–75

Chapter 38

38–54 37–45 revised
38–58 37–88 revised
38–60 37–87
38–61 37–86
38–62 37–85
38–63 37–84
38–64 37–83
38–66 37–66 revised
38–67 37–82
38–68 37–71
38–69 37–72
38–70 37–38
38–71 37–73
38–72 37–74 revised
38–73 37–75
38–74 37–76
38–75 37–77
38–76 37–3
38–77 37–78
38–78 37–79
38–79 37–80
38–80 37–44
38–81 37–81

SECTION EIGHT

PROBLEMS IN THE STUDENT SOLUTION MANUAL, IN THE STUDENT'S COMPANION, AND ON THE WILEY WEBSITE

The *Student Solution Manual* contains fully worked out solutions to about one-third of the odd numbered end-of-chapter problems and the study guide (*A Student's Companion*) contains hints for about another third. The Wiley website duplicates three or four solutions of each chapter from the *Student Solution Manual* and slightly less than half the hints in *A Student's Companion.* The *Student Solution Manual* and *A Student's Companion* are available to students as print supplements. The wiley Website is can be reached at the address given in the text. The problem solutions included in the *Student Solution Manual* and the hints in *A Student's Companion* are listed here. Those that appear on the website are underlined.

As part of each assignment you may wish to have students study a few of the solutions in the *Student Solution Manual* before attempting their own solutions to other problems. You may also wish to include in the assignment several of the problems discussed in *A Student's Companion.* These will help students develop problem-solving strategies and learn effective problem-solving techniques.

Chapter 1

Solution Manual: 1, 5, 7, 13, 19, 23, 29, 33, 37, 41, 47, 57

Study Guide: 3, 9, 11, 17, 21, 27, 31, 35, 39, 43, 45, 49, 59

Chapter 2

Solution Manual: 3, 5, 15, 19, 23, 25, 27, 39, 41, 43, 47, 49, 51, 65, 69, 75, 87, 93, 103, 107, 115

Study Guide: 7, 13, 17, 21, 31, 35, 37, 45, 53, 57, 61, 67, 73, 79, 89, 105, 109, 113, 117

Chapter 3

Solution Manual: 1, 3, 7, 11, 15, 19, 25, 31, 35, 41, 45, 47, 55, 65, 67, 77

Study Guide: 5, 13, 21, 23, 27, 29, 33, 37, 39, 43, 53, 61, 69, 73

Chapter 4

Solution Manual: 5, 9, 15, 25, 27, 31, 37, 39, 41, 53, 59, 61, 63, 69, 75, 79, 81, 87, 97, 99, 103, 107, 115, 119, 127, 135

Study Guide: 7, 19, 23, 29, 33, 35, 43, 45, 51, 57, 67, 71, 73, 83, 89, 91, 101, 105, 111, 117, 123, 129, 133

Chapter 5

Solution Manual: 5, 9, 13, 17, 23, 29, 39, 43, 49, 53, 59, 63, 67, 73, 77, 83, 89, 99

Study Guide: 3, 15, 21, 25, 35, 37, 45, 47, 55, 61, 69, 75, 81, 85, 87, 93, 97

Chapter 6

Solution Manual: 3, 5, 13, 15, 21, 25, 27, 31, 45, 47, 49, 51, 55, 61, 67, 69, 73, 77, 87, 91, 99, 105

Study Guide: 1, 7, 9, 17, 19, 23, 29, 37, 39, 43, 57, 59, 65, 75, 79, 83, 93, 101, 107

Chapter 7

Solution Manual: 3, 13, 17, 19, 29, 33, 41, 43, 45, 55, 59, 63, 65, 71, 73, 77
Study Guide: 7, 9, 11, 15, 23, 25, 35, 37, 39, 47, 51, 53, 61, 67, 69, 75

Chapter 8

Solution Manual: 1, 3, 9, 13, 17, 29, 43, 45, 51, 67, 69, 71, 83, 89, 99, 105, 115, 119, 125, 127, 129
Study Guide: 5, 11, 19, 21, 27, 35, 39, 49, 57, 65, 75, 79, 87, 91, 97, 103, 109, 121, 123, 135

Chapter 9

Solution Manual: 1, 15, 17, 23, 35, 43, 55, 59, 63, 69, 71, 73, 75, 79, 81, 89, 95, 99, 101, 105, 107, 121, 125, 131, 133, 135
Study Guide: 5, 7, 11, 13, 21, 31, 39, 49, 53, 65, 77, 83, 87, 97, 109, 113, 119, 123, 127, 137

Chapter 10

Solution Manual: 13, 21, 27, 29, 33, 35, 37, 41, 45, 47, 49, 55, 63, 69, 71, 77, 83, 85, 89, 97, 101, 107, 111, 123
Study Guide: 11, 15, 17, 23, 25, 31, 39, 51, 53, 59, 61, 65, 75, 79, 81, 87, 93, 99, 105, 109, 117

Chapter 11

Solution Manual: 3, 9, 15, 21, 27, 31, 35, 37, 41, 43, 45, 59, 67, 75, 83, 85, 95
Study Guide: 7, 9, 13, 25, 29, 39, 47, 51, 55, 61, 63, 69, 73, 81, 87

Chapter 12

Solution Manual: 5, 7, 9, 11, 19, 21, 25, 29, 31, 33, 35, 37, 39, 41, 45, 47, 51, 55, 61, 69, 75
Study Guide: 1, 3, 13, 15, 17, 23, 27, 43, 49, 53, 63, 73, 81

Chapter 13

Solution Manual: 1, 5, 11, 15, 23, 25, 31, 33, 35, 41, 53, 55, 61, 67, 69, 75, 79, 85, 97, 99, 103
Study Guide: 3, 9, 13, 19, 27, 29, 37, 39, 43, 47, 51, 59, 63, 71, 73, 83, 87, 93, 101

Chapter 14

Solution Manual: 1, 3, 7, 11, 15, 19, 21, 25, 29, 31, 35, 41, 43, 45, 49, 59, 67, 69, 71, 75, 79, 85, 89, 97
Study Guide: 5, 9, 13, 17, 23, 27, 37, 39, 47, 51, 55, 61, 73, 77, 81, 83, 87, 93

Chapter 15

Solution Manual: 1, 5, 13, 15, 21, 23, 29, 31, 39, 43, 49, 55, 59, 65, 71, 77, 79, 81, 85, 97, 113
Study Guide: 7, 9, 17, 19, 25, 27, 33, 37, 41, 47, 53, 61, 63, 67, 73, 83, 89, 99, 101

Chapter 16

Solution Manual: 13, 17, 19, 23, 29, 33, 39, 41, 45, 47, 49, 53, 55, 61, 65, 67, 77, 85, 87
Study Guide: 3, 7, 11, 15, 25, 27, 31, 37, 43, 51, 57, 63, 71, 79, 83, 91

Chapter 17

Solution Manual: 3, 5, 7, 9, 11, 21, 25, 37, 39, 43, 47, 59, 63, 77, 83, 93, 95, 107
Study Guide: 13, 15, 27, 29, 31, 33, 41, 53, 55, 61, 65, 79, 85, 99, 109

Chapter 18

Solution Manual: 11, 15, 21, 25, 27, 39, 41, 45, 49, 51, 61, 63, 77, 79, 81, 89, 93, 99
Study Guide: 7, 13, 17, 19, 29, 31, 37, 47, 55, 59, 67, 73, 75, 83, 91, 97, 101

Chapter 19

Solution Manual: 5, 13, 17, 21, 27, 31, 39, 43, 51, 59, 63, 65, 71, 75, 79, 85, 89, 91
Study Guide: 7, 9, 15, 25, 35, 37, 45, 47, 53, 57, 67, 69, 73, 77, 87

Chapter 20

Solution Manual: 1, 7, 15, 17, 21, 25, 27, 35, 37, 41, 45, 49, 51, 57, 61, 73
Study Guide: 5, 9, 11, 13, 23, 39, 43, 53, 63, 67, 71

Chapter 21

Solution Manual: 3, 5, 9, 17, 21, 29, 31, 37, 41, 43, 53, 57, 61, 71
Study Guide: 1, 7, 13, 19, 25, 27, 33, 39, 45, 49. 59, 65, 67

Chapter 22

Solution Manual: 3, 5, 7, 9, 11, 17, 21, 27, 31, 35, 39, 41, 45, 51, 55, 65, 71, 81, 87
Study Guide: 1, 13, 15, 19, 25, 29, 33, 53, 61, 67, 75, 85, 91

Chapter 23

Solution Manual: 1, 9, 11, 15, 19, 21, 27, 31, 37, 39, 41, 43, 47, 55, 57, 59, 61, 63, 77, 81, 85
Study Guide: 7, 17, 23, 25, 27, 29, 33, 41, 45, 51, 53, 65, 69, 71, 73, 79, 83

Chapter 24

Solution Manual: 1, 3, 15, 17, 23, 25, 33, 37, 41, 45, 49, 53, 55, 61, 63, 73, 75, 91, 97, 101, 107, 117
Study Guide: 5, 7, 9, 13, 19, 21, 27, 29, 35, 43, 51, 57, 59, 65, 69, 79, 81, 85, 95, 103, 105, 109

Chapter 25

Solution Manual: 3, 5, 7, 13, 17, 21, 27, 31, 35, 37, 39, 43, 45, 47, 53, 59, 63, 69, 73, 75, 81
Study Guide: 1, 9, 11, 15, 19, 23, 25, 29, 33, 41, 49, 51, 57, 61, 65, 67, 77

Chapter 26

Solution Manual: 3, 5, 15, 17, 19, 21, 23, 27, 33, 35, 39, 45, 47, 55, 57, 63, 69, 75, 79
Study Guide: 1, 7, 11, 13, 25, 29, 31, 37, 41, 43, 49, 53, 59, 65, 73, 81

Chapter 27

Solution Manual: <u>5</u>, 9, 13, 19, 21, 31, <u>33</u>, 35, 45, <u>53</u>, 55, 57, 61, 67, 73, 77, 81, 85, 91, 93, 101
Study Guide: 3, <u>7</u>, 11, <u>17</u>, 23, <u>25</u>, 29, <u>37</u>, 41, <u>43</u>, 47, <u>51</u>, 59, 69, 71, 75, 83, 99, 103, 111

Chapter 28

Solution Manual: 1, <u>9</u>, 17, <u>23</u>, 29, 33, <u>37</u>, 39, 41, 45, 47, 51, 55, 59, 61, 63, 67, 71, 75, 79, 81, 85, 89, 89
Study Guide: 5, <u>7</u>, 11, <u>13</u>, 19, <u>21</u>, 25, <u>27</u>, 31, <u>35</u>, 43, <u>53</u>, 57, 65, 69, 77, 83, 87

Chapter 29

Solution Manual: 3, <u>5</u>, 7, 13, 15, <u>17</u>, 19, 23, 27, 35, 43, <u>45</u>, 47, 49, 53, 59, 61, 75, 77, 83, 85, 89, 91, 93
Study Guide: 1, 9, <u>11</u>, 21, <u>25</u>, 29, <u>31</u>, 33, <u>37</u>, 39, <u>41</u>, 51, 57, 63, 67, 73, 79, 81, 87

Chapter 30

Solution Manual: <u>5</u>, 9, 19, <u>25</u>, 27, <u>33</u>, 35, 41, 49, 53, <u>57</u>, 61, 63, 67, 69, 71, 77, 85, 87, 89, 99
Study Guide: 3, <u>11</u>, 13, <u>15</u>, 23, <u>31</u>, 37, <u>45</u>, 47, <u>51</u>, 59, <u>65</u>, 73, 75, 79, 83, 93, 95, 97

Chapter 31

Solution Manual: 7, <u>15</u>, 19, 27, 33, 39, <u>45</u>, 47, 53, <u>55</u>, 57, 61, 65, 71, 75, 77, 89
Study Guide: 1, <u>11</u>, 13, <u>17</u>, 25, <u>29</u>, 31, <u>35</u>, 37, <u>41</u>, 43, <u>49</u>, 59, <u>63</u>, 73, 77, 83, 91, 97

Chapter 32

Solution Manual: 3, 5, 13, <u>19</u>, 21, 27, 29, <u>31</u>, 35, 37, 43, <u>49</u>, 53, 57, 63, 69, 71
Study Guide: 1, <u>9</u>, 11, 17, <u>23</u>, 25, <u>33</u>, 39, <u>41</u>, 45, <u>47</u>, 55, 61, 65, 75

Chapter 33

Solution Manual: 5, 19, <u>23</u>, 25, 27, 33, <u>43</u>, 49, <u>53</u>, 61, 65, 75, 85, 89, 93, 95, 105, 109
Study Guide: 3, <u>13</u>, 17, <u>21</u>, 29, <u>35</u>, 39, <u>47</u>, 51, <u>57</u>, 63, 67, 71, 73, 79, 83, 87, 91, 101, 107

Chapter 34

Solution Manual: <u>5</u>, 9, 11, 13, 25, 31, 35, 39, <u>43</u>, 49, 53, 57, 65, 73, 77, <u>83</u>, 89, 91, 95, 103, 105, 111
Study Guide: 3, <u>15</u>, 17, <u>21</u>, 27, <u>39</u>, 47, <u>55</u>, 61, <u>71</u>, 75, <u>85</u>, 93, <u>97</u>, 101, <u>109</u>, 115, <u>121</u>, 125, 131, 135

Chapter 35

Solution Manual: 7, 15, 17, 19, 21, <u>27</u>, 29, 39, 43, 45, 49, <u>55</u>, 61, 65, 69, 71, 75, 79, <u>81</u>, 91, 95, 107, 111, 115, 121, 123
Study Guide: 5, <u>11</u>, 23, <u>25</u>, 33, <u>37</u>, 41, <u>51</u>, 59, <u>67</u>, 73, <u>77</u>, 83, <u>87</u>, 97, <u>99</u>, 105, <u>109</u>, 113, <u>121</u>

Chapter 36

Solution Manual: 3, 5, 7, 13, 15, 17, 21, 25, 33, 39, 41, 43, 47, 51, 63, 69, 73, 79, 83, 93, 103, 105, 107, 109

Study Guide: 1, 9, 11, 19, 27, 29, 35, 37, 49, 55, 57, 59, 65, 77, 81, 89, 95

Chapter 37

Solution Manual: 1, 9, 11, 13, 17, 19, 21, 27, 29, 31, 35, 39, 41, 49, 51, 53, 55, 63, 73, 75, 81, 83, 87, 93, 99

Study Guide: 3, 5, 15, 23, 33, 37, 43, 45, 47, 59, 65, 71, 77, 89, 91

Chapter 38

Solution Manual: 9, 13, 19, 25, 29, 33, 39, 41, 45, 49, 51, 55, 57, 61, 65, 73, 83

Study Guide: 7, 15, 17, 21, 23, 31, 37, 43, 51, 59, 63, 69, 75, 79

Chapter 39

Solution Manual: 3, 9, 13, 17, 21, 25, 29, 33, 37, 41, 43, 45, 47, 53, 57

Study Guide: 5, 11, 15, 19, 23, 27, 31, 35, 39, 51, 55

Chapter 40

Solution Manual: 9, 11, 15, 17, 23, 27, 33, 35, 37, 47, 59, 63

Study Guide: 3, 7, 13, 19, 21, 25, 31, 39, 43, 45, 53, 57, 67

Chapter 41

Solution Manual: 5, 9, 15, 19, 21, 27, 31, 35, 37, 45

Study Guide: 1, 11, 13, 17, 23, 25, 29, 33, 41, 47

Chapter 42

Solution Manual: 11, 15, 19, 25, 31, 37, 41, 45, 51, 57, 65, 69, 75, 77

Study Guide: 3, 13, 23, 33, 39, 43, 49, 53, 61, 67, 79, 85, 87

Chapter 43

Solution Manual: 3, 9, 11, 15, 19, 21, 23, 27, 29, 39, 45

Study Guide: 5, 7, 13, 17, 23, 31, 35, 37, 43, 47, 51, 53

Chapter 44

Solution Manual: 7, 11, 15, 23, 33, 37, 39, 43

Study Guide: 1, 5, 13, 17, 27, 31, 35, 41

NOTES